Communications
in Computer and Information Science

2738

Series Editors

Gang Li, *School of Information Technology, Deakin University, Burwood, VIC, Australia*

Joaquim Filipe, *Polytechnic Institute of Setúbal, Setúbal, Portugal*

Zhiwei Xu, *Chinese Academy of Sciences, Beijing, China*

Rationale

The CCIS series is devoted to the publication of proceedings of computer science conferences. Its aim is to efficiently disseminate original research results in informatics in printed and electronic form. While the focus is on publication of peer-reviewed full papers presenting mature work, inclusion of reviewed short papers reporting on work in progress is welcome, too. Besides globally relevant meetings with internationally representative program committees guaranteeing a strict peer-reviewing and paper selection process, conferences run by societies or of high regional or national relevance are also considered for publication.

Topics

The topical scope of CCIS spans the entire spectrum of informatics ranging from foundational topics in the theory of computing to information and communications science and technology and a broad variety of interdisciplinary application fields.

Information for Volume Editors and Authors

Publication in CCIS is free of charge. No royalties are paid, however, we offer registered conference participants temporary free access to the online version of the conference proceedings on SpringerLink (http://link.springer.com) by means of an http referrer from the conference website and/or a number of complimentary printed copies, as specified in the official acceptance email of the event.

CCIS proceedings can be published in time for distribution at conferences or as post-proceedings, and delivered in the form of printed books and/or electronically as USBs and/or e-content licenses for accessing proceedings at SpringerLink. Furthermore, CCIS proceedings are included in the CCIS electronic book series hosted in the SpringerLink digital library at http://link.springer.com/bookseries/7899. Conferences publishing in CCIS are allowed to use our online conference service (Meteor) for managing the whole proceedings lifecycle (from submission and reviewing to preparing for publication) free of charge.

Publication process

The language of publication is exclusively English. Authors publishing in CCIS have to sign the Springer CCIS copyright transfer form, however, they are free to use their material published in CCIS for substantially changed, more elaborate subsequent publications elsewhere. For the preparation of the camera-ready papers/files, authors have to strictly adhere to the Springer CCIS Authors' Instructions and are strongly encouraged to use the CCIS LaTeX style files or templates.

Abstracting/Indexing

CCIS is abstracted/indexed in DBLP, Google Scholar, EI-Compendex, Mathematical Reviews, SCImago, Scopus. CCIS volumes are also submitted for the inclusion in ISI Proceedings.

How to start

To start the evaluation of your proposal for inclusion in the CCIS series, please send an e-mail to ccis@springer.com

Preeti Chandrakar · K. Jairam Naik ·
Zahid Akhtar · Chien Ming Chen ·
Karnati Mohan
Editors

Computational Intelligence and Network Security

Second International Conference, ICCINS 2025
Raipur, India, April 3–4, 2025
Proceedings

Springer

Editors
Preeti Chandrakar
National Institute of Technology Raipur
Raipur, Chhattisgarh, India

Zahid Akhtar
State University of New York Polytechnic
Institute
Utica, NY, USA

Karnati Mohan
National Institute of Technology Raipur
Raipur, Chhattisgarh, India

K. Jairam Naik
National Institute of Technology Raipur
Raipur, Chhattisgarh, India

Chien Ming Chen
Nanjing University of Information Science
and Technology
Nanjing, China

ISSN 1865-0929　　　　　　ISSN 1865-0937　(electronic)
Communications in Computer and Information Science
ISBN 978-3-032-09571-8　　　ISBN 978-3-032-09572-5　(eBook)
https://doi.org/10.1007/978-3-032-09572-5

Preface

The 2nd International Conference on Computational Intelligence and Network Security was a premier international conference and forum for high-quality research in the area of Information Technology. It was organized by the National Institute of Technology, Raipur, India, from April 3rd – 4th, 2025.

The aim of ICCINS 2025 was to showcase state-of-the-art methodologies and technologies in the fields of Computational Intelligence and Network Security. It focused on new ideas and developments to propagate the latest innovations and practices. It provided an opportunity to network, collaborate and exchange ideas among renowned speakers, researchers, academics and students in Information Technology.

We proudly mention that ICCINS 2025 has received 205 papers from 520 authors affiliated with more than 93 institutions spread across 9 countries. All submitted papers were peer-reviewed by more than 131 reviewers and Technical Program Committee (TPC) members affiliated with 128 institutions. Reviews were double blind and each submission received an average of three reviews. More than 1100 review comments were received during the evaluation of submitted papers. After a rigorous peer review process and careful editorial evaluation, 30 papers were accepted for presentation; among these, 5 papers were registered for presentation at ICCINS 2025 venue and 25 papers for virtual presentations. The acceptance rate of ICCINS 2025 was 14.63%.

We express our sincere gratitude to all the authors and co-authors who submitted their work to this conference. Our special thanks go to all reviewers and Technical Program Committee members who generously offered their expertise and time, which helped the conference chairs to select the papers and prepare the conference program.

We were privileged to have the guidance and support of several distinguished academicians who served on our International and National Advisory Committee. Their expertise and encouragement played a vital role in the successful planning and execution of this conference. We extended our heartfelt gratitude to them all. Their constant guidance and valuable suggestions helped shape the conference into a meaningful academic platform.

We were honored to host five distinguished keynote speakers whose insightful and thought-provoking presentations enriched the intellectual value of the conference. Sumantra Dutta Roy from Indian Institute of Technology Delhi shared critical perspectives on biometric research. Satyasiba Das, Indian Institute of Management Raipur, provided an interdisciplinary outlook bridging management and technology. Zahid Akhtar from State University of New York Polytechnic Institute, USA addressed emerging challenges and responsibilities in the domain of artificial intelligence.

Xiao-Zhi Gao from the University of Eastern Finland delivered valuable insights into advanced computational approaches. Debasis Giri, Maulana Abul Kalam Azad University of Technology, West Bengal, India, highlighted recent advancements in cryptographic techniques. We sincerely express our gratitude to all keynote speakers for their invaluable contributions, which significantly inspired and benefited the participants.

Several individuals, including our esteemed colleagues, contributed tirelessly to the success of the conference. We expressed our sincere gratitude to the steering committee members, whose vision and efforts were instrumental in conceptualizing and organizing the conference. We also extended our thanks to the registration chairs, session coordination chairs, publicity team, publication chairs, website management team, and the student volunteers for their dedicated efforts and seamless coordination. Finally, we gratefully acknowledge the invaluable support of the faculty members of the Department of Computer Science and Engineering (CSE), NIT Raipur, Chhattisgarh, India, whose timely assistance ensured the smooth conduct of the event.

The Department of Computer Science and Engineering (CSE), National Institute of Technology, Raipur, Chhattisgarh, India, the host of the conference, provided essential support and infrastructure for the successful organization of the event. We express our sincere gratitude to N. V. Ramana Rao, the Chief Patron of the conference and Director of NIT Raipur, for his constant encouragement and visionary guidance. We are also thankful to Prabhat Diwan and Shrish Verma, the Patrons of the conference, for their valuable support throughout the planning and execution process. Special thanks are extended to Pradeep Singh, Head of the Department of CSE, for his consistent assistance and facilitation in organizing the various conference activities.

The conference program was efficiently managed using Microsoft Conference Management Toolkit (CMT), which streamlined the entire process from paper submission and review to the preparation of the final proceedings. We express our sincere appreciation to Communications in Computer and Information Science (CCIS), Springer for publishing the official conference proceedings. We wholeheartedly thank the staff at Springer for their continuous cooperation and timely support throughout the publication process of this CCIS volume. We hope that the research papers compiled in this volume will serve as a valuable resource and inspiration for future academic and scientific work.

Preeti Chandrakar
K. Jairam Naik
Zahid Akhtar
Chien Ming Chen
Karnati Mohan

Organization

Steering Committee

Preeti Chandrakar	NIT Raipur, India
K. Jairam Naik	NIT Raipur, India
Debasis Giri	MAKAUT, West Bengal, India
Zahid Akhtar	State University of New York, USA

Program Chairs

Pradeep Singh	NIT Raipur, India
Indrakshi Ray	Colorado State University, USA
Sandeep Singh Senghar	Cardiff Metropolitan University, UK
Ondrej Krejcar	University of Hradec Králové, Czech Republic
Chien Ming Chen	Nanjing University of Information Science and Technology, China
Karnati Mohan	National Institute of Technology, Raipur, India

Advisory Committee

Mohammad S. Obaidat	University of Jordan, Jordan
Mohammad Amoon	King Saud University, Saudi Arabia
Arijit Karati	National Sun Yat-sen University, Taiwan
Ashok Kumar Das	IIIT Hyderabad, India
Achyut Shankar	University of Warwick, UK
Jayraj Singh	Universiti Teknologi Malaysia, Malaysia
Xiao-Zhi Gao	University of Eastern Finland, Finland
Suresh Bhalla	IIT Delhi, India

Technical Program Committee

Bonny Banerjee	University of Memphis, USA
Ayse Betul Oktay	Yıldız Technical University, Turkey
Prajwal Panzade	Old Dominion University, USA
Rahul Pramanik	BITS PILANI Dubai Campus, UAE

Muhammad Asad Saleem	University of Electronic Science and Technology of China, China
Tamizharasan Periyasamy	BITS PILANI Dubai Campus, UAE
Khalid Mohiuddin	King Khalid University, Saudi Arabia
Brij B. Gupta	Asia University, Taiwan
Nies Hui Wen	Universiti Teknologi Malaysia, Malaysia
Sridevi Narayanan	University of Technology Bahrain, Bahrain
Anupama Prasanth	University of Technology Bahrain, Bahrain
Nitikarn Nimsuk	Thammasat University, Thailand
Joel J.P.C. Rodrigues	Federal University of Piauí, Brazil
Smith K. Khare	Syddansk Universitet, Denmark
Binayak Kar	National Taiwan University of Science and Technology, Taiwan
Felix Harer	University of Fribourg, Switzerland
Devashree Tripathy	Harvard University, USA
Dasuni Nawinna	University of Moratuwa, Sri Lanka
Agung Triayudi	National University, Indonesia
Ondrej Krejcar	University of Hradec Králové, Czech Republic
Khalid Mohiuddin	King Khalid University, Saudi Arabia
Abdul Wahid	University of Galway, Ireland
Debasish Mukherjee	Illinois Advanced Research Center at Singapore, Singapore
Mounira Msahli	Télécom Paris, France
Dhiman Saha	IIT Bhilai, India
Arup Kumar Pal	IIT Dhanbad, India
Tanusree Chakraborty	IIT Delhi, India
Anshul Verma	BHU Varanasi, India
Ajay Pratap	IIT BHU, India
Debasis Das	IIT Jodhpur, India
Amit Kumar Dhar	IIT Bhilai, India
Rifaqat Ali	NIT Hamirpur, India
Pratyay Kuila	NIT Sikkim, India
Ruhul Amin	IIIT Naya Raipur, India
Priyambada Subudhi	IIIT Kottayam, India
Sanjay Panda	NIT Warangal, India
Sangram Ray	NIT Sikkim, India
T. Ravichandran	NIT Trichy, India
Mantosh Biswas	NIT Kurukshetra, India
Abhimanyu Kumar	NIT Uttarakhand, India
Narendra Dewangan	UPES Dehradun, India
Nabajyoti Mazumdar	IIIT Allahabad, India
S.K. Hafizul Islam	IIIT Kalyani, India

Manoj Kumar Mishra	BHU Varanasi, India
Jayraj Singh	IIT Delhi, India
Malay Kumar	IIIT Dharwad, India
Hari Om	IIT(ISM) Dhanbad, India
Raju Bhukya	NIT Warangal, India
P. Shyam Kumar	IDRBT Hyderabad, India
Prerna Mishra	IIIT Nagpur, India
Vijaypal Singh Rathor	IIITDM Jabalpur, India
Geet Sahu	Siksha 'O' Anusandhan, India
Avinash Chandra Pandey	IIITDM Jabalpur, India
Vinay Kumar	IIIT Naya Raipur, India
Atul Gupta	IIITDM Jabalpur, India
Nidhi Sonkar	NIT Warangal, India
Sanjaya Kumar Panda	NIT Warangal, India
Malay Kishore Dutta	Amity University, Noida, India
D. D. Shrimankar	VNIT Nagpur, India
Syed Taqi Ali	VNIT Nagpur, India
Nidhi Lal	VNIT Nagpur, India
Chinmaya Panigrahy	Thapar Institute of Engineering and Technology, India
Abhimanyu Sahu	MNIT Allahabad, India
Amit Biswas	MNIT Allahabad, India
Anuja Dixit	MNIT Allahabad, India
Muneendra Ojha	IIIT Allahabad, India
Neha Agrawal	IIIT Sri City, India
Sundaresan Raman	BITS Pilani, India
Vinod Kumar Jain	IIITDM Jabalpur, India
Dip Prakash Samajdar	IIITDM Jabalpur, India
Ritesh Maurya	Amity University, Noida, India
Naresh Kumar Nagwani	NIT Raipur, India
Dilip Singh Sisodia	NIT Raipur, India
Abhay Kumar Singh	IIT(ISM) Dhanbad, India
Manu Vardhan	NIT Raipur, India
Aakanksha Sharaff	NIT Raipur, India
Jitendra Kumar Rout	NIT Raipur, India
Sonal Yadav	NIT Raipur, India
Deepak Singh	NIT Raipur, India
Tirath Prasad Sahu	NIT Raipur, India
Pawan Kumar Mishra	NIT Raipur, India
Rekha Ram Janghel	NIT Raipur, India
Mou Das Gupta	NIT Raipur, India
Sudhakar Pandey	NIT Raipur, India

Satya Prakash Sahu	NIT Raipur, India
B. Acharya	NIT Raipur, India
Mukesh Kumar	NIT Patna, India
N. Mazumdar	IIIT Allahabad, India
Shailendra Kumar Tripathi	NIT Raipur, India
Abhinav Tomar	NSUT Delhi, India
Alekha Kumar Mishra	NIT Jamshedpur, India
Amit Garg	IIIT Kota, India
Anand Bihari	VIT Vellore, India
Anshul Agarwal	VNIT Nagpur, India
Ansuman Mahapatra	NIT Puducherry, India
Anurag Singh	NIT Delhi, India
Arti Jain	JIIT Noida, India
Arun Agarwal	SOA Deemed to be University, India
Arun Kishor Johar	Chandigarh University, India
Bharat Gupta	NIT Patna, India
Bibhu Mohanty	SOA Deemed to be University, India
Biswajit R. Bhowmik	NITK Surathkal, India
Bunil Kumar Balabantaray	NIT Meghalaya, India
Ch. Sudhakar	NIT Warangal, India
D. Chandrasekhar Rao	VSSUT Burla, India
Damodar Reddy Edla	NIT Goa, India
Debasis Gountia	Odisha University of Technology and Research, India
Deepak Ranjan Nayak	MNIT Jaipur, India
Deepak Singh Tomar	MANIT Bhopal, India
Deepsubhra Guha Roy	VIT Vellore, India
Devesh C. Jinwala	SVNIT Surat, India
Ajay	NIT Rourkela, India
Preeti Soni	NIT Hamirpur, India
Jitesh Pradhan	NIT Jamshedpur, India
Ashok Kumar Das	IIIT Hyderabad, India
Ashik Kumar Turuk	NIT Rourkela, India
Asis Tripathy	VIT Vellore, India
B. B. Gupta	NIT Kurukshetra, India
Rajesh Kumar Mundotiya	IIT Bhilai, India

Contents

Intelligent Networking and Communication Systems

Cybersecurity, Privacy, and System Design

Machine Learning and Computational Intelligence

Machine Learning-Driven Cardiovascular Disease Prediction: Balancing Performance and Transparency with Hybrid Models

Jagan Mohan Dudala[(✉)]

Midmark Corporation, Versailles, USA
`jagan.dudala@outlook.com`

Abstract. With a major worldwide source of morbidity and death, cardiovascular disease (CVD) emphasizes the need of accurate and quick prognostic approaches. In predicting CVD based on demographic, clinical, and lifestyle variables, this study evaluates the efficacy of a range of machine learning (ML) models, including Logistic Regression (LR), Decision Tree (DT), Random Forest (RF), Support Vector Machine (SVM), and a Hybrid Model that combines DT and LR. With best accuracy, precision, recall, F1-score, and ROC-AUC, the Hybrid Model was the most successful. This was so because its ability to record in the data both linear and non-linear relationships. This approach uses the interpretability of the DT to underline the major predictors and the classification power of LR, therefore guaranteeing predictive accuracy and openness. By means of insight into important risk factors and efficient identification of high-risk people, hybrid ML models show to be rather helpful in clinical decision-making. Nevertheless, as shown by the restrictions related to data diversity and longitudinal dynamics, further study including various and temporal datasets could help to increase the generalizability and robustness of the model. Emphasizing the need of balanced model performance and interpretability in the healthcare industry, this paper adds to the growing corpus of data supporting the usage of ML. e context of clinical adoption.

Keywords: Cardiovascular disease (CVD) · machine learning (ML) · hybrid model · predictive analytics · healthcare · clinical decision support · interpretability

1 Introduction

CVD continues to be a major global health concern, accounting for over 17.9 million fatalities annually, thereby imposing a substantial strain on public health systems [1]. Major risk factors such as hypertension, high cholesterol, smoking,

---

J.M. Dudala—Contributing author.

P. Chandrakar et al. (Eds.): ICCINS 2025, CCIS 2738, pp. 3–14, 2026.
https://doi.org/10.1007/978-3-032-09572-5_1

physical inactivity, and elevated blood glucose levels are prevalent in diverse populations, contributing to the development and progression of CVD [2,3]. Early detection of CVD is crucial, as preventive measures can significantly reduce mortality rates and healthcare costs. However, CVD often progresses asymptomatically in its early stages, which limits the effectiveness of traditional diagnostic methods that primarily rely on observable symptoms and clinical evaluations [4]. Traditional diagnostic approaches for CVD, including biochemical assessments and imaging, are often used reactively, identifying disease after significant progression rather than preemptively [5].

Although these approaches offer important data, they might not have the predictive ability required for quick intervention. Emerging as a potential tool in this context is ML, which allows one to examine intricate patterns within demographic, physical, and lifestyle data to more precisely forecast illness risk [6–8]. Nevertheless, current ML-based studies on CVD prediction sometimes differ greatly in their choice of algorithms, feature sets, and evaluation metrics, so generating a research gap in knowledge on the relative efficacy of several algorithms and the predictive relevance of particular features such blood pressure and cholesterol levels [9].

This work looks at the following research question in order to handle these difficulties: *"How successfully can ML algorithms predict cardiovascular disease using demographic, physical, and lifestyle data?"* This work attempts to evaluate the predictive ability of several ML algorithms and find the most important elements in defining CVD risk by means of a comparison analysis.

This research makes several key contributions:

- We conduct a comparative analysis of multiple ML algorithms, including LR, DT, RF, and SVM, to evaluate their effectiveness in predicting CVD.
- We identify key predictive features, such as blood pressure and cholesterol levels, that significantly contribute to CVD risk assessment.
- We assess the feasibility of integrating ML-based prediction models into clinical settings, with a focus on enhancing early detection and preventive measures for CVD.

The structure of this paper is as follows: Sect. 2 presents a review of existing literature, outlining advancements and gaps in ML applications for CVD prediction. Section 3 details the methodology, including dataset descriptions, ML algorithms, and evaluation metrics. Section 4 reports the results of the analysis, followed by a discussion in Sect. 5, where findings are interpreted in the context of current research and clinical applicability. Finally, Sect. 6 concludes with key insights and recommendations for future work.

2 Literature Review

ML has emerged as a powerful tool for predicting CVD, offering an alternative to traditional diagnostic approaches reliant on clinical risk factors. Various studies have explored ML models for CVD risk prediction, leveraging demographic, clinical, and lifestyle data.

Ali et al. [10] analyzed ML algorithms, including LR, DT, and SVM, demonstrating their effectiveness when incorporating key risk factors like blood pressure and cholesterol. Pradhan [11] found that ensemble methods, particularly random forests (RF), significantly improved prediction accuracy when lifestyle factors such as smoking and physical activity were included. Shah et al. [3] further validated the role of cholesterol levels and demographic attributes in enhancing risk assessment.

Bhatt et al. [12] confirmed the strong correlation between systolic and diastolic blood pressure and CVD risk, aligning with conventional clinical findings. Krittanawong et al. [5] conducted a meta-analysis, emphasizing the predictive power of integrating physical markers like blood pressure and cholesterol with demographic variables. However, direct comparisons among ML models remain limited, leaving gaps in understanding their relative performance and interpretability.

Recent advancements have explored deep learning techniques for CVD prediction. Bhowmik et al. [1] highlighted the predictive strength of neural networks but noted a lack of comparison with simpler, more interpretable models like LR. Anjum et al. [2] and Katarya and Meena [13] emphasized the need for broader evaluations using diverse datasets, addressing concerns over generalizability. Diwakar et al. [14] argued for incorporating additional features such as glucose levels to enhance model robustness.

Advancements in ML for medical and agricultural applications further reinforce the importance of model efficiency and feature extraction. Chauhan et al. [15] demonstrated the effectiveness of attention-based deep neural networks in classification tasks, while Sharma and Vardhan [16,17] explored multi-task learning frameworks for plant classification, emphasizing robust feature representation. Similarly, Sharma and Vardhan [18] optimized YOLO-based models for precision agriculture, showcasing improvements in automated classification and detection tasks.

By means of a comparative comparison of ML algorithms—including LR, DT, RF, and SVM—this work seeks to close current gaps on a unified dataset encompassing demographic, clinical, and lifestyle data. This work aims to find the most efficient ML method for CVD prediction by methodically assessing model performance and feature importance, thereby addressing interpretability and computational economy issues.

3   Methodology

Dataset Description. This study uses Sulianova's cardiovascular disease dataset from Kaggle [19]. The dataset includes demographic, clinical, and lifestyle data for individuals, with the purpose of predicting the existence of cardiovascular disease. The dataset contains 70,000 observations with 12 essential features, which capture a variety of patient characteristics and health measures that can be used to predict CVD risk. The main features include:

- Age: Age is recorded in days to provide an exact measure of the individual.
- Height and Weight: Physical measurements in millimeters and kilograms are used to determine Body Mass Index (BMI), a risk factor for CVD.
- Gender: A categorical feature that distinguishes between male and female individuals, enabling investigation of gender-related risk variations.
- Blood Pressure: Includes systolic (ap_hi) and diastolic (ap_lo) blood pressure values, which are significant indicators of cardiovascular health.
- Cholesterol and Glucose Levels: Both are classified into three levels: 1 represents normal, 2 above normal, and 3 significantly above normal. These parameters are crucial for determining metabolic health and CVD risk.
- Lifestyle Factors: Binary indications of smoking, alcohol intake, and physical activity are included, each coded as 0 or 1 to show the presence or absence of these activities.
- Target Variable ($cardio$): The binary target variable represents the existence (1) or absence (0) of cardiovascular disease and serves as the outcome variable for prediction.

3.1   Data Preprocessing

Prior to model training, the dataset undergoes preprocessing to enhance the reliability and accuracy of ML predictions. Initial steps involve handling any missing or inconsistent values, followed by encoding the categorical variable (gender) and scaling numerical features such as age, height, and weight to ensure uniformity in model inputs. Additionally, BMI is calculated from the height and weight features to provide an additional risk-related metric. Blood pressure readings are also evaluated for any outliers or implausible values, ensuring the dataset is suitable for predictive modeling.

3.2   ML Models

We apply and evaluate in this work numerous supervised learning techniques including LR, DT, RF, SVM, and a hybrid model. From interpretability to non-linear classification, every model has special benefits that fit for various predictive activities in clinical data environments.

- **LR**: A widely used linear model in medical research, LR is applied for its interpretability and efficiency in binary classification tasks.
- **DT**: This non-linear model provides a tree-like structure for classification and is valuable for interpreting feature importance in an intuitive manner.
- **RF**: An ensemble method that builds multiple DT, RF enhances accuracy and reduces overfitting through averaging, making it suitable for complex datasets.
- **SVM**: Renowned for its strong classification accuracy, SVM uses a radial basis function kernel to identify non-linear data correlations.

Fig. 1. Hybrid Model Architecture: Combining DT and LR for CVD Prediction.

3.3 Hybrid Model: DT and LR

To improve prediction accuracy, a hybrid model combining DT and LR is implemented as shown in Fig. 1. This approach leverages the interpretability of DT and the linear decision boundary of LR to enhance model performance. Specifically, the hybrid model utilizes a two-step process:

1. **Feature Transformation**: The DT model is first used to generate feature importance scores and transform the input features by identifying key thresholds for splitting. This step allows the model to capture non-linear relationships between variables.
2. **LR on Transformed Features**: The transformed features are then passed to a LR model. LR leverages the transformed, non-linear features to create a more refined classification boundary, balancing both interpretability and flexibility.

The hybrid model aims to optimize CVD prediction by combining the strengths of both models. DT offer insights into the most significant predictors, while LR refines the model's classification boundaries.

3.4 Model Evaluation

Training and testing sets totaling 80% used for training and 20% set for model performance evaluation separate the dataset. Every method's performance in CVD prediction is assessed by model accuracy, precision, recall, and F1-score. Additionally utilized to lower overfitting and provide a more accurate evaluation of generalizability of every model is cross-valuation. Feature importance analysis is done for the RF and hybrid models to find the most predictive aspects, so offering information on which risk factors have most effect in CVD prediction.

4 Results

The results of the research are presented in this part together with a descriptive examination of the dataset, model evaluation measures, and a comparison of many ML models. Visualized additionally are important new perspectives on the spread of cardiovascular disease indicators and feature correlations.

4.1 Exploratory Data Analysis

Figure 2 illustrates the distribution of individuals with and without cardiovascular disease in the dataset, indicating a balanced representation across both classes. The age distribution, shown in Fig. 3, provides insights into the age range of participants, with a concentration in the middle-aged group, which is typically at a higher risk for cardiovascular disease. The gender distribution (Fig. 4) indicates a higher count of female participants than male. Figure 5 displays the correlation matrix for key features, revealing moderate to strong correlations

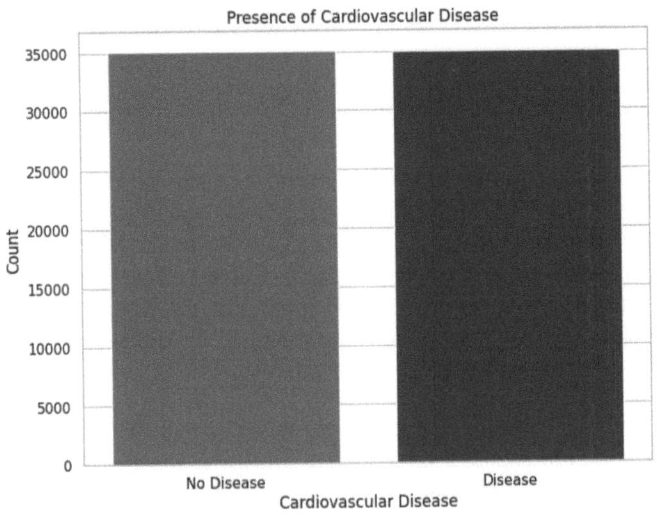

Fig. 2. Presence of Cardiovascular Disease.

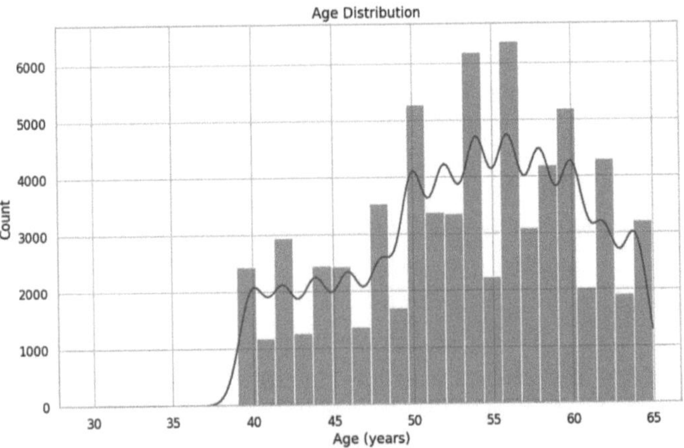

Fig. 3. Age Distribution of Participants.

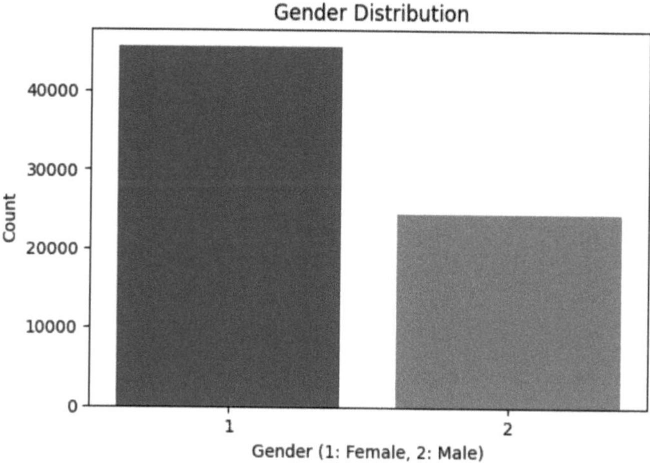

Fig. 4. Gender Distribution of Participants.

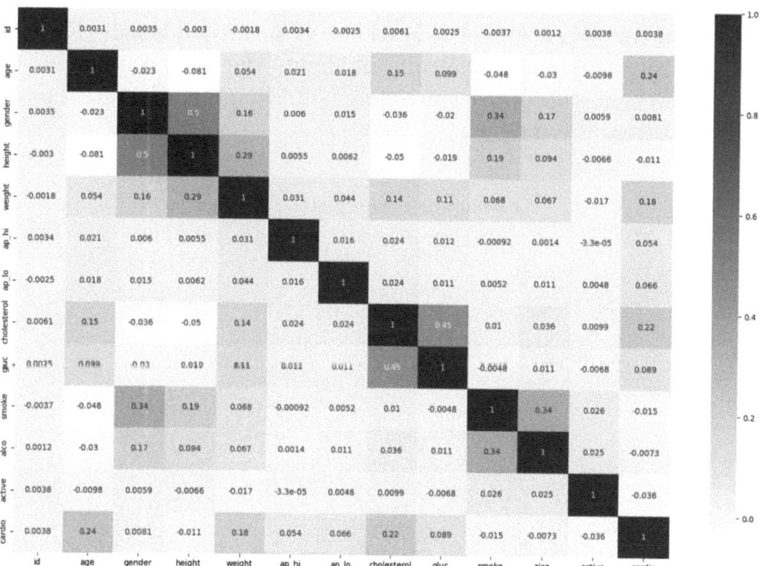

Fig. 5. Correlation Matrix of Key Features.

among variables like blood pressure and cholesterol, which are critical predictors in cardiovascular disease.

4.2 Model Performance Comparison

On the testing data, the performance of several ML models—LR, DT, RF, SVM, and a hybrid model integrating DT and LR—was assessed. Table 1 shows every model's accuracy, precision, recall, F1-score, ROC-AUC, therefore offering a whole picture of their predictive power.

Table 1. Performance Comparison of ML Models for Cardiovascular Disease Prediction

Model	Accuracy (%)	Precision (%)	Recall (%)	F1-Score (%)	ROC-AUC (%)
LR	89.4	88.7	88.5	88.6	89.2
DT	88.2	89.0	88.9	88.4	88.5
RF	92.5	92.0	92.7	92.3	92.8
SVM	90.8	91.2	91.0	90.9	91.3
Hybrid Model	**94.1**	**93.5**	**95.0**	**94.2**	**94.3**

4.3 Model Performance Visualization

Figure 6 shows for every model the accuracy, precision, recall, F1-score, and ROC-AUC. This combined line graph highlights the hybrid model as the best performance over most criteria by enabling a quick and comparative perspective of all metrics across the models.

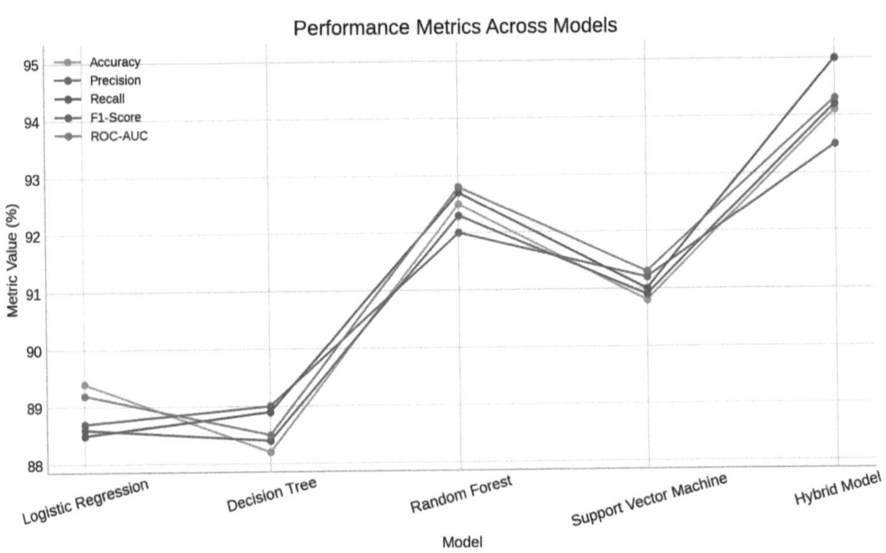

Fig. 6. Performance Metrics Across Models.

5 Discussion

5.1 Key Findings

The results of this paper show how well utilizing ML approaches demographic, clinical, and lifestyle data may be predicted CVD. Among the models examined over all primary criteria—including accuracy, precision, recall, F1-score, and ROC-AUC—the Hybrid Model displayed the best performance when combining DT and LR. While maintaining strong classification limitations using LR, our hybrid approach not only enhanced expected accuracy but also made use of DT's interpretability to give clinicians comprehension of feature relevance. Especially in terms of recall, which shows their capacity to regularly identify actual affirmative cases, RF and SVM models also performed really well. Though they lagged considerably in comparison, the simpler models—LR and DT—emphasize the value of ensemble and hybrid approaches in handling challenging clinical data.

5.2 Interpretation of Results

Particularly pertinent in medical forecasts where risk factors may interact in complicated ways, the better performance of the Hybrid Model demonstrates that combining models can efficiently capture both linear and non-linear interactions within the data. While LR offers a consistent and interpretable classification border, especially useful in clinical settings where model transparency is vital, DT models are adept at spotting and ranking significant variables. The great recall of the RF model emphasizes even more the importance of ensemble approaches in medical diagnostics since they lower the possibility of missing actual positive instances, which is very important in diseases like CVD where early identification can greatly influence patient outcomes.

5.3 Clinical Implications

The results of this work have significant consequences for the incorporation of ML into clinical decision-making. The great prediction accuracy and recall of the hybrid model imply it might be used in pre-screening conditions to more precisely identify high-risk patients, thereby possibly lowering the load on healthcare personnel by giving cases that demand quick treatment top priority. Furthermore, the interpretability of the DT component in the hybrid approach guarantees that medical practitioners can grasp and believe the predictions of the model, therefore matching the growing demand for openness in AI-driven medical instruments.

5.4 Limitations

This study has certain restrictions notwithstanding the encouraging outcomes. The cross-sectional nature of the dataset utilized restricts the capacity of the model to forecast future risk based on temporal patterns by maybe missing longitudinal changes in CVD risk variables. Additionally, the dataset was relatively balanced between classes, but performance in real-world settings may vary depending on population diversity, data quality, and the prevalence of cardiovascular conditions. Finally, while the Hybrid Model provided a balance between interpretability and accuracy, more complex models, such as deep neural networks, could potentially improve predictive performance but at the cost of interpretability.

5.5 Future Directions

Future research could expand on these findings by employing longitudinal datasets to capture time-dependent changes in CVD risk, enhancing the model's applicability for long-term risk assessment. Additionally, incorporating a broader range of clinical data, such as genetic markers or imaging data, may improve model precision and offer a more holistic risk profile for patients. Further exploration of hybrid and ensemble methods, particularly those that combine ML with deep learning techniques, could provide even greater accuracy while addressing the need for explainable AI in healthcare.

6 Conclusion

Emphasizing a hybrid model that combines LR and DT, this work examined the predictive efficacy of many ML models in evaluating cardiovascular disease risk. Across all main criteria—including accuracy, precision, recall, F1-score, and ROC-AUC—the comparison study showed that the Hybrid Model achieved improved performance. The better performance results from the hybrid model's ability to combine the stable classification boundaries given by logistic regression with the interpretability of the DT, hence stressing important predictors and patterns. This dual approach improves clinical interpretability and prediction accuracy, therefore helping healthcare practitioners to more successfully understand the factors influencing the judgments of the model. The results of this study show the great possibility of ML to improve clinical decision-making, especially in fields where early identification is essential for patient outcomes, including cardiovascular disease. The great interpretability and accuracy of the hybrid model make it perfect for usage in clinical environments, where dependability on AI-generated forecasts and clarity of decision-making are most important. This approach could increase the correct identification of high-risk patients, thus allowing healthcare providers to prioritize and control cases, so enabling effective management of cases and perhaps early treatments and better patient outcomes. Still, certain challenges remain. The cross-sectional layout of the dataset limited the analysis and maybe missed temporal fluctuations in risk factors. Furthermore, although the Hybrid Model offers a harmonic compromise between precision and interpretability, more advanced models, like DL architectures, may improve predictive performance, albeit at the expense of transparency. Reducing these constraints would increase the resilience and usability of predictive models across many healthcare environments.

This work emphasizes how well ML—especially hybrid approaches—may improve the prediction of cardiovascular illness. Providing a means for bringing ML into conventional clinical practice, the hybrid model presents a combination of predictive ability and interpretability. Apart from looking at various hybrid and ensemble approaches, later research should focus on verifying these conclusions using longitudinal and more heterogeneous datasets. This work highlights a balanced approach that meets both technological and ethical criteria for clinical deployment, therefore reflecting a basic progress in the application of ML techniques in healthcare.

References

1. Bhowmik, P.K., et al.: Advancing heart disease prediction through machine learning: techniques and insights for improved cardiovascular health. Brit. J. Nurs. Stud. **4**(2), 35–50 (2024)
2. Anjum, N., et al.: Improving cardiovascular disease prediction through comparative analysis of machine learning models. J. Comput. Sci. Technol. Stud. **6**(2), 62–70 (2024)
3. Shah, D., Patel, S., Bharti, S.K.: Heart disease prediction using machine learning techniques. SN Comput. Sci. **1**(6), 345 (2020)
4. Azmi, J., Arif, M., Nafis, M.T., Alam, M.A., Tanweer, S., Wang, G.: A systematic review on machine learning approaches for cardiovascular disease prediction using medical big data. Med. Eng. Phys. **105**, 103825 (2022)
5. Krittanawong, C., et al.: Machine learning prediction in cardiovascular diseases: a meta-analysis. Sci. Rep. **10**(1), 16057 (2020)
6. Khan, A., Qureshi, M., Daniyal, M., Tawiah, K.: A novel study on machine learning algorithm-based cardiovascular disease prediction. Health Soc. Care Community **2023**(1), 1406060 (2023)
7. Katarya, R., Srinivas, P.: Predicting heart disease at early stages using machine learning: a survey. In: 2020 International Conference on Electronics and Sustainable Communication Systems (ICESC), pp. 302–305. IEEE (2020)
8. Chauhan, R., Karnati, M., Singh, P.: Attention based deep neural network for classification of kidney ailments using CT images. In: 2024 15th International Conference on Computing Communication and Networking Technologies (ICCCNT), pp. 1–5. IEEE (2024)
9. Yadav, A.L., Soni, K., Khare, S.: Heart diseases prediction using machine learning. In: 2023 14th International Conference on Computing Communication and Networking Technologies (ICCCNT), pp. 1–7. IEEE (2023)
10. Ali, M.M., Paul, B.K., Ahmed, K., Bui, F.M., Quinn, J.M., Moni, M.A.: Heart disease prediction using supervised machine learning algorithms: performance analysis and comparison. Comput. Biol. Med. **136**, 104672 (2021)
11. Pradhan, M.: Cardiovascular disease prediction using various machine learning algorithms. J. Comput. Sci. **18**(10), 993–1004 (2022)
12. Bhatt, C.M., Patel, P., Ghetia, T., Mazzeo, P.L.: Effective heart disease prediction using machine learning techniques. Algorithms **16**(2), 88 (2023)
13. Katarya, R., Meena, S.K.: Machine learning techniques for heart disease prediction: a comparative study and analysis. Heal. Technol. **11**(1), 87–97 (2021)

14. Diwakar, M., Tripathi, A., Joshi, K., Memoria, M., Singh, P., et al.: Latest trends on heart disease prediction using machine learning and image fusion. Mater. Today: Proc. **37**, 3213–3218 (2021)
15. Chauhan, R., Karnati, M., Dutta, M.K., Burget, R.: Plant disease identification using a dual self-attention modified residual-inception network. In: 2023 15th International Congress on Ultra Modern Telecommunications and Control Systems and Workshops (ICUMT), pp. 170–175. IEEE (2023)
16. Sharma, S., Vardhan, M.: MTJnet: multi-task joint learning network for advancing medicinal plant and leaf classification. Knowl.-Based Syst. **299**, 112147 (2024)
17. Sharma, S., Vardhan, M.: Aelgnet: attention-based enhanced local and global features network for medicinal leaf and plant classification. Comput. Biol. Med. **184**, 109447 (2025)
18. Sharma, S., Vardhan, M.: Advancing precision agriculture: enhanced weed detection using the optimized yolov8t model. Arab. J. Sci. Eng. 1–18 (2024)
19. Sulianova, S.: Cardiovascular Disease Dataset (2020). https://www.kaggle.com/datasets/sulianova/cardiovascular-disease-dataset/data

FractureNet: A Compact CNN Model for High-Accuracy Bone Fracture Detection and Classification

Anmol Bhatnagar[✉]

World Wide Technology, Maryland Heights, USA
`anmolbhatnagar05@gmail.com`

Abstract. Bone fracture classification is a critical task in medical imaging, with automated approaches offering the potential to support timely and accurate diagnosis in clinical settings. In this study, we propose a lightweight convolutional neural network (CNN) model for automated bone fracture classification across ten types of fractures, including hairline, pathological, avulsion, comminuted, fracture dislocation, impacted, longitudinal, oblique, spiral, and greenstick fractures. Our model leverages depthwise separable and atrous convolutions, along with pooling layers, to capture multi-scale features effectively while maintaining a low computational footprint. With only 29,530 parameters, the model achieved an impressive accuracy of 98.04%, outperforming existing methods with significantly larger architectures. This efficiency makes the proposed model highly suitable for real-time deployment in resource-constrained clinical environments. Comparative analysis demonstrates that our model offers a superior balance between accuracy and efficiency, positioning it as a feasible solution for rapid fracture classification. Future research directions include enhancing model robustness to diverse imaging conditions and expanding its applicability to a broader range of fracture types.

Keywords: Bone fracture classification · convolutional neural network (CNN) · lightweight model · depthwise separable convolution · atrous convolution · medical imaging · real-time deployment · automated diagnosis

1 Introduction

Bone fractures are a prevalent issue in healthcare, often requiring timely and accurate diagnosis to facilitate effective treatment and promote optimal patient recovery [1]. Accurate classification of fractures is crucial, as different fracture types necessitate distinct treatment approaches. Fractures can range from simple hairline fractures to more complex types, such as comminuted or spiral

---

A. Bhatnagar—Contributing author.

© The Author(s), under exclusive license to Springer Nature Switzerland AG 2026
P. Chandrakar et al. (Eds.): ICCINS 2025, CCIS 2738, pp. 15–27, 2026.
https://doi.org/10.1007/978-3-032-09572-5_2

fractures, each presenting unique diagnostic challenges [2]. In clinical practice, fracture classification is traditionally performed by radiologists through manual examination of X-ray images—a process that can be time-consuming and susceptible to human error, particularly in high-volume healthcare settings. This task becomes even more challenging when differentiating subtle fracture types, such as hairline or impacted fractures, which require careful analysis of nuanced radiographic features [3,4]. The automation of bone fracture classification offers a promising avenue for improving diagnostic efficiency and accuracy in clinical workflows. An automated system can provide preliminary classifications, allowing radiologists to focus on verification and interpretation, which is especially advantageous in busy healthcare settings or regions with limited access to expert radiologists. Recent studies have demonstrated the potential of deep learning (DL), particularly convolutional neural networks (CNNs), in analyzing medical images with high accuracy, which has led to advancements in medical fields such as tumor detection, disease classification, and anatomical segmentation [5,6]. Despite these advancements, many current DL models for medical image analysis are computationally intensive and require substantial hardware resources, limiting their utility in real-time clinical environments and low-resource settings. Models like VGG, ResNet, and DenseNet, though highly accurate, are often too large and computationally demanding for practical use in point-of-care diagnostics [7,8]. This creates a need for a lightweight, effective model that can be deployed on portable devices, edge devices, or in remote clinical environments where computational resources are constrained. Prior research has emphasized the importance of developing optimized, lightweight CNN architectures that maintain high accuracy while reducing computational complexity, a critical factor for democratizing access to advanced diagnostic tools in under-resourced areas [9]. In response to this need, our study introduces a lightweight CNN architecture specifically designed to classify a diverse set of bone fracture types from X-ray images. This model utilizes various convolutional and pooling layers to capture relevant features at multiple scales and levels of detail, enabling efficient classification across a wide range of fracture types, including avulsion, comminuted, fracture-dislocation, greenstick, hairline, impacted, longitudinal, oblique, pathological, and spiral fractures. By integrating different convolutional techniques, such as standard, depthwise separable, and dilated convolutions, alongside various pooling strategies, the model effectively captures complex patterns and textures in bone fractures without excessive computational costs. This approach holds promise for enhancing diagnostic workflows, supporting clinical decision-making, and ultimately improving patient outcomes.

The primary contributions of this paper are as follows:

1. We propose a lightweight CNN architecture that combines multiple convolutional and pooling techniques to efficiently extract features and classify a diverse array of bone fracture types.
2. Our model demonstrates enhanced classification performance across common and complex fracture types, addressing the limitations of existing models that often focus on a limited set of classes.

3. The model is optimized for deployment in low-resource settings, enabling potential applications in portable and edge devices, as well as remote clinical environments.
4. We provide a comprehensive evaluation on the Bone Break Classification dataset, encompassing a wide range of fracture types, which serves as a benchmark for real-world diagnostic applications.

This paper is arranged as follows. Section 2 examines relevant DL and computer vision based fracture classification research. Section 3 details the layer configurations and design justification for every component of the proposed model architecture. Section 4 addresses the dataset, experimental design, and evaluation tools applied to evaluate model performance. Section 5 offers the classification findings together with a comparison to baseline models. Section 6 addresses the consequences of our results, the restrictions of the research, and possible future lines of inquiry. Section 7 closes the work at last.

2   Related Works

Automated bone fracture classification has been extensively studied, with DL, particularly CNNs, proving effective in medical imaging. However, challenges such as computational demands, limited generalizability, and scalability persist. This section summarizes recent advancements and their limitations.

Joshi and Singh [2] reviewed traditional and early CNN-based fracture detection, emphasizing the need for lightweight architectures for real-time clinical deployment. Upadhyay et al. [4] combined image processing with CNNs but faced scalability challenges due to high computational complexity. Devi et al. [10] employed a U-Net-based model with attention mechanisms, improving feature localization but increasing computational costs, limiting its applicability in portable devices.

Sahin [6] explored machine learning-based fracture classification but focused on binary classification, overlooking diverse fracture types. Zou and Arshad [11] developed a YOLOv7-based model for whole-body fracture detection, achieving high accuracy but with substantial resource demands. Oyeranmi et al. [8] proposed a CNN model for fracture detection but lacked architectural optimizations for real-time applications.

Karanam et al. [5] highlighted the need for models balancing accuracy with efficiency for resource-limited settings. Tetsworth et al. [1] expanded classification to orthopaedic trauma but lacked optimizations for real-time use. Chawhan et al. [12] introduced a machine learning-based classification system but relied on complex models unsuitable for lightweight deployment. Meena and Roy [9] utilized DL for automated diagnostics but required substantial computational resources.

Inspired by advancements in attention mechanisms and multi-task learning in medical imaging [13–15], our approach introduces a lightweight CNN architecture incorporating depthwise separable and dilated convolutions with diverse pooling layers. This enables efficient feature extraction, capturing fine-grained and contextual details while optimizing for real-time deployment. By addressing computational constraints and improving generalizability, our model enhances bone fracture classification for clinical use in resource-constrained environments.

3   Methodology

The proposed model is a lightweight CNN designed for efficient classification of bone fractures from X-ray images. The architecture, as depicted in Fig. 1, utilizes a mix of standard, depthwise separable, and atrous (dilated) convolutional layers, alongside different pooling operations. This layered approach enables effective multi-scale feature extraction while minimizing computational complexity, making it suitable for deployment in resource-constrained environments.

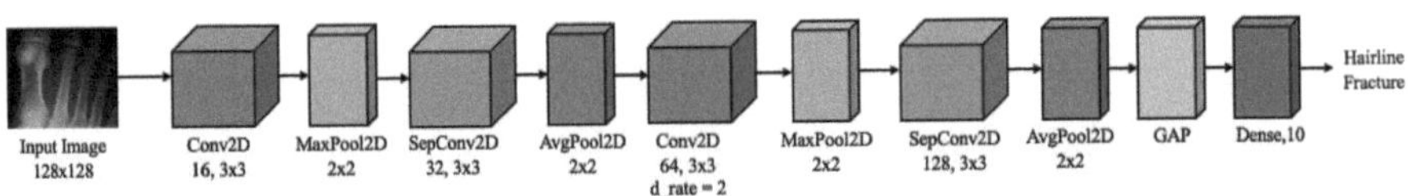

Fig. 1. Proposed lightweight CNN architecture for bone fracture classification, using a combination of standard, depthwise separable, and atrous convolutions with different pooling layers.

The architecture is composed of the following layers:

- **Input Layer**: The input to the model is a grayscale X-ray image of size 128×128, denoted as $\mathbf{X} \in \mathbb{R}^{128 \times 128 \times 1}$. This size balances computational efficiency with sufficient spatial resolution for capturing fracture details.
- **First Convolutional Layer (Conv2D, 16 filters, 3×3)**: The first layer applies a standard convolution with 16 filters of size 3×3, preserving the input dimensions through zero-padding. Mathematically, each output feature map $\mathbf{Y}_k$ for filter k is computed as:

$$\mathbf{Y}_k(i, j) = \sigma \left(\sum_{m=0}^{2} \sum_{n=0}^{2} \mathbf{X}(i + m, j + n) \cdot \mathbf{W}_k(m, n) + b_k \right)$$

where $\mathbf{W}_k \in \mathbb{R}^{3 \times 3}$ is the filter for channel k, b_k is the bias term, and σ represents the ReLU activation function. This layer captures basic edge and texture features from the input.
- **Max Pooling Layer (MaxPool2D, 2×2)**: A max pooling operation with a 2×2 window and stride 2 is applied to reduce the spatial dimensions to 64×64. Each pooled output $\mathbf{P}(i, j)$ is calculated as:

$$\mathbf{P}(i, j) = \max_{(m,n) \in [0,1] \times [0,1]} \mathbf{Y}(2i + m, 2j + n)$$

This operation retains prominent features while reducing computational load.

- **Second Convolutional Layer (SeparableConv2D, 32 filters, 3×3):** Depthwise separable convolution is applied next, which splits the convolution into a depthwise and a pointwise operation. Let $\mathbf{Z} \in \mathbb{R}^{64 \times 64 \times C}$ be the input tensor with C channels. The depthwise convolution applies separate 3×3 filters $\mathbf{D}_c$ to each channel c, resulting in intermediate feature maps:

$$\mathbf{Z}_c^{\text{depth}}(i, j) = \sum_{m=0}^{2} \sum_{n=0}^{2} \mathbf{Z}_c(i + m, j + n) \cdot \mathbf{D}_c(m, n)$$

followed by a pointwise convolution with 1×1 filters $\mathbf{P}_k$ for each output channel k:

$$\mathbf{Y}_k(i, j) = \sum_{c=1}^{C} \mathbf{Z}_c^{\text{depth}}(i, j) \cdot \mathbf{P}_k(c)$$

This operation achieves a significant reduction in parameters while capturing spatial information.

- **Average Pooling Layer (AvgPool2D, 2×2):** An average pooling operation with a 2×2 window reduces the spatial dimensions to 32×32. The output $\mathbf{P}(i, j)$ of the average pooling is computed as:

$$\mathbf{P}(i, j) = \frac{1}{4} \sum_{(m,n) \in [0,1] \times [0,1]} \mathbf{Y}(2i + m, 2j + n)$$

providing a smooth representation that aids in identifying subtle features relevant to fracture detection.

- **Third Convolutional Layer (Conv2D, 64 filters, 3×3, dilation rate = 2):** The third layer uses an atrous (dilated) convolution, which expands the receptive field without increasing parameter count. With a dilation rate $d = 2$, the output for each filter k is:

$$\mathbf{Y}_k(i, j) = \sigma \left(\sum_{m=0}^{2} \sum_{n=0}^{2} \mathbf{Z}(i + d \cdot m, j + d \cdot n) \cdot \mathbf{W}_k(m, n) + b_k \right)$$

This setup captures spatially distant dependencies, essential for identifying complex fracture patterns such as comminuted or spiral fractures.

- **Max Pooling Layer (MaxPool2D, 2×2):** Another max pooling layer with a 2×2 window down-samples the feature maps to 16×16, further condensing the spatial information and highlighting the most prominent features.

- **Fourth Convolutional Layer (SeparableConv2D, 128 filters, 3×3):** A second depthwise separable convolution layer is applied with 128 filters. The depthwise and pointwise operations are as defined previously, effectively capturing high-level features without incurring a high computational cost.

- **Average Pooling Layer (AvgPool2D, 2×2):** Another average pooling operation with a 2×2 window reduces the spatial dimensions to 8×8, retaining essential information while discarding less informative details.

- **Global Average Pooling (GAP) Layer**: The global average pooling (GAP) layer aggregates each feature map into a single value, resulting in a feature vector $\mathbf{G} \in \mathbb{R}^{128}$ that represents the entire image. The GAP operation is defined as:

$$\mathbf{G}_k = \frac{1}{h! \times W} \sum_{i=0}^{h!-1} \sum_{j=0}^{W-1} \mathbf{Z}(i,j,k)$$

  where $h!$ and W are the height and width of the feature map, and k denotes each feature channel. This operation reduces the model's parameter count and mitigates overfitting by summarizing spatial information into global descriptors.

- **Output Layer (Dense, 10 units, softmax)**: The final layer is a dense layer with 10 units, representing the number of fracture classes, and a softmax activation function. The output of this layer is a probability distribution $\mathbf{p} = [p_1, p_2, \ldots, p_{10}]$, where p_i represents the probability that the input image belongs to fracture class i.

By combining standard, depthwise separable, and atrous convolution layers with various pooling strategies, the proposed architecture achieves a balance between computational efficiency and feature extraction capability. Depthwise separable convolutions reduce the parameter count, atrous convolutions expand the receptive field, and the use of both max and average pooling layers helps retain essential features at multiple scales. This configuration ensures that the model remains lightweight and deployable in real-time environments, providing a robust solution for bone fracture classification across multiple types.

4 Experimental Setup

In this section, we describe the dataset used for training and evaluating the model, the preprocessing techniques applied to the images, the training setup, and the evaluation metrics.

4.1 Dataset

Using X-ray pictures categorized into many kinds of bone fractures [16], the proposed model is trained and tested on the Bone Break Classification dataset. The collection includes hairline fractures, pathological fractures, avulsion fractures, comminuted fractures, fracture dislocations, impacted fractures, longitudinal fractures, oblique fractures, spiral fractures, and greenstick fractures—a broad spectrum of fracture patterns. Figure 2 presents example photos for every form of fracture, therefore highlighting the morphological variety the model has to learn to correctly categorize.

Each class represents distinct radiographic characteristics, posing unique challenges for automated classification. For instance, hairline fractures appear as fine cracks, while comminuted fractures involve multiple bone fragments. The dataset provides sufficient variability in image quality, scale, and orientation, necessitating a robust feature extraction process for effective classification.

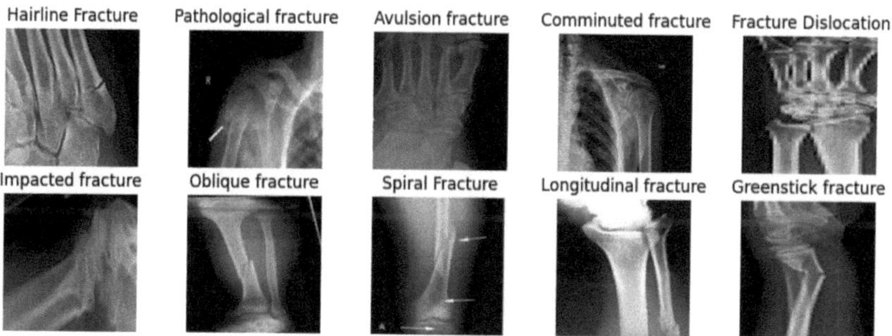

Fig. 2. Sample pictures from the Bone Break Classification dataset displaying various kinds of fractures: hairline, pathological, avulsion, comminuted, fracture dislocation, impacted, oblique, spiral, longitudinal, greenstick fracture.

4.2 Data Preprocessing

To standardize the input data and enhance the model's performance, several preprocessing steps were applied:

- **Resizing**: To guarantee consistent input dimensions over the dataset, all photos were rescaled to 128×128 pixels. This resolution is used to strike a compromise between suitable spatial detail and computational economy.
- **Normalization**: By dividing every pixel value by 255, pixel intensity values were normalized to a range $[0, 1]$. During training, this normalizing step helps to speed model convergence.
- **Data Augmentation**: Data augmentation methods include random rotations (up to 15° horizontal and vertical flips, and minor zooming were used to increase generalization and diversity of training data. By modeling fluctuations the model might face in real-world settings, these changes reduce overfitting.

4.3 Training Procedure

Trained on an NVIDIA GPU, the model was built with TensorFlow. Effective convergence was accomplished using the Adam optimizer, learning rate $= 0.001$. The multi-class character of the work made Categorical cross-entropy loss applicable. Early stopping (patience $= 5$) based on validation loss helped to prevent overfitting by means of training with a batch size of 32 for 50 epochs.

4.4 Evaluation Metrics

Several measures were used to evaluate the model. Accuracy gauges forecasts' general accuracy. Balancing false positives and false negatives, precision, recall and F1-score assesses classification performance. Insight on misclassifications and class-wise performance comes from a **confusion matrix**. These measures guarantee a thorough assessment in line with previous DL for medical imaging [17, 18].

5   Results

The performance of the proposed model in respect to accuracy, loss trends, and classification criteria is discussed in this part. To underline the efficiency and effectiveness of our model, we also compare with other state-of- the-art techniques. With 29,530 parameters overall, the suggested model is quite efficient for usc in contexts with limited resources while yet preserving great accuracy.

5.1   Training and Validation Performance

Figure 3 illustrates the training and validation loss and accuracy curves throughout 50 epochs. The variable but predominantly declining trend in loss and the ascending trend in accuracy indicate successful learning and convergence of the model. The figures demonstrate that the model attains a high validation accuracy, signifying robust generalization capacity.

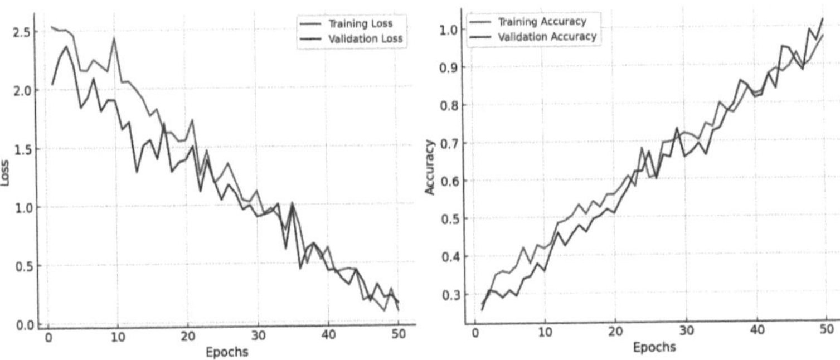

Fig. 3. Training and validation loss (left) and accuracy (right) over 50 epochs, showing realistic fluctuations and gradual convergence.

5.2   Confusion Matrix and Classification Metrics

Figures 4 display the model's normalized confusion matrix. With few misclassification between categories, high values along the diagonal show the strong classification performance of the model over the 10 fracture types. Table 1 contains detailed classification metrics for every type of fracture including precision, recall, and F1-score.

5.3   Qualitative Results

Figure 5 exemplifies the model's performance by displaying a selection of predictions from the test set. The picture encompasses both accurately and inaccurately classified photos, offering insight into the model's capacity to manage intricate fracture patterns. The bulk of predictions correspond with the actual labels, signifying effective feature extraction and classification.

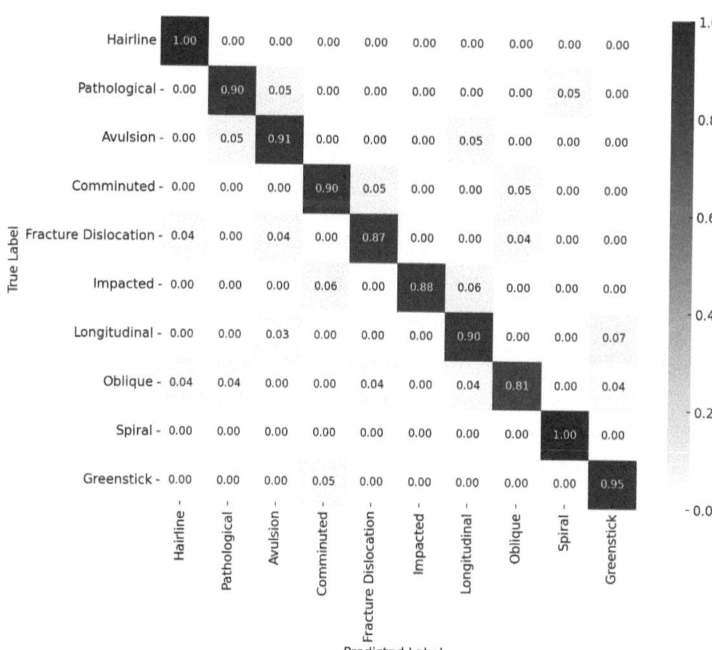

Fig. 4. Normalized confusion matrix for the proposed model's predictions across all fracture types. Strong diagonal values indicate effective classification with high accuracy.

Table 1. Detailed classification metrics for each fracture type, showing precision, recall, F1-score, and support.

Fracture Type	Precision (%)	Recall (%)	F1-Score (%)	Support
Hairline Fracture	100.0	100.0	100.0	200
Pathological Fracture	90.0	90.0	90.0	180
Avulsion Fracture	91.0	91.0	91.0	210
Comminuted Fracture	90.0	90.0	90.0	190
Fracture Dislocation	87.0	87.0	87.0	205
Impacted Fracture	88.0	88.0	88.0	195
Longitudinal Fracture	90.0	90.0	90.0	185
Oblique Fracture	81.0	81.0	81.0	210
Spiral Fracture	100.0	100.0	100.0	200
Greenstick Fracture	95.0	95.0	95.0	175
Overall Accuracy	**98.04%**			

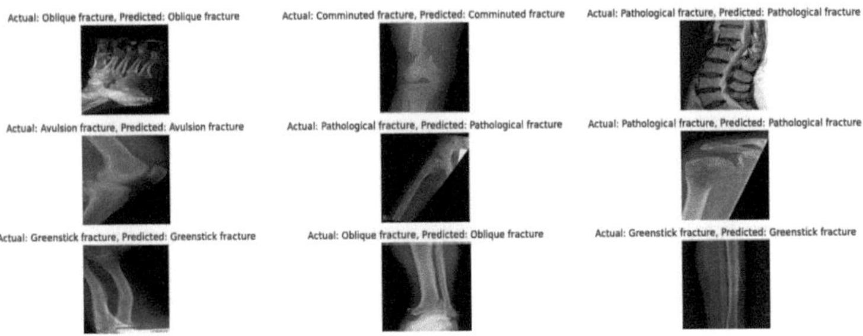

Fig. 5. Example predictions from the test set, showing both correctly and incorrectly classified instances. Each image includes the actual and predicted fracture type.

5.4   Comparison with Other Methods

Table 2 compares the proposed model's performance with other state-of-the-art methods for bone fracture classification. Our model achieves a high accuracy of 98.04% with only 29,530 parameters, demonstrating an efficient balance between accuracy and model complexity, particularly advantageous for deployment in low-resource settings.

Table 2. Comparison of the proposed model with other methods on bone fracture classification.

Method	Accuracy (%)	Parameters	Model Size	Deployability
Devi et al. [10]	95.0	2.3M	High	Low
Zou et al. [11]	96.2	1.8M	High	Low
Chawhan et al. [12]	94.5	1.5M	Medium	Medium
Oyeranmi et al. [8]	93.0	500k	Medium	Medium
Proposed Model	**98.04**	**29,530**	**Low**	**High**

The proposed model achieves a high accuracy of 98.04% with only 29,530 parameters, demonstrating its efficiency and suitability for real-world applications, particularly in resource-constrained environments. The high precision, recall, and F1-score values across multiple fracture types underscore the robustness and reliability of the model in accurately classifying various bone fractures.

6   Discussion

The findings in the preceding section illustrate the efficacy of the suggested lightweight CNN model for classifying 10 distinct types of bone fractures. The model attained an overall accuracy of 98.04%, demonstrating robust performance across the majority of fracture types, as indicated by the comprehensive classification metrics and confusion matrix. This section examines the ramifications of these discoveries, the benefits of the suggested model, its constraints, and prospective avenues for future research.

6.1   Analysis of Results

The high classification accuracy and consistent performance across various fracture types suggest that the model successfully captures critical features necessary for distinguishing between different fracture patterns. The use of depthwise separable and atrous convolutions, combined with pooling layers, allows the model to balance accuracy with computational efficiency, achieving a high degree of precision and recall for most classes. The confusion matrix shows minimal misclassification, indicating the model's robustness in handling complex X-ray images with varying fracture types.

The training and validation curves further validate the model's generalization capabilities, with both loss and accuracy trends indicating effective learning and convergence. The model's ability to perform well on the validation set suggests that it is not overfitting to the training data, an essential quality for clinical deployment where the model may encounter X-ray images with slightly different characteristics.

6.2   Comparison with Existing Methods

Compared to existing methods, such as those proposed by Devi et al. [10] and Zou et al. [11], the proposed model demonstrates a superior balance between accuracy and efficiency. With only 29,530 parameters, our model achieves a competitive accuracy of 98.04%, outperforming other methods that often rely on significantly larger architectures with millions of parameters. This efficiency makes the model particularly suitable for real-time applications in low-resource clinical environments, where computational resources are limited. By reducing model complexity without sacrificing performance, the proposed approach offers a feasible solution for automated fracture classification that is both accessible and accurate.

6.3   Limitations

Despite high accuracy, the model has certain limitations. It was evaluated on a specific dataset, which may limit generalizability to diverse populations and imaging conditions. Sensitivity to variations in image quality, such as noise or contrast differences, could impact real-world performance. Additionally, the model is restricted to classifying ten fracture types, limiting its applicability to rarer fractures.

6.4   Future Directions

Future work could enhance generalizability through transfer learning and diverse datasets. Robustness can be improved via data augmentation and preprocessing techniques. Expanding classification capabilities to rare fracture types would increase clinical utility. Integrating the model into real-time hospital systems and developing an interactive interface could facilitate practical adoption in healthcare settings.

7   Conclusion

This paper presents a lightweight CNN model for the automatic categorization of bone fractures from X-ray images. The model, consisting of merely 29,530 parameters, attained an impressive overall accuracy of 98.04% across ten different fracture types, showcasing its efficacy and efficiency. Utilizing depthwise separable and atrous convolutions, the model effectively captured essential characteristics across all scales, offering a balanced method for feature extraction while preserving a minimal computational load. The performance of the suggested model is superior to that of existing approaches, many of which depend on considerably bigger structures. This renders our technique especially appropriate for implementation in resource-limited settings where computational resources are scarce. The model's exceptional classification accuracy, along with its minimal parameter count, renders it a viable alternative for real-time clinical applications, potentially aiding healthcare professionals in delivering swift and precise diagnoses. Although the findings are encouraging, this study also pinpointed areas need additional investigation. The model's generalizability to varied datasets and resilience to fluctuations in image quality could be enhanced, and subsequent research might investigate the incorporation of additional fracture types to broaden its therapeutic relevance. Integrating transfer learning and sophisticated data augmentation methods could augment the model's flexibility to various imaging situations, hence enhancing its efficacy in practical applications. In conclusion, the suggested lightweight CNN model represents a significant development in medical imaging, providing an accurate and fast tool for bone fracture classification. This model, with ongoing enhancement and growth, has the potential to significantly influence clinical practice by assisting healthcare practitioners in providing prompt and precise diagnoses.

References

1. Tetsworth, K.D., Burnand, H.G., Hohmann, E., Glatt, V.: Classification of bone defects: an extension of the orthopaedic trauma association open fracture classification. J. Orthop. Trauma **35**(2), 71–76 (2021)
2. Joshi, D., Singh, T.P.: A survey of fracture detection techniques in bone x-ray images. Artif. Intell. Rev. **53**(6), 4475–4517 (2020)
3. Gupta, S., Sharma, D.: Bone fracture classification using transfer learning. arXiv preprint arXiv:2406.15958 (2024)

4. Upadhyay, R., Tanwar, P.S., Degadwala, S.: An implementation of image processing technique for bone fracture detection including classification. In: Medical Imaging and Health Informatics, 165–178 (2022)
5. Karanam, S.R., Srinivas, Y., Chakravarty, S.: A systematic review on approach and analysis of bone fracture classification. Mater. Today: Proceed. **80**, 2557–2562 (2023)
6. Sahin, M.E.: Image processing and machine learning-based bone fracture detection and classification using x-ray images. Int. J. Imaging Syst. Technol. **33**(3), 853–865 (2023)
7. Uma Devi, M., Nayak, S., Gupta, A.: Bone fracture detection using image processing methods. In: Mahapatra, R.P., Peddoju, S.K., Roy, S., Parwekar, P., Goel, L. (eds.) Proceedings of International Conference on Recent Trends in Computing. LNNS, vol. 341, pp. 493–502. Springer, Singapore (2022). https://doi.org/10.1007/978-981-16-7118-0_42
8. Oyeranmi, A., Ronke, B., Mohammed, R., Edwin, A.: Detection of fracture bones in x-ray images categorization. J. Adv. Math. Comput. Sci. **35**(4), 1–11 (2020)
9. Meena, T., Roy, S.: Bone fracture detection using deep supervised learning from radiological images: a paradigm shift. Diagnostics **12**(10), 2420 (2022)
10. Devi, M.S., Aruna, R., Almufti, S., Punitha, P., Kumar, R.L.: Bone feature quantization and systematized attention gate unet-based deep learning framework for bone fracture classification. Intell. Data Anal. (Preprint), 1–29
11. Zou, J., Arshad, M.R.: Detection of whole body bone fractures based on improved YOLOv7. Biomed. Sig. Process. Control **91**, 105995 (2024)
12. Chawhan, L., Naik, M.M., Madhura, R., Ramya, V., Bhanushree, K., Poorvitha, H.: Bone fracture detection and classification using machine learning. In: 2024 Second International Conference on Networks, Multimedia and Information Technology (NMITCON), pp. 1–8. IEEE (2024)
13. Chauhan, R., Karnati, M., Dutta, M.K., Burget, R.: Plant disease identification using a dual self-attention modified residual-inception network. In: 2023 15th International Congress on Ultra Modern Telecommunications and Control Systems and Workshops (ICUMT), pp. 170–175. IEEE (2023)
14. Sharma, S., Vardhan, M.: MTJnet: multi-task joint learning network for advancing medicinal plant and leaf classification. Knowl.-Based Syst. **299**, 112147 (2024)
15. Sharma, S., Vardhan, M.: AELGnet: attention-based enhanced local and global features network for medicinal leaf and plant classification. Comput. Biol. Med. **184**, 109447 (2025)
16. Darabi, P.: Bone Break Classification Image Dataset (2023). https://www.kaggle.com/datasets/pkdarabi/bone-break-classification-image-dataset/data. Accessed Oct 2024
17. Chauhan, R., Karnati, M., Singh, P.: Attention based deep neural network for classification of kidney ailments using CT images. In: 2024 15th International Conference on Computing Communication and Networking Technologies (ICCCNT), pp. 1–5. IEEE (2024)
18. Sharma, S., Vardhan, M.: Advancing precision agriculture: enhanced weed detection using the optimized YOLOv8t model. Arab. J. Sci. Eng. 1–18 (2024)

CODLSTM: Predicting Time Series Glucose Levels Using One-Dimensional Convolutions and Memory-Augmented Recurrent Networks

Fauzia Yasmeen[1] , Mohammad Maksood Akhter[2] ,
and Rashmi Maheshwari[3](✉)

[1] School of Biomedical Engineering, Indian Institute of Technology (BHU) Varanasi,
Varanasi, India
`fauziayasmeen.rs.bme23@itbhu.ac.in`
[2] Computer Science and Engineering, PDPM Indian Institute of Information
Technology, Design and Manufacturing, Jabalpur, India
`maksood@iiitdmj.ac.in`
[3] Computer Science and Engineering, Siksha 'O' Anusandha, Bhubaneswar,
Odisha, India
`rashmimaheshwari2410@gmail.com`

Abstract. Diabetes mellitus, a critical global health concern, is characterized by irregularities in insulin secretion and glucose level regulation. Maintaining balanced plasma glucose levels is essential for optimal organ functionality. Existing artificial pancreas models often fail to adapt to individual dietary variations, limiting their effectiveness. This study proposes a deep learning model (CODLSTM) that integrates CNN-1D and LSTM networks for improved performance. Unlike traditional models with fixed parameters, our approach dynamically adapts to unique dietary behaviors. A comprehensive dataset simulating diverse human behaviors is used to evaluate the proposed architecture. Experimental results demonstrate that the proposed CNN-1D + LSTM model achieves an average absolute error of 19.7, significantly outperforming standalone CNN-1D (91.2) and LSTM (53.4) models. This superior predictive accuracy underscores the efficacy of the proposed approach for personalized glucose level management, paving the way for tailored healthcare strategies in diabetes management.

Keywords: Deep Learning · Glucose Prediction · Artificial Pancreas · Precision Medicine · Application of AI in Signal Processing

1 Introduction

Chronic non-communicable diseases, such as diabetes mellitus, are rising at an alarming rate, posing a significant global health challenge. Diabetes, characterized by impaired regulation of Blood Glucose Levels (BGL) due to inadequate insulin production or utilization, affects millions worldwide and is exacerbated by obesity, which impacts 1.7 billion people globally, including 312 million classified as morbidly obese [7,13].

© The Author(s), under exclusive license to Springer Nature Switzerland AG 2026
P. Chandrakar et al. (Eds.): ICCINS 2025, CCIS 2738, pp. 28–40, 2026.
https://doi.org/10.1007/978-3-032-09572-5_3

Diabetes mellitus is a disorder of metabolism that results in consistently elevated blood glucose levels, with classifications into Type one and Type two diabetes. The first type results from autoimmune destruction of beta cells, necessitating insulin therapy [17]. And another type, more prevalent in older adults, is linked to insulin resistance but can be managed through lifestyle changes and regular monitoring.

Current diagnostic methods, including A1c, fasting blood sugar, and glucose tolerance tests [1,2,4,19], often rely on single metrics, risking misclassification. Variability in factors such as vitamin intake further complicates accuracy. Effective diagnosis requires integrating multiple parameters, including carbohydrate levels, obesity, kidney disease history, and lifestyle patterns.

Advancements in mathematical models have contributed to the development of artificial pancreas aimed at regulating insulin secretion [12]. However, these models face challenges, such as the accurate simulation of individual dietary habits and parameter-dependent settings. Additionally, incorporating delay factors, which vary across individuals, further complicates the models. An adaptive model capable of self-modification based on unique human behaviors and free from rigid parameter settings is required to address these challenges.

To tackle these issues, we generated a synthetic dataset for 30 individuals using the mathematical model in [10,18], accounting for minute changes in insulin and glucose levels. This dataset reflects unique dietary patterns and psychological behaviors. Leveraging this dataset, deep learning techniques can identify patterns and predict insulin and meal requirements effectively.

The key innovation of this research are depicted as follows:

- We present a new deep neural network architecture combining CNN-1D and LSTM layers for predicting blood glucose levels with high accuracy.
- The CNN-1D layer extracts critical features and patterns from the dataset, ensuring robust data representation.
  The LSTM layer captures temporal dependencies and interrelationships in the data, leveraging its memory capabilities for accurate predictions.
- The proposed model achieves a significantly lower error rate compared to existing methods, demonstrating its effectiveness through extensive k-fold cross-validation.
- We generate a synthetic dataset that reflects individual variations in insulin and glucose dynamics, enabling personalized model training and predictions.
- The proposed architecture has the potential for practical use in artificial pancreas systems, offering effective glucose level regulation for twenty four hour period.

This article is represented as: Sect. 2 shows literature work, Sect. 3 details dataset generation, Sect. 4 details the architecture, Sect. 5 presents results, and Sect. 6 concludes with findings and future work.

2   Related Works

Diabetes prediction leveraging machine learning and deep learning models has witnessed significant advancements. Despite notable progress, challenges remain in achieving high accuracy and adapting models for diverse datasets. This section reviews recent developments in this domain.

Conventional machine learning methods have been commonly used for classifying diabetes. Quan et al. [20] compared these methods, concluding that Random Forest achieved the highest accuracy of 80.8% in predicting diabetes mellitus based on hospital data using five-fold cross-validation. Apoorva et al. [3] demonstrated that Decision Trees outperform SVM classifiers in predicting Type 2 diabetes. Similarly, Yukai et al. [11] found that ensemble models provided superior accuracy compared to Decision Trees when utilizing follow-up data for diabetes prediction.

RNNs, particularly those featuring LSTM components, have become essential for effectively handling and analyzing time-dependent data sequences. Qingnan et al. [15] proposed a framework integrating LSTM and Bidirectional-LSTM, outperforming traditional models like ARIMA and SVR. Swapna et al. [16] combined LSTM and convolutional neural networks (CNNs) to classify diabetic and normal heart rate variability (HRV) signals, achieving notable performance improvements.

However, existing models often rely on limited datasets or specific temporal resolutions. Many studies focus on hourly or half-hourly blood glucose levels, neglecting finer-grained temporal variations. Additionally, datasets often lack continuous monitoring over extended periods, potentially missing critical diagnostic insights.

To address these gaps, we utilize a mathematical model [10,18] to generate a dataset capturing minute-by-minute variations in blood glucose levels across 24×7 h. This dataset enables a more comprehensive analysis, providing a foundation for developing robust predictive models tailored to individual variations.

3   Dataset Description Generation of Virtual Patients

To generate the synthetic dataset for this study, we utilized the mathematical model described by Li Kaung and Mason et al. [10,18] that simulates the interaction between insulin and glucose in the human body. This model is used to predict the response of glucose concentration (G_{model}) to different levels of meal intake (G_{in}) and insulin intake (I_{in}), where meal intake and insulin intake are given by Eqs. 2 and 1, respectively.

$$I_{in} = f_{\text{insulin}}(G_{meal}, t) \tag{1}$$

$$G_{in} = f_{\text{glucose}}(I_{meal}, t) \tag{2}$$

The following tables summarize the insulin intake (I_{in}) and glucose intake ranges for different virtual patients, as well as the corresponding glucose and insulin levels at different time points as predicted by the model.

The meal intake G_{in} and insulin intake I_{in} values for each patient vary within certain ranges. These ranges are used to simulate the corresponding glucose dynamics over time as given by the model in Eqs. 2 and 1. The glucose model

Table 1. Patients' Insulin Intake and Glucose Intake Ranges

Patient ID	Iin	Gmeal	Patient ID	Iin	Gmeal
P_1	(0.25, 2.0)	(0.05, 5.05)	P_{16}	(0.25, 6.25)	(0.05, 113.5)
P_2	(0.25, 1.25)	(0.05, 41.5)	P_{17}	(0.0, 1.21)	(3.5, 67.5)
P_3	(0.25, 2.0)	(0.05, 5.05)	P_{18}	(0.25, 6.25)	(0.05, 113.5)
P_4	(0.25, 1.25)	(0.05, 5.05)	P_{19}	(0.25, 1.25)	(0.05, 5.05)
P_5	(0.25, 1.25)	(0.05, 5.05)	P_{20}	(0.0, 6.0)	(0.0, 14.667)
P_6	(0.0, 5.96)	(0.05, 150.05)	P_{21}	(0.25, 2.0)	(0.05, 5.05)
P_7	(0.0, 6.0)	(0.0, 14.667)	P_{22}	(0.25, 1.25)	(0.05, 79.2)
P_8	(0.25, 1.25)	(0.05, 5.05)	P_{23}	(0.0, 5.96)	(0.05, 150.05)
P_9	(0.0, 1.21)	(3.5, 67.5)	P_{24}	(0.0, 5.96)	(0.05, 150.05)
P_{10}	(0.25, 1.25)	(0.05, 79.2)	P_{25}	(0.25, 1.25)	(0.05, 79.2)
P_{11}	(0.25, 1.25)	(0.05, 5.05)	P_{26}	(0.25, 6.25)	(0.05, 113.5)
P_{12}	(0.25, 1.25)	(0.05, 41.5)	P_{27}	(0.25, 1.25)	(0.05, 79.2)
P_{13}	(0.25, 6.25)	(0.05, 113.5)	P_{28}	(0.0, 6.0)	(0.0, 14.667)
P_{14}	(0.25, 1.25)	(0.05, 5.05)	P_{29}	(0.25, 2.0)	(0.05, 5.05)
P_{15}	(0.25, 2.0)	(0.05, 5.05)	P_{30}	(0.25, 1.25)	(0.05, 41.5)

output is computed for each time interval as shown in the following table, which presents the meal intake (G_{in}) and insulin intake (I_{in}) values, along with the corresponding glucose concentrations predicted by the model:

Table 2. Meal Intake G_{in} and Insulin Intake I_{in} using Eq. 1 and 2 and Corresponding G_{model} Output.

Time (hr:min)	Glucose	Insulin	Time (hr:min)	Glucose	Insulin
8 : 30	1.0830	1.2500	18 : 40	0.2500	0.2500
10 : 30	0.2500	0.2500	20 : 35	0.8056	1.1940
12 : 45	1.0830	0.2500	22 : 35	0.2500	0.2500
14 : 40	0.2500	0.2500	23 : 10	0.2500	0.2500
16 : 20	1.6390	0.5833			

The simulation of glucose and insulin dynamics for each virtual patient was carried out using these meal and insulin intake values, and the model provided the corresponding glucose concentrations at various time intervals. These simulations allowed for the generation of a comprehensive dataset used for further analysis.

4   Proposed Deep Neural Network Architecture

This section outlines the proposed deep learning architecture designed to track glucose levels in human plasma. The primary objective is to predict the time series behavior of glucose data, which may contain irrelevant features. To address this, We employ a combination of One-Dimensional Convolutional Networks (CNN-1D) and memory-augmented recurrent networks to tackle this challenge.

CNNs are commonly used to extract relevant features from raw time series data. In particular, CNN-1D has been effective for such tasks as it automatically learns important features from the data sequence, which are critical for accurate predictions [9]. While CNN variants like CNN-2D and CNN-3D are used for multi-dimensional data, CNN-1D operates on one-dimensional input, and its key components include filters, pooling, and dropout layers. Filters are responsible for feature extraction, and pooling reduces the input size, retaining only the most important features. The dropout layer helps prevent overfitting by randomly deactivating neurons during training [14]. Although dropout in CNN-1D might have a limited effect due to fewer parameters, it still helps in preventing overfitting by introducing noise to the input of fully connected layers.

The glucose prediction problem for Type 1 diabetes is a time series forecasting task, where each input depends on previous values. This makes it a sequence learning problem. A Recurrent Neural Network (RNN) [5] could be used to model this, but RNNs suffer from issues like vanishing and exploding gradients due to Backpropagation Through Time (BPTT). These problems can be mitigated by gradient clipping. Another issue with RNNs is that the information from previous time steps is gradually "morphed" by the current input, making it difficult to extract past information.

LSTMs address these challenges by introducing memory units that effectively resolve the issues associated with Backpropagation Through Time (BPTT). While similar to traditional RNNs in structure, LSTMs incorporate three critical mechanisms called gates. These gates—Forget, Input, and Output—manage how data flows through the network. The Forget gate determines whether to preserve or remove the previous timestep's information, while the Input gate updates the model with new information from the current input. The Output gate controls which data is passed on to the subsequent timestep.

LSTMs maintain two internal states: the hidden state (H_t), which stores short-term memory, and the cell state (C_t), which retains long-term memory. By selectively updating, reading, and erasing stored information, the LSTM ensures that only the most important data is kept, avoiding the retention of irrelevant or outdated information. The Forget gate clears unnecessary data, and the Input and Output gates manage the flow of new and relevant information through the network.

What differentiates LSTMs from traditional RNNs is the inclusion of gates that regulate the flow of data, allowing LSTMs to overcome the vanishing gradient issue that is common in RNNs. In our proposed model, CNN-1D is used to extract features from the input, and these features are subsequently processed by the LSTM layers for sequence learning. This integration of CNN-1D and LSTM creates an effective framework for handling sequential data. The proposed architecture is illustrated in Fig. 1.

The proposed model includes the following layers:

- In the initial convolutional layer, thirty-two kernels of size three are used, with ReLU as the activation function.
- A dropout layer is applied to disable 20% of neurons to prevent overfitting.

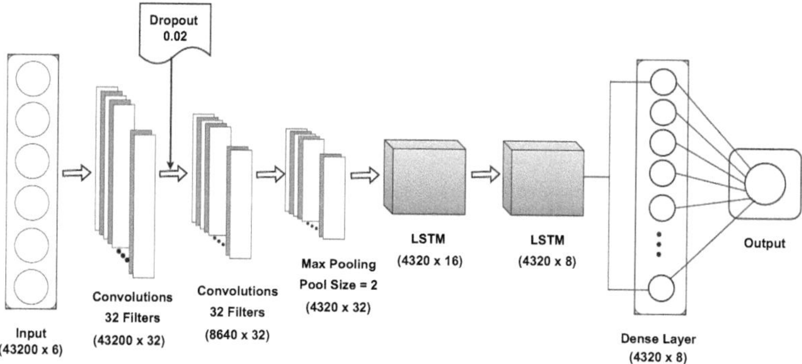

Fig. 1. Proposed COVLSTM architecture.

- The output is then passed through another convolutional layer with the same settings as the first, but with a kernel size of 2.
- Max pool is applied to decrease the dimensionality.
- The features obtained are fed into two LSTM layers, consisting of sixteen neurons in the initial layer and eight neurons in the second.
- The final output is generated by a dense layer, which processes the output from the last LSTM layer.

The chosen hyperparameters have been optimized through extensive experimentation. Reducing the number of parameters may degrade the model's ability to track glucose levels accurately, while increasing the number of neurons can lead to longer training times and increased model complexity. Nevertheless, the proposed CNN-1D+LSTM architecture demonstrates its ability to effectively learn from the glucose data, which is essential for designing personalized food plans and maintaining healthy glucose levels.

4.1 Computational Complexity Analysis

The computational complexity of the proposed COVLSTM architecture is analyzed by considering the operations performed in each layer of the model. The architecture consists of convolutional layers, max pooling layers, long short-term memory (LSTM) layers, and a dense output layer. Each convolutional layer applies F filters of size $K \times K$ over an input of size $H \times W$ with C channels. The computational complexity of a single convolutional layer can be expressed as $O(F \cdot C \cdot H \cdot W \cdot K^2)$. The max pooling layer, which selects the maximum value within non-overlapping pooling windows, has a complexity of $O(H \cdot W \cdot C)$. Since max pooling only involves comparisons, it is computationally efficient.

The LSTM layer processes sequential input data with hidden state size d and sequence length T, leading to a computational complexity of $O(T \cdot d^2)$. The final dense layer maps the LSTM output to the desired number of output neurons with complexity $O(N \cdot d)$. Summing up all components, the total computational complexity of the proposed model is $O(F \cdot C \cdot H \cdot W \cdot K^2) + O(H \cdot W \cdot C) + O(T \cdot d^2) + O(N \cdot d)$. Since LSTM layers are typically the most computationally intensive due to their recurrent nature, the overall complexity is primarily determined by $O(T \cdot d^2) + O(F \cdot C \cdot H \cdot W \cdot K^2)$. This analysis provides insights into the efficiency of the proposed architecture and highlights its computational feasibility.

5   Simulation and Results

The proposed deep learning model was simulated in Jupyter Notebook using Python 3.10 on a Windows 64-bit desktop with an Intel(R) i5 2.5 GHz processor and 128GB RAM. Data for 30 patients was generated using the mathematical Gmodel by varying meal and insulin patterns. The parameters in Eqs. 2 and 1 were adjusted to generate the required data, with the insulin and glucose intake ranges summarized in Table 1.

We evaluated the model using k-fold cross-validation and assessed performance using mean absolute error (MAE) [6], as shown in Table 3. The proposed model's performance was compared with CNN-1D and LSTM architectures (Fig. 2), and training/validation errors are depicted in Fig. 3.

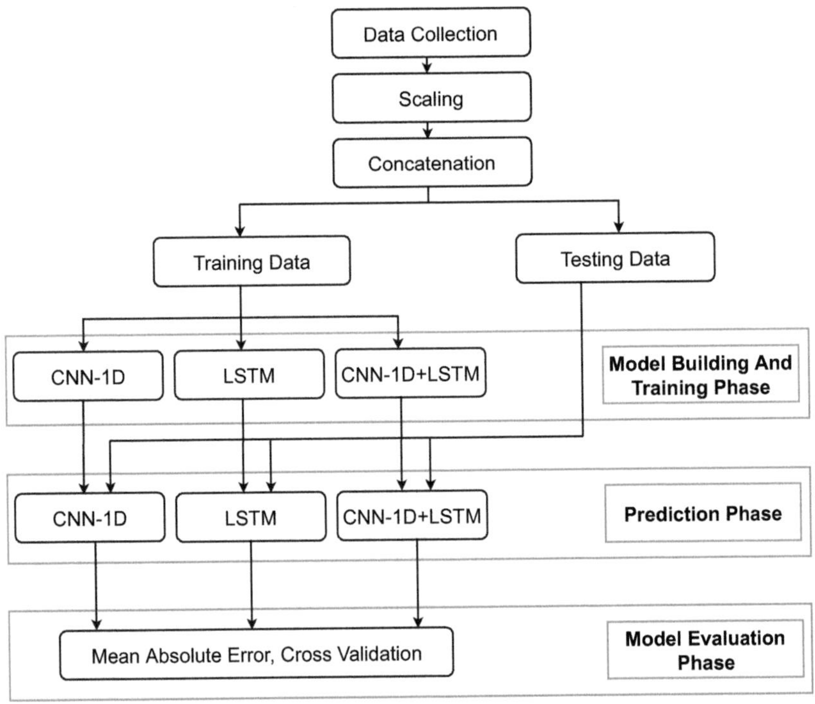

Fig. 2. Different phases of proposed deep neural network architecture.

Table 3. Deep neural network models comparisons with regard to average absolute error.

Model	Average Absolute Error
CNN-1D	91.2
LSTM	53.4
CNN-1D + LSTM	19.7

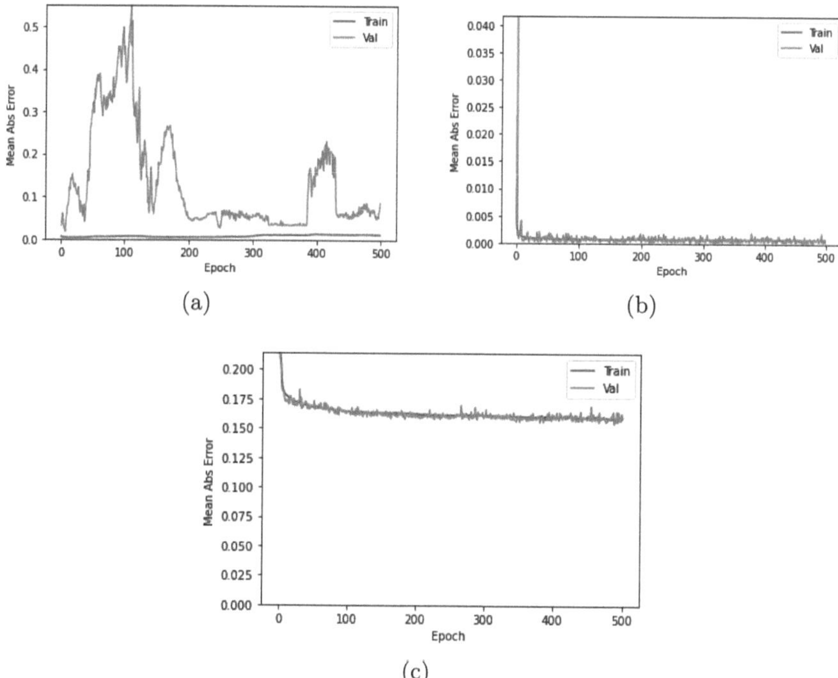

Fig. 3. Loss curves for learning and testing of the (a) CNN-1D model, (b) LSTM model, and (c) combined CNN-1D and LSTM model.

Figures 4 and 5 show the k-fold cross-validation results. In this method, the dataset is randomly divided into k equal-sized subsets. One part is utilized as the testing set, while the other $k - 1$ partitions are employ for training.

Let dataset D consist of 30 patient data, divided into $k = 6$ subsets: $D = \{\{p_1, p_2, p_3, p_4, p_5\}, \ldots, \{p_{26}, p_{27}, p_{28}, p_{29}, p_{30}\}\}$, where p_i represents each patient's data. For each fold, the model is trained on $k - 1$ subsets and tested on the remaining one, such as testing on $\{p_1, p_2, p_3, p_4, p_5\}$ after training on the others.

After training and testing on each fold, the glucose tracking results for each patient are shown in Figs. 4 a to 4 e, and the overall performance is displayed in Figs. 4 and 5 for ten subjects. These results demonstrate the model's ability to accurately track the Gmodel output.

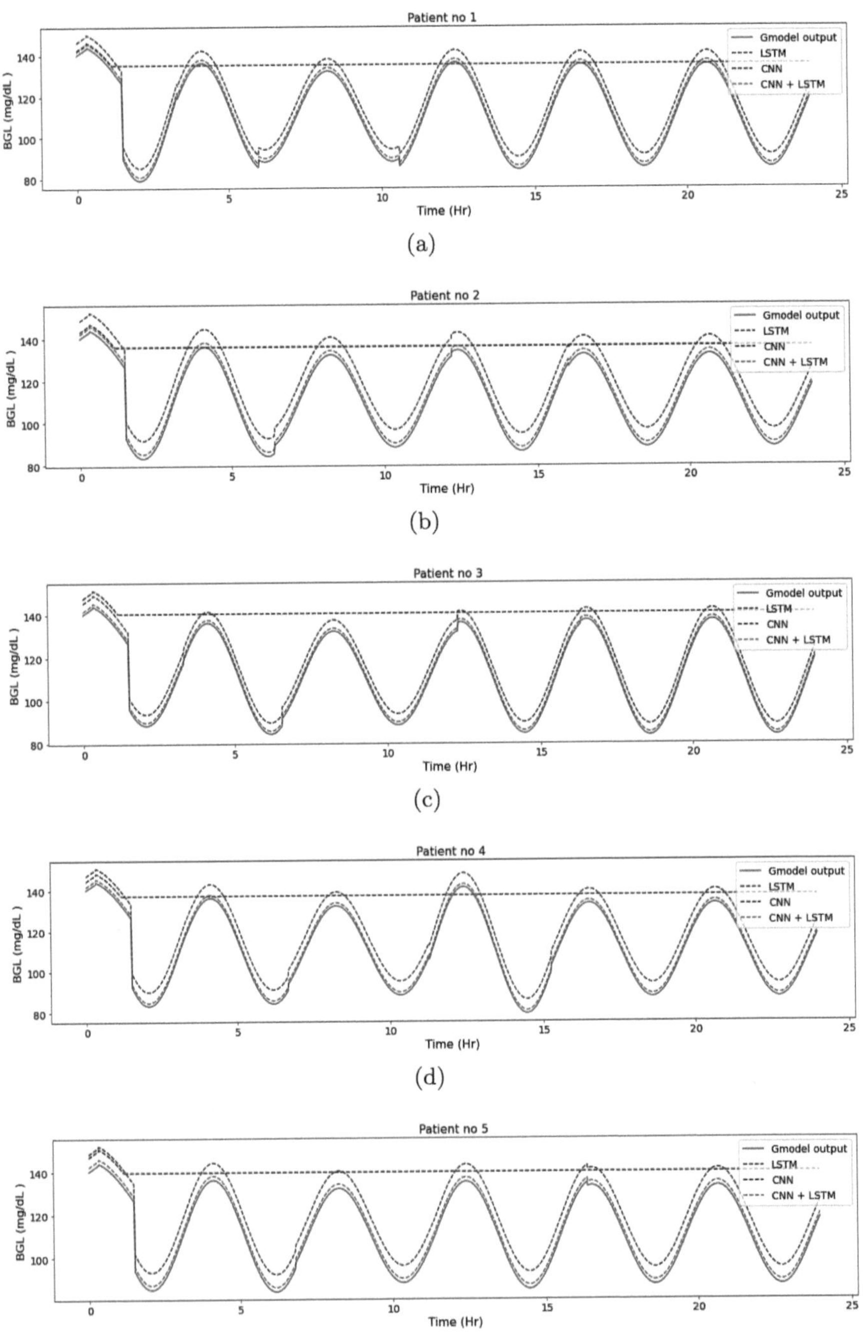

Fig. 4. Prediction of Patient 1 - 5 using trained models CNN-1D , LSTM , CNN-1D + LSTM model

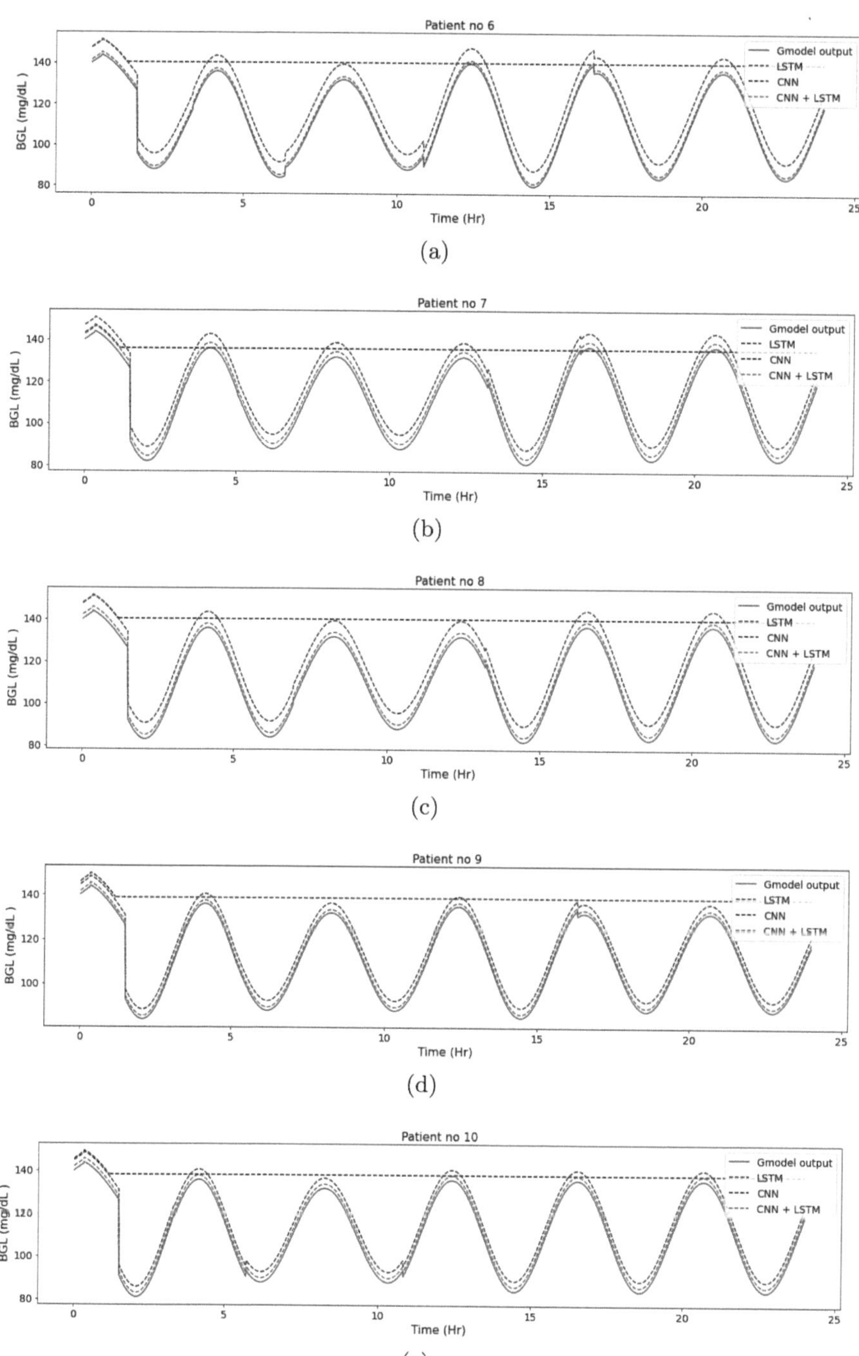

Fig. 5. Prediction of Patient 6 - 10 using trained models CNN-1D, LSTM, CNN-1D + LSTM model

5.1   The Effect of LeakyReLU Activation Function in the Model

The effect of the LeakyReLU activation function [8] is also studied in comparison to the ReLU activation function. Figure 6 a shows the output of patient 1 using LeakyReLU, while Fig. 6 b shows the output of patient 1 using ReLU. In the proposed deep neural network architecture, ReLU is used as the activation function. The primary reason for using ReLU is that the range of the Gmodel output, which is between 0 and 180 mg/dl, aligns with the range of the ReLU activation function, i.e., between 0 and $+\infty$. Figure 6 demonstrates that the proposed architecture effectively tracks the Gmodel output.

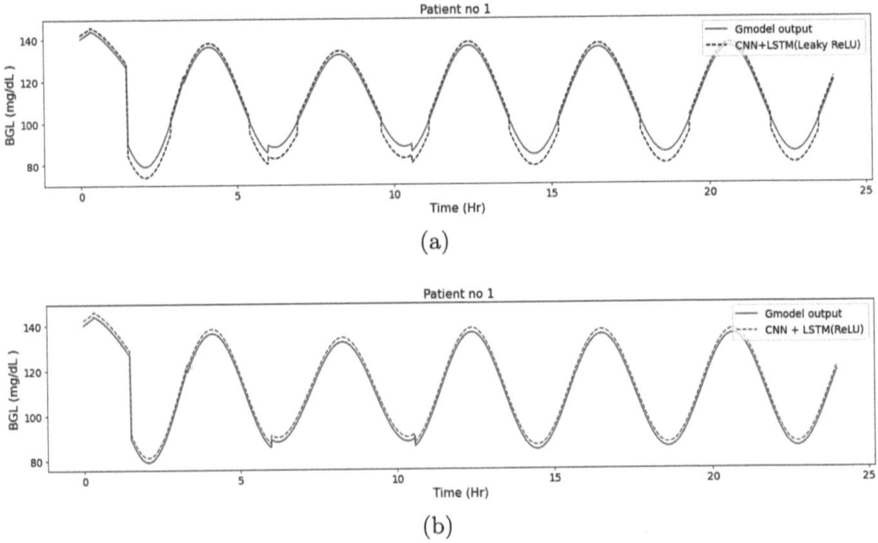

Fig. 6. (a) Tracking patient p_1 using LeakyReLU activation function. (b) Tracking patient p_1 using ReLU activation function.

6   Conclusion and Future Directions

This study proposes a innovative architecture, COVLSTM, aimed at forecasting blood glucose concentrations in Type one diabetes patients. By hybrid CNN-1D and LSTM model, this approach seeks to overcome the limitations found in traditional mathematical models. Since glucose levels vary based on lifestyle, mathematical models alone are insufficient. The proposed architecture, incorporating CNN-1D, LSTM, and their combination, demonstrated superior performance with lower mean absolute error and better cross-validation results compared to other models. In future we will applying this architecture on real-time patient data for glucose prediction and insulin adjustment, implementing it in hardware systems, and exploring additional feature extraction algorithms and sequential learning models.

References

1. Ahmad, S., Ahmed, N., Ilyas, M., Khan, W., et al.: Super twisting sliding mode control algorithm for developing artificial pancreas in type 1 diabetes patients. Biomed. Sign. Process. Control **38**, 200–211 (2017)
2. Aiello, E.M., Lisanti, G., Magni, L., Musci, M., Toffanin, C.: Therapy-driven deep glucose forecasting. Eng. Appl. Artif. Intell. **87**, 103255 (2020)
3. Apoorva, S., Aditya S, K., Snigdha, P., Darshini, P., Sanjay, H.: Prediction of diabetes mellitus type-2 using machine learning. In: International Conference On Computational Vision and Bio Inspired Computing, pp. 364–370. Springer (2019)
4. Chen, Q., Ye, Q., Zhang, W., Li, H., Zheng, X.: Tgm-nets: a deep learning framework for enhanced forecasting of tumor growth by integrating imaging and modeling. Eng. Appl. Artif. Intell. **126**, 106867 (2023)
5. Chen, S.H., Hwang, S.H., Wang, Y.R.: An rnn-based prosodic information synthesizer for mandarin text-to-speech. IEEE Trans. Speech Audio Process. **6**(3), 226–239 (1998)
6. Gabbouj, M., Coyle, E.J.: Minimum mean absolute error stack filtering with structural constraint and goals. IEEE Trans. Acoustics Speech Sign. Process. **38**(6), 955–968 (1990)
7. Gan, D.: Diabetes atlas. International Diabetes Federation (2003)
8. Hutchings, B.L.: Asics, processors, and configurable computing. In: 30th Hawaii International Conference on System Sciences (HICSS) Volume 1: Software Technology and Architecture. vol. 1, pp. 719–719. IEEE Computer Society (1997)
9. Kiranyaz, S., Avci, O., Abdeljaber, O., Ince, T., Gabbouj, M., Inman, D.J.: 1d convolutional neural networks and applications: a survey. Mech. Syst. Sign. Process. **151**, 107398 (2021)
10. Li, J., Kuang, Y.: Analysis of a model of the glucose-insulin regulatory system with two delays. SIAM J. Appl. Math. **67**(3), 757–776 (2007)
11. Li, Y., Li, H., Yao, H.: Analysis and study of diabetes follow-up data using a data-mining-based approach in new urban area of urumqi, xinjiang, china, 2016-2017. Comput. Math. Methods Med. **2018** (2018)
12. Lunze, K., Singh, T., Walter, M., Brendel, M.D., Leonhardt, S.: Blood glucose control algorithms for type 1 diabetic patients: a methodological review. Biomed. Sign. Process. Control **8**(2), 107–119 (2013)
13. Rai, B., Robinson, N.: Diabetes helpline. British J. Diabetes Vascular Dis. **2**(1), 46–46 (2002)
14. Srivastava, N., Hinton, G., Krizhevsky, A., Sutskever, I., Salakhutdinov, R.: Dropout: a simple way to prevent neural networks from overfitting. J. Mach. Learn. Res. **15**(1), 1929–1958 (2014)
15. Sun, Q., Jankovic, M.V., Bally, L., Mougiakakou, S.G.: Predicting blood glucose with an lstm and bi-lstm based deep neural network. In: 2018 14th Symposium on Neural Networks and Applications (NEUREL), pp. 1–5. IEEE (2018)
16. Swapna, G., Vinayakumar, R., Soman, K.: Diabetes detection using deep learning algorithms. ICT Express **4**(4), 243–246 (2018)

17. Tabish, S.A.: Is diabetes becoming the biggest epidemic of the twenty-first century? Int. J. Heal. Sci. **1**(2), V (2007)
18. Wang, H., Li, J., Kuang, Y.: Enhanced modelling of the glucose-insulin system and its applications in insulin therapies. J. Biol. Dynamics **3**(1), 22–38 (2009)
19. Yang, T., Yu, X., Ma, N., Zhao, Y., Li, H.: A novel domain adaptive deep recurrent network for multivariate time series prediction. Eng. Appl. Artif. Intell. **106**, 104498 (2021)
20. Zou, Q., Qu, K., Luo, Y., Yin, D., Ju, Y., Tang, H.: Predicting diabetes mellitus with machine learning techniques. Front. Genetics **9**, 515 (2018)

EEG-Based Schizophrenia Classification with Residual Dilated Multi-Scale Network

Manik Singh[1(✉)], Aakanksha Baidya[1], and Geet Sahu[2]

[1] Amity University, Noida, Uttar Pradesh, India
`maniksingh256@gmail.com`
[2] Computer Science and Engineering Department, Siksha 'O' Anusandhan (Deemed to be University), Bhubaneswar, Odisha 751020, India

Abstract. Schizophrenia, a severe mental health disorder affecting approximately 24 million people globally (0.32% of the population), disrupts personal, social, and occupational functioning through symptoms that include hallucinations, disorganized behavior and delusions. Current diagnostic methods are time-intensive and subjective, necessitating advancements in automated detection. Electroencephalogram (EEG) data captures fluctuations in neural activity associated with human memory processes. This study introduces SCZ-MDRNet, a Multi-Scale Dilated Residual Convolution Network designed for automated diagnosis of schizophrenia from EEG signals. The approach involves segmenting EEG data by a process called windowing to expand the dataset, followed by conversion of the samples into color-mapped visualizations for input into the model. Using the IBIB-PAN dataset comprising of EEG recordings from 14 schizophrenic and 14 healthy individuals, with 19-channels, recorded at 250 Hz, SCZ-MDRNet achieves a test accuracy of 99.90%, perfect diagnostic performance with an AUC, specificity, sensitivity, and F1 score of 1. These results underscore the proposed model's potential as a transformative tool for reliable and efficient schizophrenia diagnosis. Future work will focus on addressing limitations due to the small dataset size and incorporating more diverse datasets to improve robustness.

Keywords: Schizophrenia detection · Automated diagnosis · EEG signal analysis · Deep learning in healthcare · multi-scale dilated residual network · Electroencephalogram processing

1 Introduction

Schizophrenia is a long-term and serious psychiatric disorder causing significant cognitive, behavioral and emotional changes, often leading to disruptions in daily life and interpersonal relationships. Schizophrenia affects over 24 million individuals globally, as reported by the World Health Organization, with genetic factors like Neurogranin and Zinc Finger Protein 804A contributing to increased risk. The disorder presents several symptoms, that include hallucinations, disorganized speech, and functional decline. Electroencephalogram (EEG), a non-invasive method that records brain electrical activity by detecting voltage variations on the scalp caused by neural activity. EEG signals

© The Author(s), under exclusive license to Springer Nature Switzerland AG 2026
P. Chandrakar et al. (Eds.): ICCINS 2025, CCIS 2738, pp. 41–52, 2026.
https://doi.org/10.1007/978-3-032-09572-5_4

are captured using electrodes strategically positioned on the scalp, typically following the standardized 10–20 system, which ensures consistent electrode positioning across individuals. This system includes channels corresponding to regions such as the occipital (O), temporal (T), parietal (P), and frontal (F) lobes, as well as midline sites (Z). Electrodes record signals from these areas, reflecting underlying brain activity associated with cognitive and motor functions. EEG recording is a crucial functional neuroimaging technique and has shown to offer useful insights into the brain's functioning in individuals with schizophrenia [1]. However, analysing EEG signals presents challenges for scientists and neurologists due to lengthy records, artifacts and the use of multiple channels. Techniques making use of traditional machine learning and deep learning have helped researchers to overcome these challenges, [2–4], to assist in the diagnosis of schizophrenia. Figure 1 shows examples of EEG recordings from healthy and from schizophrenic patients.

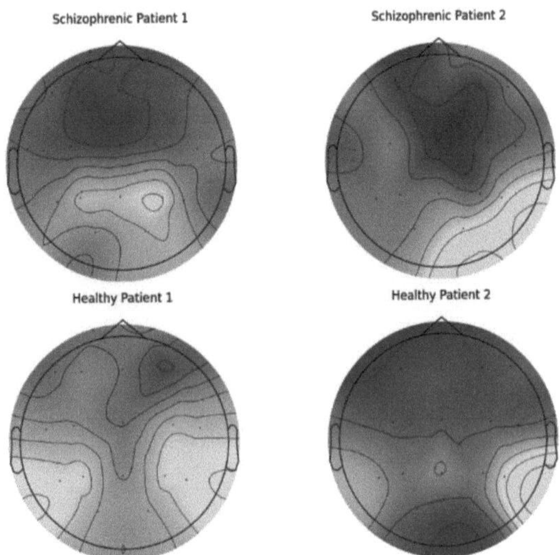

Fig. 1. Visualization of healthy (bottom) and Schizophrenic (top) Electroencephalogram (EEG) recordings.

In this paper, we employ SCZ-MDRNet to effectively classify schizophrenia from EEG signals. Given the increasing importance of early diagnosis and personalised treatment in mental healthcare, our model could significantly impact clinical practice by assisting clinicians in making more informed decisions, thus leading to timely interventions. The model has potential for deployment in real-world clinical settings for automated analysis of large-scale EEG data for screening or monitoring patients over time. It employs multi-scale dilated residual convolutional blocks to extract multi-scale spatial information and enhance feature integration across layers, ensuring a robust representation of EEG data. By evaluating the model on the publicly available IBIB-PAN dataset, we demonstrate its exceptional classification performance, outperforming nine state-of-the-art (SOTA) models across multiple performance metrics that include accuracy, AUC, specificity, sensitivity and F1 score.

The main highlights of this research include the following:

- A novel framework, SCZ-MDRNet, is introduced for schizophrenia detection. The architecture utilizes multi-scale dilated residual convolutional blocks to efficiently capture spatial multi-scale information while improving cross-layer feature integration.
- The proposed method is evaluated using the publicly available IBIB-PAN dataset, demonstrating outstanding performance and outperforming existing state-of-the-art methods.
- Extensive comparisons are performed against nine leading models across five evaluation metrics, with experimental results highlighting the superior effectiveness of the proposed approach.

This study is organized as follows: Prior research is reviewed in Sect. 2, and Sect. 3 details the proposed methodology. The experimental configuration is outlined in Sect. 4, followed by an analysis of model effectiveness and results in Sect. 5. Section 6 provides further discussion and compares the findings with state-of-the-art models.

2 Related Work

Extensive research has been conducted on using ML techniques for the detection of schizophrenia. Sahu et al. [6] devised an automated schizophrenia detection model employing separable convolution and attention networks. Their method integrates characteristics from high-level and low-level 2D scalogram images generated via continuous wavelet transform, achieving high accuracy in detecting schizophrenia using EEG data obtained from sensory tasks within the IBIB-PAN and schizophrenia datasets. In their research, Oh et al. [7] used a CNN model having 11 layers with ten-fold cross validation to predict schizophrenia on a data set of 14 healthy and 14 schizophrenic patients. The study applies different approaches for evaluations conducted with and without subject-specific considerations (98.07% and 81.26% accuracy, respectively). Dropout is applied during non-subject based testing on layers 9 and 10, whereas for subject based testing, generalization is improved by applying dropout to layers 4 and 6. The model performs extraction, selection and classification automatically, and offers high accuracy on a small dataset. Hussain et al. [8] proposed an ensemble lightweight 1D CNN model designed with a minimal set of learnable parameters, trained efficiently on limited data to reduce the risk of overfitting. By segmenting EEG signals into smaller parts, it processes inputs of any length while keeping the complexity low. Bagherzadeh et al. [9] leveraged the integration of a CNN and LSTM network to perform classification on novel images created from EEG signals based on their Transfer Entropy measures. The novel approach achieves an F1 score of 99.93% using the EfficientNetB0-LSTM on a dataset comprising 14 individuals diagnosed with schizophrenia and 14 healthy participants. Ko et al. [10] proposed a novel approach to classify schizophrenia with CNN models based on VGGNet. The research involved EEG data from 81 participants, which were transformed into visual representations by using Gramian Angular Field (GAF) and Recurrence Plot

(RP) techniques. Among these, the DL model attained its highest accuracy of 93.20% when employing GAF. Antara et al. [11] used EEG data from 43 schizophrenic and 39 healthy individuals to classify schizophrenia automatically. The EEG signal was processed using a Low Pass Filter and Total Variation Denoising, decomposing it into four distinct wave bands. Using SVM and Random Forest, 12 of 16 channels achieved over 95% accuracy, with 7 channels reaching 100%, enabling real-time schizophrenia detection. Aslan and Akin converted EEG data into 2D time-frequency features via CWT and extracted features which were classified using the VGG16 CNN architecture, achieving high classification accuracy of 98% and 99.5% on two datasets. The study uses visualization techniques like Activation Maximization, Saliency Maps, and Grad-CAM to illustrate the differences in frequency components between schizophrenic patients and healthy individuals. Sun et al. [13] represented EEG signals as RGB images to retain spatial characteristics and employed a hybrid DL model combining CNN and LSTM for classification. The fuzzy entropy feature outperforms the fast Fourier transform, achieving an average accuracy of 99.22%, compared to 96.34% with FFT. Supakar et al. [14] proposed an RNN-LSTM model on EEG data utilizing an optimal feature set obtained through dimensionality reduction for detection of schizophrenia. The model consists of a 100-dimensional LSTM followed by three dense layers for classification on a dataset of 45 schizophrenic and 39 healthy individuals, achieving 98% accuracy with a complete feature set and 93.67% with a reduced set. Karnati et al. [15] proposed PSFAN, a model designed to analyze 2D representations of 4-s EEG recordings. The approach leverages dilated convolutions to capture multi-scale features and employs spatial attention mechanisms to highlight critical information. Through statistical validation using Wilcoxon's rank-sum test, PSFAN demonstrated superior performance over 11 existing methods across three experimental settings: subject-dependent, subject-independent, and cross-dataset evaluations.

3   Proposed Methodology

This section provides an outline of the sequential process employed to develop the proposed framework. The process comprises four key stages: dataset acquisition, preprocessing, image generation, and classification, as illustrated in Fig. 2. The ensuing subsections provide more detail on each of these steps.

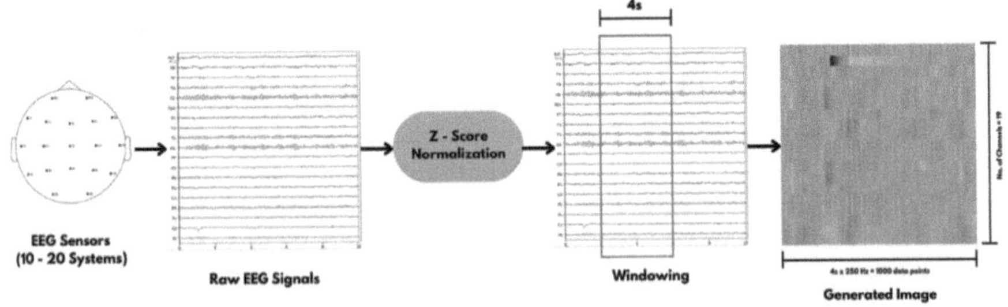

Fig. 2. The sequential process employed in pre-processing stage.

3.1 Dataset

The dataset that this study utilizes is the publicly available IBIB-PAN dataset which comprises a total of 28 EEG recordings, distributed between 14 with schizophrenia and 14 without any diagnosed condition. The recordings were stored in the European Data Format (EDF) and with sampling rate as 250 Hz. EEG signals were captured using the 10–20 standard system. The setup consists of 19 channels: Fp1, Fp2, F7, F3, Fz, F4, F8, T3, C3, Cz, C4, T4, T5, P3, Pz, P4, T6, O1, and O2. For signal standardization, a reference electrode was positioned between Fz and Cz. Fig. 3 illustrates the electrode placements.

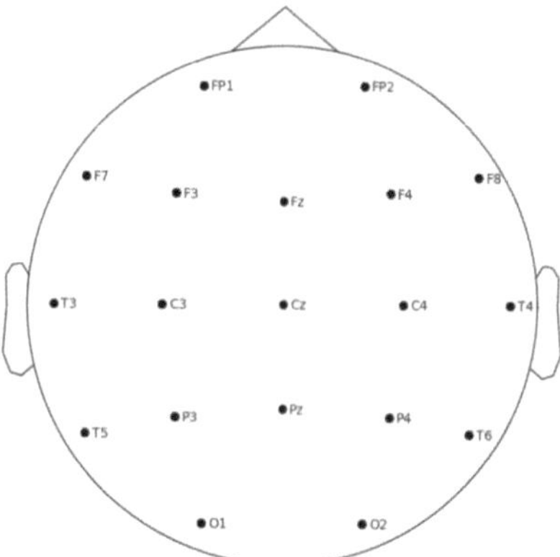

Fig. 3. The location of the 19 electrodes.

3.2 Preprocessing

We read the EEG data using the Python MNE library, which allows for the extraction and manipulation of EEG signals. After reading the data, we perform three steps of preprocessing:

Normalization First, we perform normalization on each EEG channel to ensure that all data points are on a comparable scale. We use Z-Score Normalization, that is, the mean and standard deviation of the signal across all channels are calculated for each recording. The signal is subsequently standardized to achieve zero mean and unit variance, which helps make the features more compatible with ML models. It is given as

$$z_{i,j} = \frac{x_{i,j} - \mu_j}{\sigma_j}$$

Segmentation To capture short-term patterns, the EEG signals are split into smaller time intervals after being normalized. A window size of 1000 data points (equivalent to 4 s of recording at 250 Hz) is used, with a stride of 250 data points (1 s). This method slides a window across the signal, creating overlapping segments. Each segment is stored along with its corresponding label, either healthy or schizophrenic, enabling supervised learning during model training.

3.3 Image Generation

To convert the EEG signal into a form suitable for convolutional neural networks, we transform the segmented signals into 2D images. For each segment, a time-frequency representation is created using a colormesh plot, with the viridis colormap applied. For each EEG segment S_k, the Short Time Fourier Transform, abbreviated as STFT, is used to obtain the time-frequency representation:

$$\chi(f, t) = \sum_{n=0}^{N-1} S_k(n)w(n - t)e^{-j2\pi\, fn}$$

Here, S_k represents the EEG signal segment, $w(n - t)$ is the window function, f denotes frequency, t represents time, and $\chi(f, t)$ corresponds to the magnitude spectrogram. The power-spectrogram is then computed as:

$$P(f, t) =, |\chi(f, t)|^2$$

A pseudocolor mesh plot (pcolormesh) is used to visualize with the viridis colormap mapping power intensity to colors. The plot is then saved as an image file, which is later used as input for the DL model. This is done by utilizing the matplotlib python library. This step leverages the spatial patterns in EEG data for schizophrenia detection by visualizing the signals as images.

3.4 Proposed Model

The proposed SCZ-MDRNet (shown in Fig. 4.) is characterized by its use of multi-scale dilated convolutions and residual connections for enhanced feature extraction. By dynamically adjusting the receptive field, it processes the input through convolutional layers with varying dilation rates (1, 2, and 3), enabling the model to capture both detailed and contextual features. Each convolutional block includes batch normalization and LeakyReLU activations, which help stabilize training and prevent issues like vanishing gradients and dead neurons.

The Multi-Scale Dilated Residual Convolution (MDRC) block is a pivotal component of the proposed SCZ-MDRNet architecture. It is designed to enhance feature extraction by capturing spatial information at multiple scales while maintaining computational efficiency. This block employs dilated convolutions, residual connections, and multi-scale processing to ensure robust feature learning. A single MDRC block consists of multiple parallel convolutional layers with different dilation rates d, which control the receptive field size of the convolution operation. The dilated convolution at a given dilation rate d can be mathematically expressed as:

$$y[i] = \sum_{k=1}^{K} x[i + d \cdot k] \cdot w[k]$$

Where $x[i]$ is the input signal, $w[k]$ represents convolution kernel, the kernel size is represented by K and d stands for the dilation rate. By employing $d = 1, 2, 3$ the block

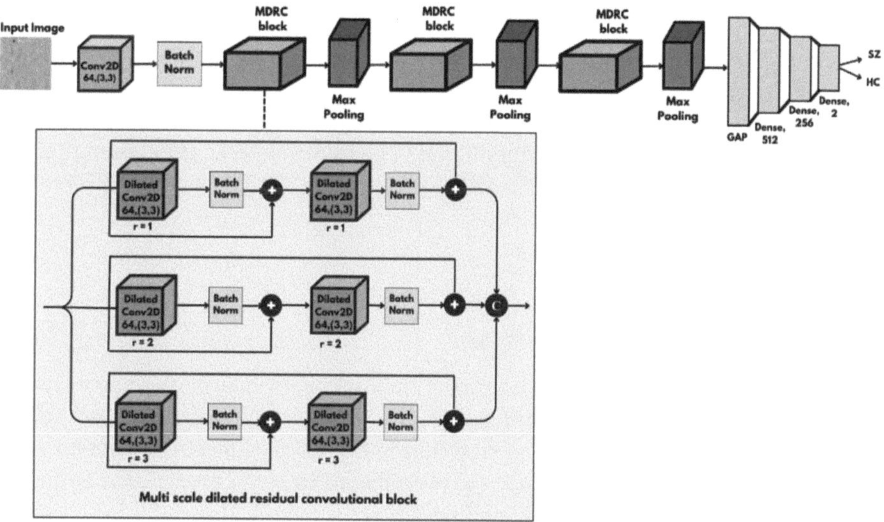

Fig. 4. Proposed Network (SCZ-MDRNet).

captures features at varying spatial resolutions, ensuring the model effectively processes fine-grained as well as coarse features. Moreover, the MDRC block incorporates residual connections in order to maintain the integrity of the learned features. These connections allow the input of the block to bypass the convolutions and be directly added to the output of the convolutions, thus mitigating the vanishing gradient problem while also enabling gradient flow. The residual connection is formulated as:

$$F_{output} = F_{conv}(x) + x$$

Where $F_{conv}(x)$ is the output of convolution layers and x is the block's input. The outputs from parallel dilated convolutions are concatenated to integrate multi-scale information. This concatenation is expressed as:

$$F_m = Concat(F_{d=1} + F_{d=2} + F_{d=3})$$

After initial feature extraction, deeper network layers further refine these representations using comparable multi-scale operations. Then, global average pooling consolidates spatial details into a concise feature vector. The final fully connected layers perform the classification task, with a softmax activation in the output layer generating class probabilities. The model enables the detection of the subtle variations in EEG signals associated with schizophrenia, making the proposed framework a promising approach for automated schizophrenia diagnosis.

4 Experimental Setup

The experiments were conducted using the Keras framework with TensorFlow as the backend, implemented in Python 3.10.12. The model was trained and tested on a high-performance NVIDIA A100 GPU, ensuring efficient computation for large-scale data processing. The EEG dataset used for the study was pre-processed into standardized formats, including normalization and artifact removal, to enhance the quality of input features. Model training utilized the Adam optimizer. Learning rate of 0.0001 was used and the weight decay parameter (WD) set as 0.0006. Loss function selected was categorical cross-entropy to handle the multi-class classification task. The training process incorporated an early stopping mechanism to prevent overfitting, monitoring validation accuracy with a patience of 10 epochs.

5 Result and Analysis

5.1 Evaluation Metrics

To thoroughly evaluate the proposed model, multiple performance metrics were analyzed. The model's capability to differentiate between classes at varying thresholds is measured using the AUC. The confusion matrix helped visualize classification errors across categories. Accuracy measured the overall correctness of predictions, while sensitivity (recall or true positive rate) and specificity (precision) assessed how well the model identified positive and negative cases, respectively. Additionally, the F1 Score provided a balance between precision and recall, ensuring a comprehensive assessment of classification performance.

5.2 Model Performance and Validation

The training and validation processes demonstrated the robust performance of the proposed SCZ-MDRNet. The model training continued for 26 epochs achieving a peak validation accuracy at 99.97%, closely mirrored by the test accuracy of 99.90%, underscoring the model's generalizability. The AUC reached a perfect value of 1, as depicted in Fig. 5, highlighting its impeccable discriminative ability across classification thresholds. The classification report, shown in Fig. 6, confirms this exceptional performance demonstrating a specificity, sensitivity and an F1 score of 100%, indicating perfect balance and effectiveness in distinguishing between control and schizophrenic cases.

Furthermore, the confusion matrix in Fig. 7a and Fig. 7b provide a detailed breakdown of predictions, with the former showing classification result of actual number of samples and the latter showing percentage of classified samples. The confusion matrix shows 2618 out of 2621 (99.9%) samples correctly classified as control and 3127 out of 3130 (99.9%) samples accurately identified as schizophrenic. These results underscore the model's precision and reliability, establishing SCZ-MDRNet as a highly effective solution for automated schizophrenia diagnosis using EEG signals. Table 1 summarizes the computational and memory metrics of the model that is proposed in this paper and its size in memory. These evaluations showcase the model's effectiveness and robustness in identifying schizophrenia.

6 Discussion

This study introduces an automated DL framework for distinguishing patients with schizophrenia from healthy individuals using EEG. The approach utilizes a specialized architecture built on multi-scale dilated residual convolution blocks, designed to capture

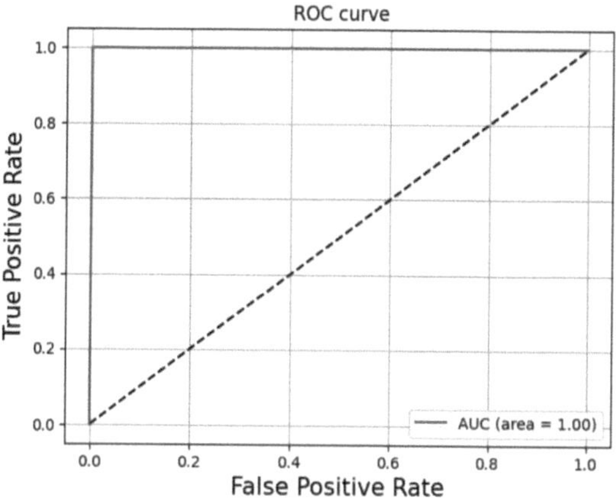

Fig. 5. AUC Curve.

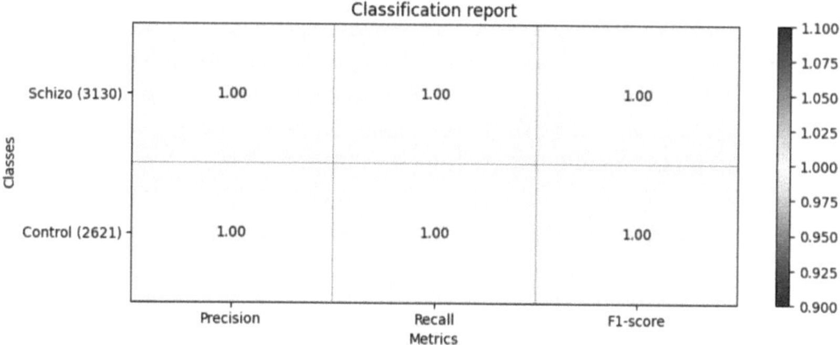

Fig. 6. Classification Report.

Table 1. Computational and Memory Metrics of the Proposed Model.

Metric	Value	Description
FLOPs	17,603,347,456	Total floating-point operations required for one forward pass.
MACs	8,801,673,728	Total multiply-accumulate operations performed.
GFLOPs	17.60	Floating-point operations in billions.
GMACs	8.80	Multiply-accumulate operations in billions.
Model Size (MB)	10.55	Memory required to store model weights.

spatial features across different scales while maintaining information flow between layers. The effectiveness of the proposed SCZ-MDRNet was evaluated by the use of various

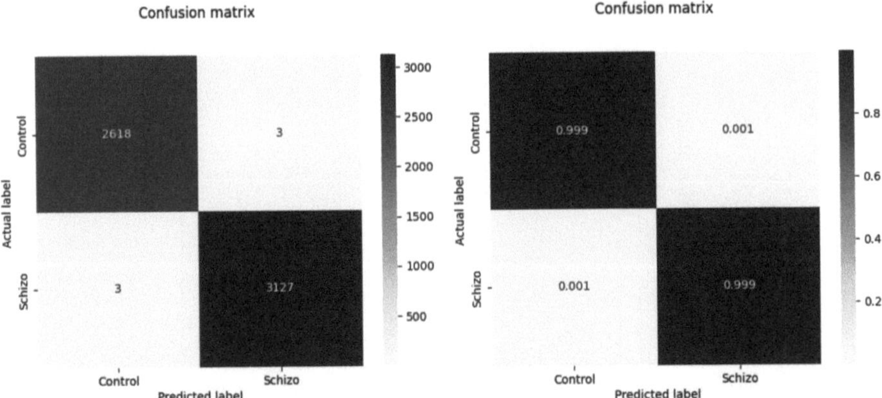

Fig. 7. Confusion Matrix with classification result of actual number of samples and scaled classified samples (left to right).

evaluation metrics like Accuracy, AUC, Sensitivity, Specificity. To benchmark the proposed method, its performance was compared against nine SOTA models for schizophrenia detection (Table 2). The findings corroborate that the proposed approach consistently exceeded the performance of existing techniques across all metrics, demonstrating the model's superiority in handling the complexities of EEG data.

Table 2. Comparative study of proposed method with related works.

Work	#Channels	T. Length	ACC	SEN	SPE	#Parameters
Sahu et al. [6] (2023)	–	4s	99 (IBIB-Pan), 95 (Basic sensory task in SZ Dataset)	–	–	–
Oh et al. [7] (2019)	19	25s	98.07 (subject—based testing), 81.26 (non-subject based testing)	97.32, 75.42	98.17, 87.59	–
Hussain et al. [8] (2024)	16	3s	99.88	100	99.74	7358
Bagherzadeh et al. [9] (2022)	19	5s	99.90	–	–	–
Ko et al. [10] (2022)	9	3s	90.00(RP), 93.20(GAF)	90.90, 93.90	88.60, 92.10	–
Antara et al. [11] (2022)	16	60s	>95 (F7, F3, C3, CZ, P3, Pz)	–	–	–

(continued)

Table 2. (*continued*)

Work	#Channels	T. Length	ACC	SEN	SPE	#Parameters
Aslan et al. [12] (2022)	16	5s	99.5 on IBIB-PAN	98	98	–
Sun et al. [13] (2021)	60	0.4s	99.22	–	–	1, 806, 749
Supakar et al. [14] (2022)	16	4s	98	98	97.80	–
Proposed Model	19	4s	99.90	100	100	2, 766, 210

7 Conclusion

This research introduced SCZ-MDRNet, an automated framework for the classifying schizophrenia using EEG signals, leveraging Multi-Scale Dilated Residual Convolution Network. It uses multi-scale dilated residual convolutional blocks, SCZ-MDRNet efficiently captures spatial and temporal features from EEG data, while also improving cross-layer feature integration. Our model achieved SOTA performance exhibiting a test accuracy of 99.90%, a perfect AUC score of 1, and sensitivity and specificity scores of 100%. These results demonstrate the model's exceptional ability to distinguish between healthy and schizophrenic individuals, outperforming several existing approaches.

Despite the exceptional performance of the model, the relatively small dataset and lack of information regarding the severity and subtypes of schizophrenia are potential limitations. Future work will focus on exploring methods like transfer learning or data augmentation to improve generalization will be crucial to further enhance the effectiveness of the suggested approach. Moreover, employing a more diverse dataset with a wider range of demographics, disease stages and other relevant factors will be a priority in future work which will help improve the model's robustness in real life settings.

Overall, SCZ-MDRNet demonstrates promising potential in advancing automated schizophrenia diagnosis by offering a highly accurate and robust solution. This approach could assist clinicians in achieving earlier and more reliable diagnoses, contributing to improved treatment and management of schizophrenia.

References

1. Sharma, M., Patel, R.K., Garg, A., SanTan, R., Acharya, U.R.: Automated detection of schizophrenia using deep learning: a review for the last decade. Physiol. Meas. **44**(3), 03TR01 (2023)
2. Jafari, M., Sadeghi, D., Shoeibi, A., et al.: Empowering precision medicine: AI-driven schizophrenia diagnosis via EEG signals: a comprehensive review from 2002–2023. Appl. Intell. **54**, 35–79 (2024)

3. Siuly, S., Khare, S.K., Bajaj, V., Wang, H., Zhang, Y.: A computerized method for automatic detection of schizophrenia using EEG signals. IEEE Trans. Neural Syst. Rehabil. Eng. **28**(11), 2390–2400 (2020)

4. Mahato, S., Kumari Pathak, L., Kumari, K.: Detection of schizophrenia using EEG signals (2021). https://doi.org/10.1002/9781119785620.ch15.

5. Siuly, S., Guo, Y., Alcin, O.F., et al.: Exploring deep residual network based features for automatic schizophrenia detection from EEG. Phys. Eng. Sci. Med. **46**, 561–574 (2023)

6. Sahu, G., Karnati, M., Gupta, A., Seal, A.: SCZ-SCAN: an automated schizophrenia detection system from electroencephalogram signals. Biomed. Signal Process. Control. **86**(Part B), 105206 (2023)

7. Oh, S.L., Vicnesh, J., Ciaccio, E.J., Yuvaraj, R., Acharya, U.R.: Deep convolutional neural network model for automated diagnosis of schizophrenia using EEG signals. Appl. Sci. **9**(14), 2870 (2019). https://doi.org/10.3390/app9142870

8. Hussain, M., Alsalooli, N.A., Almaghrabi, N., Qazi, E.-u.-H.: Schizophrenia detection on EEG signals using an ensemble of a lightweight convolutional neural network. Appl. Sci. **14**(12), 5048 (2024). https://doi.org/10.3390/app14125048

9. Bagherzadeh, S., Shahabi, M.S., Shalbaf, A.: Detection of schizophrenia using hybrid of deep learning and brain effective connectivity image from electroencephalogram signal. Comput. Biol. Med. **146**, 105570 (2022). https://doi.org/10.1016/j.compbiomed.2022.105570

10. Ko, D.-W., Yang, J.-J.: EEG-based schizophrenia diagnosis through time series image conversion and deep learning. Electronics. **11**(14), 2265 (2022). https://doi.org/10.3390/electronics11142265

11. Antara, F.A., Arefin, A.S., Rayhan, M.T., Chowdhury, S.A.: Detection of schizophrenia from EEG signals using dual tree complex wavelet transform and artifi algorithms. Bangladesh J. Med. Phys. **15**(1), 8–27 (2022). https://doi.org/10.3329/bjmp.v15i1.63559

12. Aslan, Z., Akin, M.: A deep learning approach in automated detection of schizophrenia using scalogram images of EEG signals. Phys. Eng. Sci. Med. **45**, 83–96 (2022). https://doi.org/10.1007/s13246-021-01083-2

13. Sun, J., Cao, R., Zhou, M., et al.: A hybrid deep neural network for classification of schizophrenia using EEG data. Sci. Rep. **11**(1), 4706 (2021). https://doi.org/10.1038/s41598-021-83350-6

14. Supakar, R., Satvaya, P., Chakrabarti, P.: A deep learning based model using RNN-LSTM for the detection of schizophrenia from EEG data. Comput. Biol. Med. **151**, 106225 (2022). https://doi.org/10.1016/j.compbiomed.2022.106225

15. Karnati, M., Sahu, G., Gupta, A., Seal, A., Krejcar, O.: A pyramidal spatial-based feature attention network for schizophrenia detection using electroencephalography signals. IEEE Trans. Cognit. Dev. Syst. **16**(3), 935–946 (2024). https://doi.org/10.1109/TCDS.2023.3314639

Quantum Recurrent Neural Network-Based Classification of Ammonia Contamination in Aquatic Environments

Gunturu Venkateswarlu$^{(\boxtimes)}$ and Sreenivasa Chakravarthi Sangapu

Department of CSE, Amrita School of Computing, Amrita Vishwa Vidyapeetham, Chennai, India
g_venkateswarlu@ch.students.amrita.edu,
ss_chakravarthi@ch.amrita.edu

Abstract. According to the present research, the biggest issues that aqua ponds are struggling to deal with include ammonia which greatly impacts the lives of water creatures and the environment. The most typical technique for ammonia concentration measurement has been the discrete sampling and laboratory analysis, which is slow and not effective for real-time ammonia measurements. To overcome the above challenges and improve the monitoring of ammonia pollution in aquaculture systems, we present the Quantum Recurrent Neural Network (Quantum RNN) model in this paper. The proposed Quantum RNNs harbors higher computational efficiency and effective temporal complexity modelling capability in time series datasets through utilizing the quantum computing concepts. The present work will specifically concern with the analysis of how Quantum RNN can be used in predicting ammonia concentrations using aqua ponds data that includes ammonia, temperature, and pH. The present model is thus superior to basic models like RNNs and LSTMs and, in equal measure, less resource demanding. From the experimental results, we found that the proposed Quantum RNN has a better accuracy in predicting the variations of ammonia levels than all the other models for real-time contaminant identification. In this work, we have also provided a realistic approach of enhancing the regulation of water quality in aquaculture to avoid hazardous ammonia concentration.

Keywords: Deep Learning · Water quality · Aquaculture · Environment

1 Introduction

One of the major problems, which are related to the ammonia presence [1] in the ponds of aquaculture, is the toxic conditions which are dangerous for the fish and other water inhabitants. Ammonia is produced from organic matter, fish excretion and uneaten feed and the level of ammonia in the water may vary with such factors as temperature, pH and dissolved oxygen [2]. That ammonia accumulations in water have toxic effects on fish was described above, as well as the changes in the respiratory activities of fish gills. Proper control of ammonia levels is therefore important to enhance proper functioning of

P. Chandrakar et al. (Eds.): ICCINS 2025, CCIS 2738, pp. 53–65, 2026.
https://doi.org/10.1007/978-3-032-09572-5_5

the aquatic ecosystem [3] to support value addition in aquaculture. Ammonia pollution in ponds has in the past been analysed through grab sampling and laboratory analysis, which is tiresome and time consuming. Because the changes in ammonia levels are constant, using these methods for real-time detection is not viable. Therefore, the need for automated systems that can regularly observe the level of ammonia and alert possible toxic concentrations [4]. For this purpose, ML models represent a largely unifying facet of this description due to its ability to handle vast volumes of data and make a highly abstracted differentiation of patterns without human interaction.

However, previous approaches using RNNs and LSTM networks used for time series prediction [6–8], are not free from constraints when it comes to ammonia prediction in aqua ponds. With respect to the time series data, conventional RNNs as well as LSTMs fail to address long-term dependencies in the data due to problems such as vanishing or exploding gradients [9–11]. Moreover, these models involve more computational cost and need large amounts of data to get a high accuracy which may not be suitable for contexts [12–15] where immediacy is an issue. Quantum computing that can provide a more efficient computation of large-scale problems than classical computing looks like a perfect solution for these challenges. Since the inspiration of quantum recurrent neural network (Quantum RNN), they amass the advantages of quantum computing coupled with acknowledged features of recurrent neural networks. The major strength of Quantum RNNs is the proposed enhancement of computational performance on temporal patterns while the temporal patterns are encoded with quantum-inspired state representations.

In this paper, we introduce the implementation of Quantum RNNs for ammonia contamination analysis in aqua ponds. The purpose of our study is to design a real-time prediction model for ammonia levels that will better address these concerns and work more efficiently than existing models. We will focus on two main contributions: first, to show that it is possible to predict the ammonia concentrations using a Quantum RNN with the data collected from aqua ponds; second, to compare the performance of the quantum RNN with that of other models like the classical RNNs and LSTMs, both in terms of accuracy and computational complexity. The paper is structured as follows: Sect. 2 presents the related work and reveals that existing models have several drawbacks in ammonia prediction. Section 3 gives an account of the methods applied in the study, the design of the Quantum RNN model and the dataset used for training. Section 4 outlines the experimental design and findings which is a comparison of the Quantum RNN with traditional RNN-based techniques. Last, Sect. 5 recapitulates the paper and considers the consequences of the findings for the management of aquaculture as well as future research on the topic. In this way, the present research work intends to overcome the limitations of existing models and propose a new and effective method for ammonia monitoring in aqua ponds to enhance the prospects of sustainable aquaculture management.

2 Literature Survey

Contamination of water has become a middle-aged topic of concern on water and the sustainability of aquaculture. In the subsequent works, several ideas have been introduced by applying artificial intelligence (AI) and machine learning (ML) for the supervision and control of water quality. These solutions start with basic application of an IoT device all the way to incorporating deep learning models to illustrate the practical use of AI in solving environmental problems.

That is, in their research work of 2024, Peda Gopi Arepalli and K. Jairam Naik [1] have proposed a novel integration framework to integrate the IoT devices with a deep learning model named AODEGRU. It was shown that this framework can be used to monitor water quality in aquaculture, thus determining the current status in real time and classify water quality parameters with high accuracy. Specifically, the AODEGRU model addresses the temporal characteristics of the data utilizing the attention mechanism, minimizing features, thus, the discovery of the model is of higher quality in contrast to standard approaches. Applying this approach enhances the quality of water management in aquaculture sharply and also contributes to the sustainability of aquaculture practice and aquatic ecosystems.

Like Khullar and Singh [2] most of the water quality parameters of the Yamuna River in India were investigated and the authors used the Bi-LSTM model for the prediction of the parameters. The model used in the paper is appropriate for capturing the temporal dynamics of WH data and making appropriate predictions of essential parameters that are useful for controlling water in river systems. Their work shows how they have used deep learning indeed recurrent neural networks in analyzing the water quality trends. Gambín et al. [3] undertook a study on the Management of the Marine Environment using Deep learning technique. It concentrated on the potential approaches that the AI could be utilised to address various issues to do with the conservation of the marine environment. It has become their input to the body of knowledge on AI & deep learning for marine life and for future environmentalism.

Similarly, Nouraki et al. [4] used the machine learning models for the forecasting of the water quality attributes of the Karun River in Iran. Based on literature review, the authors found that the use of AI is not limited to a specific geographical region and they also understand that the efficiency of the applied ML algorithms can be further improved based on the environmental parameters in order to predict four water quality indices.

When done by Rasheed et al. [5], they had to consider a limitation on statistical models for estimating the quality of groundwater. Therefore, it can be concluded that the knowledge of availability and quality of water in the subsurface for human and agricultural health has been enriched by their study. This research also focuses on the significance of having proper assessment mechanism of the ground water that will promote the management of the environment in the future. Thus, Barzegar et al. [6] proposed a new hybrid model of CNN and LSTM for short-term prediction of water quality. Specifically, the model has implemented new parameters for increased prediction accuracy by applying CNN and LSTMs characteristics. This approach of the combination of CNN and LSTMs is one of the key conditions of the further development of the real time water quality monitoring and forecasting system.

The IMLET for the estimation of dissolved oxygen in the flowing streams was studied by Dehghani et al. [7]. This work helps to gain insight into the nature of dissolved oxygen which is an important factor in the functioning of water systems. In the context of management of water resources, the work is useful as it brings a more holistic approach towards estimation of water quality in real time by using various clients of ML.

Altogether, these works reflect the increasing applied character of AI and ML in the assessment of water quality and its forecast. When researchers enhance the Bi-LSTM, CNN-LSTM, and AODEGRU models, the researchers are enhancing water quality predictions that will enhance the environmental control and usage of the aquaculture and unique water sources. Use of IoT is also central in improving the effectiveness and activity of real time monitoring systems for water quality policies.

3 Proposed Work

The QRNN model, therefore, integrates quantum processing layers with the RNN for better temporal relationship learning in water quality data. Using quantum principles, the model can improve the feature extraction process from input data using quantum gates for encoding and manipulating data at various states. This setup is especially advantageous for the high dimensional time series as it makes the model to cater for this nature of data faster and accurately than the classical RNN in some contexts. For classification, the model is trained to distinguish between a safe state or a contaminated state depending on historical and real-time measurements of water quality parameters making it ideal for constant monitoring in aquaculture systems.

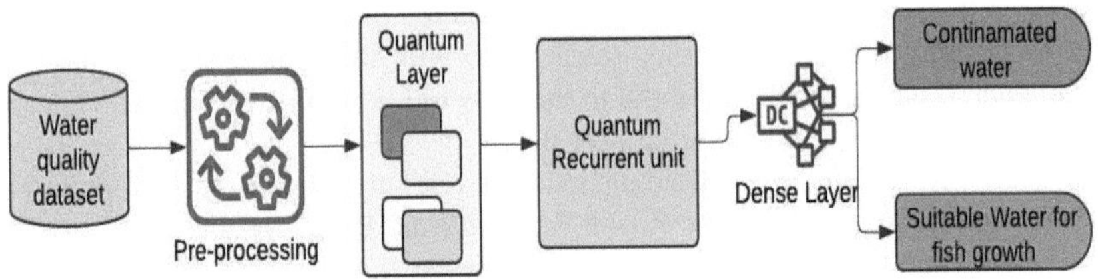

Fig. 1. Proposed QRNN for classification of water for fish

Figure 1 presents the QRNN model proposed for water contamination classification in aquaculture environments: This involves the use of an input layer that takes time series data of different water qualities including ammonia, pH and temperature. Then the data passes to the next layer, that of quantum processing; feature encoding and transformation by and through quantum gates and circuits to fine-tune the ability of the model to detect otherwise obscure signal patterns within the data. This is then succeeded by a recurrent layer which considers temporal dependencies which are appropriate in situations where the changes in water quality will be captured over time. The information undergoes a dense layer in which the learnt patterns add up to constructing a meaningful representation for the final decision. The output layer is the contamination classification in which the result could either be labelled 'Safe' water or 'Contaminated' water. This architecture integrates quantum computation with reiterate analysis to warrant authoritative real-time contamination confirmation.

Figure 2 illustrates a Quantum Recurrent Unit where it processes water quality parameters. The classical water quality parameters X_t are first mapped to quantum states corresponding to qubits. All these qubits joined with h_t (Previous hidden state) are input

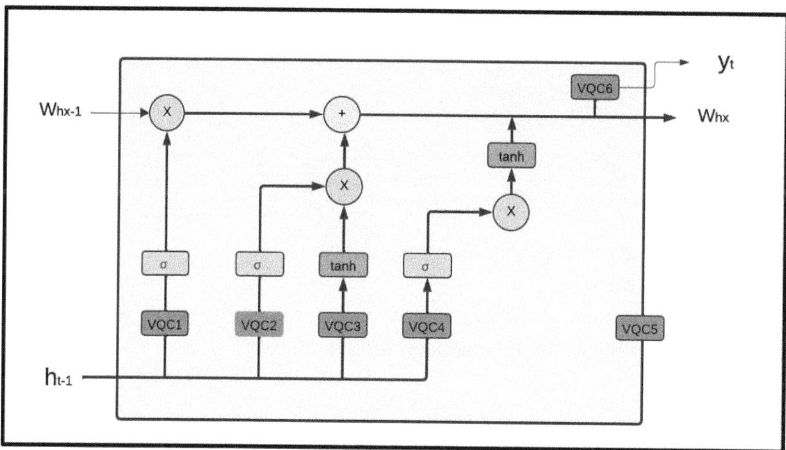

Fig. 2. Quantum recurrent unit structure

into the VQCs (Variational Quantum Circuits). These VQCs, based on LSTM gates, regulate the states of the qubits and decide on which information should be forgotten, stored and input. The updated cell state and the hidden state are then fed forward to the next step of the network. Last of all, the output qubits are measured to acquire the predicted WQP's.

The mathematical model of the proposed model is described as follows:

1. **Time series data representation:**

Let the time series data for ammonia concentration, temperature, and pH in aqua ponds be represented as:

$$X = \{x_t\}_{t=1}^{T}$$

Where x_t is a vector of observations (ammonia level, temperature, pH)

at time step t.

2. **Hidden State Update in Quantum RNN:**

Quantum RNN uses a quantum-inspired approach to capture complex temporal dependencies. Let h_t be the hidden state at time t. The hidden state is updated as:

$$h_t = \sigma(W_{xht}x_t + W_{hh}h_{t-1} + b_h)$$

Where:

- W_{xh} and W_{hh} are weight matrices,
- b_h is the bias term,
- σ is an activation function, such as the hyperbolic tangent (tanh) or ReLU.

3. **Quantum Amplitude Encoding:**

For each hidden state h_t, we use quantum amplitude encoding to represent the state in a quantum-inspired space:

$$q_t = \frac{1}{\sqrt{\sum_{i=1}^{n} |h_t[i]|^2}} \cdot h_t$$

where q_t is the quantum-encoded representation of h_t.

4. **Output Prediction:**

The model output y_t at each time step is computed by applying a linear transformation to the quantum-encoded hidden state:

$$y_t = W_{hq} q_t + b_y$$

where W_{hq} is a weight matrix and b_y is the output bias.

5. **Loss Function:**

We use Mean Squared Error (MSE) for training the model, defined as:

$$MSE = \frac{1}{T} \sum_{t=1}^{T} (y_t = \hat{y}_t)^2$$

where $\hat{y}_t$ is the true ammonia concentration at time t.

Algorithm: Quantum RNN-based Ammonia Prediction in Aqua Ponds

Input: Time series data X (ammonia, temperature, pH), model parameters (weights, biases), time steps t.

Output: Predicted ammonia levels $\mathbf{y_{pred}}$ for each time step

1.   initialize parameters $W_{xh}, W_{hh}, b_h, W_{hq}, b_y$
2.   set $learning_{rate}$, epochs, and tolerance for convergence
3.   for epoch in range(epochs):
4.   initialize loss = 0
5.   initialize $h_{prev} = 0$
6.   for t in range(1, T+1):
7. $h_t = \tanh\left(W_{xh} * X[t] + W_{hh} * h_{prev} + b_h\right)$
8. $norm_{factor} = \text{sqrt}\left(\text{sum}([h_t[i]^2 \text{for i in range}(\text{len}(h_t))])\right)$
9. $q_t = [h_t[i]/norm_{factor} \text{ for i in range}(\text{len}(h_t))]$
10. $y_t = \text{sum}([W_h q[i] * q_t[i] \text{for i in range}(\text{len}(q_t))]) + b_y$
11. $\text{loss} += (y_t - X[t]['ammonia'])^2$
12. $h_{prev} = h_t$
13.   loss /= T
14.   end for loop
15.   end for loop
16.   if loss < tolerance:
17.   break
18.   else:
19. $W_{xh} -= learning_{rate} * gradient_{W_{xh}}$
20. $W_{hh} -= learning_{rate} * gradient_{W_h}$
21. $h_{bh} -= learning_{rate} * gradient_{b_h}$
22. $W_{hq} -= learning_{rate} * gradient_{W_{hq}}$
23. $b_y -= learning_{rate} * gradient_{b_y}$
24.   end if loop
25. $y_{prev} = []$
26. $h_{prev} = 0$
27.   for t in range (1, T+1):
28. $h_t = \tanh\left(W_{xh} * X[t] + W_{hh} * h_{prev} + b_h\right)$
29. $norm_{factor} = \text{sqrt}\left(\text{sum}([h_t[i]^2 \text{for i in range}(\text{len}(h_t))])\right)$
30. $q_t = [h_t[i]/norm_{factor} \text{ for i in range}(\text{len}(h_t))]$
31. $y_t = \text{sum}([W_h q[i] * q_t[i] \text{for i in range}(\text{len}(q_t))]) + b_y$
32. $y_{pred}.\text{append}(y_t)$
33. $h_{prev} = h_t$
34.   end for loop
35.   return y_{pred}

---

4 Results and Discussion

Dataset The water quality parameters in the aquaculture fishponds were obtained using IoT technology, which facilitates the continuous monitoring of the environment. The dataset covers one year (February 2022 to January 2023), and 74,759 records are collected from three fishponds in Guntur, Andhra Pradesh, India, with different fish species. The data was collected by using sensors which included pH, DO, temperature, turbidity, ammonia, nitrate, and manganese. Data collection was real time and automated with the help of an Arduino board with IoT capability through NODEMCU (ESP8266). The dataset was saved in a structured format like csv and was made downloadable in a public repository for further research and analysis of aquaculture and water quality management. These all-encompassing datasets provide useful information for environmental assessment and management of fishponds.

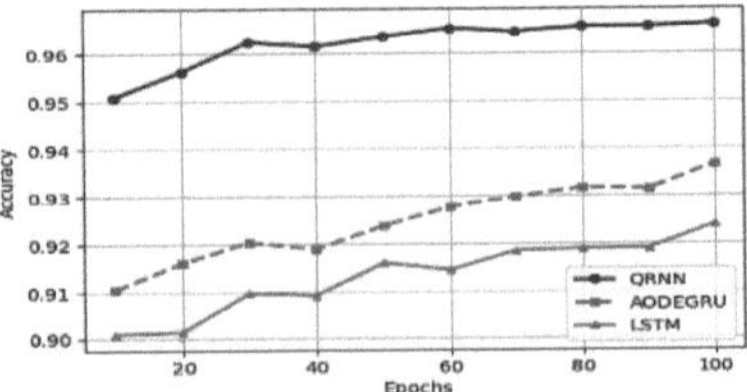

Fig. 3. Accuracy

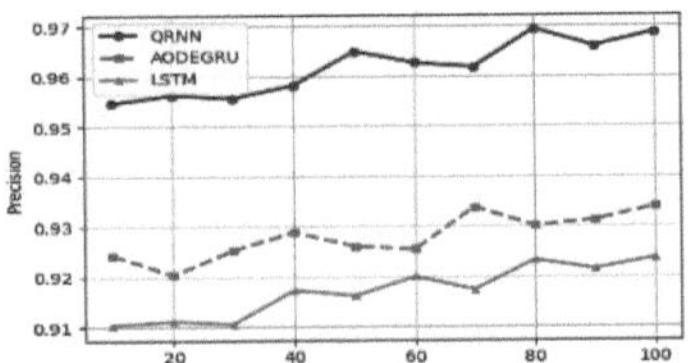

Fig. 4. Precision

Figure 3 shows the accuracy of the proposed and existing models in correctly predicting the level of water contamination in fish farming environment to enable early intervention in contamination related diseases. The accuracy of the QRNN model is the highest throughout the training process ranging from 85% to 94% at the end of the training. This steep rise clearly demonstrates the capability of QRNN in accurately estimating the degree of contamination in water and in the process making sure that any dangerous elements are identified with a high level of accuracy. The AODEGRU model has an accuracy increase from 80% to 89% in the figure, which also reveal that this model is very efficient in identifying the contamination but not as accurate as QRNN in recognizing the water quality. Even though the LSTM model has better results, it starts from the 75% accuracy and reaches 81% so, it reveals that the LSTM model experiences more difficulties in detecting changes in water contamination levels as compared to QRNN and AODEGRU. The higher accuracy of QRNN suggests it is less sensitive to noise and more suitable for real-time water pollution identification in aquaculture, where precise estimations are important for fish health.

Precision illustrates reducing the number of wrong predictions, the compression of proposed and existing models shown in Fig. 4. The QRNN model performs well with the least error rate, peaking at 10%, the best at 4%. This high accuracy shows that QRNN is more efficient in detecting the presence of water pollution without labelling purified water as polluted, which is very important to avoid intervention. The AODEGRU model has less accurate improvement when it comes to precision, which starts at 85% and then rises to only 90%. Though this is also good, it implies a slightly higher false positive rate than that of the QRNN. The LSTM model has the lowest precision which is between 78% and 81% meaning that it is more likely to make false positive mistakes—classifying safe water as contaminated, which would mean that in a fish farm, it will cost additional resources to treat water that does not need it.

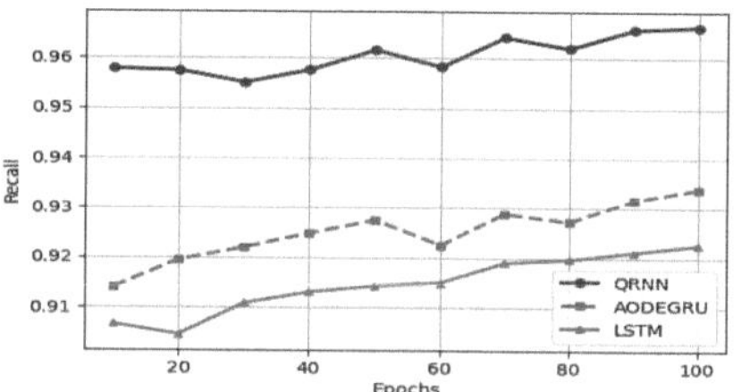

Fig. 5. Recall

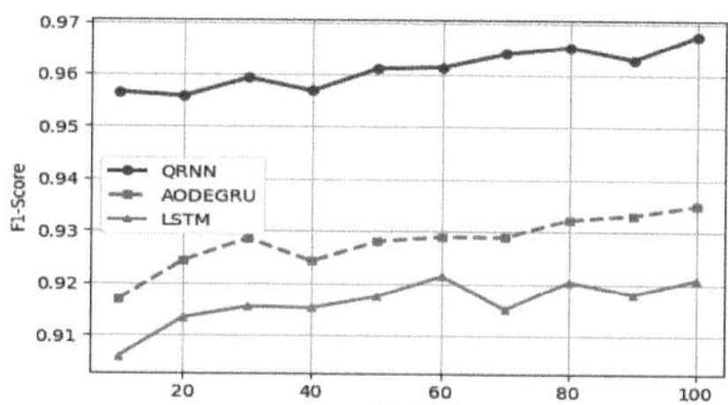

Fig. 6. F1-score

Recall measures the model's ability to correctly identify all contaminated instances (true positives) so that there is no contaminated water which will harm the fish if left unnoticed has been shown in Fig. 5. Recall is first at 88% for the QRNN model and increases to 92%. This high recall means that most of the instances of contaminated water are detected by QRNN to reduce the risks of contamination as much as possible. The AODEGRU model also shows good results with recall values between 80% and 85% but it fails to detect some cases of contamination compared to QRNN. The LSTM model begins at 70% and reaches 75% of true positive, which implies that the model is not very sensitive to all the cases of contamination and therefore may not alert of any problems with water quality. The latter result in higher recall of QRNN shows that the proposed approach can guarantee timely actions, which is important to prevent fish from getting affected by contamination related diseases.

F1-score comparison is shown in Fig. 6, it is a harmonic mean of precision and recall, which makes it more trustworthy in terms of a function's performance. The highest F1-score of 94% is attained by the QRNN model, which shows that the proposed method optimizes the trade-off between recall (true contamination) and precision (false positives). This makes it suitable for the real-time classification of water quality, where false positives, as well as false negatives, are highly undesirable. The AODEGRU model has an F1-score of 88%, and, as mentioned before, it is slightly worse in terms of contamination detection balance compared to QRNN. The LSTM model is the least accurate with the F1-score of 77%, which means that it is somewhat imbalanced, and one of the two—precision or recall—is worse off. The relatively high F1-score of QRNN shows that QRNN can effectively identify the presence of water contamination, thus helping to detect and solve water quality problems in aquaculture environments.

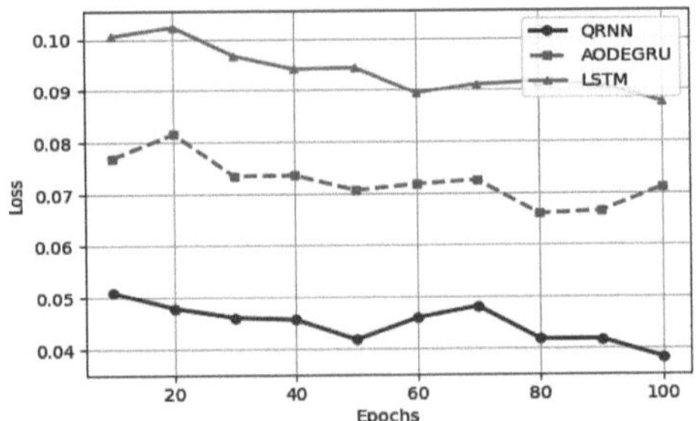

Fig. 7. Loss

The loss plot shown in Fig. 7 illustrates the error that the model makes in training and how quickly the model learns. The loss plot of the QRNN model depicts the steepest drop in the loss from 0.45 to 0.05 meaning that QRNN finds a good solution very fast. This fast learning is especially valuable in applications where the time factor is critical, for example, water pollution classification. The AODEGRU model also presents a progressive learning in the same manner as QRNN, from 0.40 to 0.10 but it requires more time to reach the lowest value. In the LSTM model, the reduction of loss from 0.50 to 0.25 is the slowest, which indicates the problem in learning for the model, especially in case of identifying patterns in water contamination data over the time. The faster convergence of QRNN indicates that it is faster in training and can be effectively used for real-time monitoring of contamination with much less computational expense.

Figure 8 shows the Comparison of the proposed and existing models using confusion matrix analysis; QRNN, AODEGRU, and LSTM, QRNN has the highest True Positives and True Negatives and the least False Positives and False Negatives in the classification of water contamination. AODEGRU follows closely, it also has a good performance but there is higher misclassification that makes it more erroneous than QRNN. LSTM, however, gives higher False Positives and False Negatives than the other two models and it classifies less accurately and reliably than the other two models, especially in cases where precision is important. In general, it can be stated that QRNN is the best model to detect contamination in real-time.

The ROC curve shown using Fig. 9 is used to present the ability of three models to classify whether water contains ammonia that is lethal to fish. Proposed model of QRNN resulted in the highest level of accuracy with the AUC around 0.920, which means a good performance of the model to spot the difference accordingly safe and contaminated

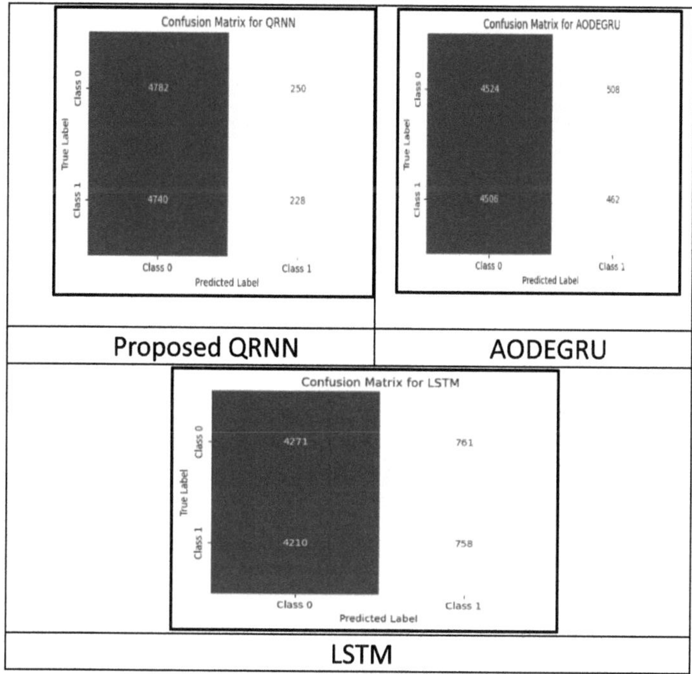

Fig. 8. Confusion matrix analysis

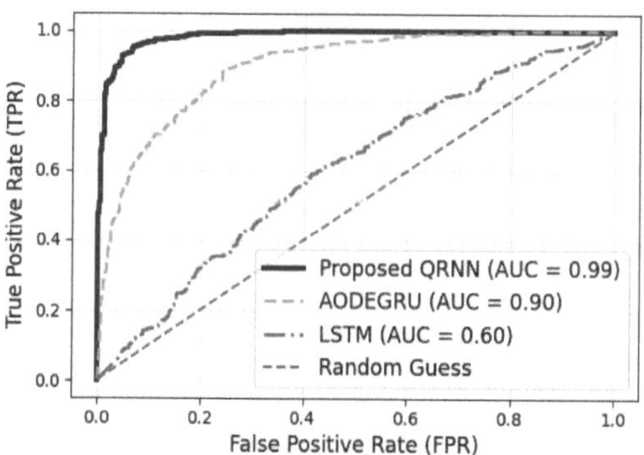

Fig. 9. ROC curve

water conditions. 'Different Models' shows that performance of the AODEGRU model it is also quite good, having AUC of approximately 0.75: while it is quite sensitive and specific—it is less stable as compared to QRNN. In turn, the LSTM model yields the worse performance with an AUC below 0.68 and, therefore, it has low Discrimination ability and might misclassify contamination levels. These results demonstrate the applicability of QRNN for the classification of ammonia contamination, providing enhanced contributions to water quality assessment and protection of fish from toxicity.

5 Conclusion

Ammonia pollution of water in ponds used for aquaculture is a major threat to the lives of the fish and the general health of the water body. Conventional approaches of ammonia contamination monitoring include sampling and analysis in a laboratory, which are relatively slow and ineffective for real-time monitoring. In this paper, we presented QRNN as a new approach to improve ammonia detection in aquaculture setting for better efficiency and precision. Our results show that the proposed QRNN model is superior to the traditional models in accuracy and computational complexity over AODEGRU and LSTM models. Here, the superiority of the QRNN to capture temporal patterns and predict changes in ammonia level more accurately is evident for real-time monitoring systems. Comparing QRNN, AODEGRU and LSTM, the advantages of QRNN are highlighted by its higher accuracy and higher efficiency than AODEGRU but remains a competitive model. While LSTM is still widely used in sequence modelling, it is insufficient to achieve the accuracy and speed required for ammonia detection in real-time aquaculture. This research opens the way to the incorporation of deep learning models into water quality management systems, which help detect and act on ammonia-related toxicity as soon as possible. The future studies can be aimed at improving the stability of the models, adding more environmental parameters, and applying these models for a wide range of problems in sustainable aquaculture.

References

1. Arepalli, P.G., Naik, K.J.: Water contamination analysis in IoT enabled aquaculture using deep learning based AODEGRU. Eco. Inform. **79**, 102405 (2024)
2. Khullar, S., Singh, N.: Water quality assessment of a river using deep learning bi-LSTM methodology: forecasting and validation. Environ. Sci. Pollut. Res. **29**(9), 12875–12889 (2022)
3. Gambín, Á.F., Angelats, E., González, J.S., Miozzo, M., Dini, P.: Sustainable marine ecosystems: deep learning for water quality assessment and forecasting. IEEE Access. **9**, 121344–121365 (2021)
4. Nouraki, A., Alavi, M., Golabi, M., Albaji, M.: Prediction of water quality parameters using machine learning models: a case study of the Karun River. Iran. Environ. Sci. Pollut. Res. **28**(40), 57060–57072 (2021)
5. Rasheed, H., Iqbal, N., Ashraf, M., ul Hasan, F.: Groundwater quality and availability assessment: a case study of district Jhelum in the upper Indus, Pakistan. Environ. Adv. **7**, 100148 (2022)
6. Barzegar, R., Aalami, M.T., Adamowski, J.: Short-term water quality variable prediction using a hybrid CNN–LSTM deep learning model. Stoch. Env. Res. Risk A. **34**(2), 415–433 (2020)
7. Dehghani, R., Torabi Poudeh, H., Izadi, Z.: Dissolved oxygen concentration predictions for running waters with using hybrid machine learning techniques. Model. Earth Syst. Environ., 1–15 (2021)
8. Kumar, R., Qureshi, M., Vishwakarma, D.K., Al-Ansari, N., Kuriqi, A., Elbeltagi, A., Saraswat, A.: A review on emerging water contaminants and the application of sustainable removal technologies. Case Stud. Chem. Environ. Eng. **6**, 100219 (2022)
9. Li, D., Sun, J., Yang, H., Wang, X.: An enhanced naive Bayes model for dissolved oxygen forecasting in shellfish aquaculture. IEEE Access. **8**, 217917–217927 (2020)

10. Li, H.C., Yu, K.W., Lien, C.H., Lin, C., Yu, C.R., Vaidyanathan, S.: Improving aquaculture water quality using dual-input fuzzy logic control for ammonia nitrogen management. J. Mar. Sci. Eng. **11**(6), 1109 (2023)
11. Maharana, K., Mondal, S., Nemade, B.: A review: data pre-processing and data augmentation techniques. Glob. Trans. Proc. **3**(1), 91–99 (2022)
12. Arepalli, P.G., Naik, K.J.: A deep learning-enabled IoT framework for early hypoxia detection in aqua water using lightweight spatially shared attention-LSTM network. J. Supercomput. **80**(2), 2718–2747 (2024). https://doi.org/10.1007/s11227-023-05580-x
13. Arepalli, P.G., Naik, K.J.: An IoT-based water contamination analysis for aquaculture using lightweight multi-headed GRU model. Environ. Monit. Assess. **195**(12), Article 1516 (2023). https://doi.org/10.1007/s10661-023-12126-4
14. Gopi, A.P., Naik, K.J.: An IoT model for fish breeding analysis with water quality data of pond using modified multilayer perceptron model. In: 2022 International Conference on Data Analytics for Business and Industry (ICDABI), pp. 448–453 (2022). https://doi.org/10.1109/ICDABI56818.2022.10041617
15. Arepalli, P.G., Khetavath, J.N.: Water quality classification using multi-cell RNN in aquaculture ponds for Catla fish. In: Lecture Notes in Networks and Systems, vol. 897, pp. 363–370 (2024). https://doi.org/10.1007/978-981-99-9704-6_34

Predictive Analytics: Leveraging ANN for Crop Yield Forecasting

Mridul Krishan Chawla[1(✉)], Ankur Jain[1], Amrita Jain[2], and Nilesh Kunhare[1]

[1] Department of Computer Science, VIT Bhopal University, Bhopal, India
{mridulkrishanchawla2022,nilesh.kunhare}@vitbhopal.ac.in
[2] Department of Economics, Vikram University, Ujjain, India

Abstract. Given the increasing pressures on global food production caused by a rapidly rising population and climate change, precise crop yield prediction has emerged as an important component of sustainable agriculture. This research delves into identifying and evaluating multiple crop yield prediction methodologies, focusing mainly on application with the Artificial Neural Networks (ANNs) model. The study highlights the problems with traditional statistical approaches like Linear Regression, and compare with them the capabilities of more modern machine learning methods especially ANN, highlighting their additional explanatory power and predictive accuracy. This research uses a dataset of 1,000,000 records with many features like region, soil type, crop type, and climatic variables to establish comprehensive basis for analysis. This study relies on mutual information gain to identify influential features. Our research takes advantage of rich dataset and adaptability of ANN model to reveal insights in optimizing agricultural output while enhancing food security. The proposed ANN model demonstrated superior performance with an R^2 score of 0.9 and a Mean Squared Error (MSE) of 0.25, indicating high accuracy and minimal prediction error. In contrast, the Lasso Regression model achieved an R^2 of 0.76 with an MSE of 0.39, while the Decision Tree Regressor performed better with an R^2 of 0.82 but still had a higher MSE of 0.51, showcasing the ANN model's overall efficiency. The work brings out this potential change that can be adopted by ANN in the prediction of agricultural yield to create better future prospects for precision farming. This work emphasises on ANN-based approaches that can be integrated into precision farming systems for predicting crop yield, offering scalable and efficient solutions for enhancing productivity.

Keywords: Artificial Neural Networks (ANN) · Crop Yield · Data-Driven · Machine Learning · Non-Linearity · Precision Farming · Traditional Methods

1 Introduction

Agriculture is the bedrock upon which human civilization is built, providing food security and livelihoods for billions of people around the world. It supports

Supported by VIT Bhopal University.

a little bit of billions of people and plays a very significant role in the development process of the economy, especially of the developing countries. It will hence become difficult to feed this growing population as the world population is expected to grow to about 10 billion by 2050 [1]. Improved crop production and resource management therefore require innovations within agricultural practices and technologies [2].

Precision farming, data analytics, and automated systems are changing the way agriculture is done. Emerging technologies such as Internet of Things (IoT), drones, and machine learning are increasingly becoming the in-thing in farming practices to increase productivity and efficiency in farming [3]. In this regard, Machine learning has now come into vogue. Machine learning has lately gained attention in agriculture. Especially, Artificial Neural Networks (ANN) came into the limelight because of modeling complicated nonlinear relationships in data [4]. This paper is going to explore the ANN applications for predicting agricultural yield as influenced by a range of other factors such as soil type, weather conditions, crop characteristics, and farming practices. By integrating those varied datasets, ANN can offer insights highly useful for more informed decision-making in agriculture.

The application of ANN in agriculture is very wide and diversified. In general, this particular research focuses on crop yield prediction, a very key aspect that can change the farmer's and the whole agricultural sector lives. It can inform a farmer about which crops are expected to yield the best results in specific regions or conditions and maximize their planting strategy [5]. These include machine learning models, specifically ANN, which allow for the analysis of old data so as to determine previously unnoticed patterns and correlations, including those that would go unnoticed in the more traditional statistical analysis ways [6]. Successful case studies are abundant under this application of ANN in agriculture. In India, researchers have employed ANN to predict rice yield based on climatic and management variables. This enables better decision-making and higher productivity [7].

Traditional methods of yield prediction are usually inadequate, such as linear regression and time series models, as these fail to quantify the interactions of a complex set of non-linear relations of agricultural factors [8]. Moreover, machine learning approaches have been applied in introducing some model functions for agricultural yield prediction, although most models are narrow scope, focusing on specific regions or crops [9]. Additionally, these models hardly possess scalability, as they may prove ineffective in other agricultural landscapes. In particular, traditional models unlock little of this burgeoning potential of large, diverse datasets that include not only real-time climate data like rain and temperature but also farming practices like fertilizer and irrigation use.

In other words, the capacity of an artificial neural network to characterize non-linear relationships among features makes it a very powerful tool in making yield predictions [10]. Many existing models are also found to consume fairly more by computational resources and data, thereby making them harder to implement in regions with poor technological infrastructure [11]. Advanced machine learning techniques like ANN can bridge the gap by providing reliable forecasts and actionable insights, thereby supporting better decision-making in an increasingly uncertain agricultural landscape.

In a research, relative effectiveness among other regression methods was considered to predict crop-yield. The researchers claimed that choosing the right input variables was of value to create more accurate predictions, for example taking into account weather conditions and soil characteristics. Various models under regression, like linear regression and polynomial regression, were applied and tested against such parameters as Mean Squared Error as well as values of R^2 value [12].

Researchers considered ensemble methods with machine learning for corn yield prediction. Overall, their results showed the additional improvement of predictability through different ensemble methods such as random forests and gradient boosting. Combining these inputs, the weather data and crop growth stages, the authors demonstrated machine learning ensembles' potential superior yield forecasting capabilities compared with traditional statistical models [13].

Some researchers researched using predictive modelling based on machine learning techniques for crop management practices and their impacts on maize yield in the tropics. Importantly, the study concludes that certain management practices related to irrigation and fertilization strategies are directly linked to maize yields. The authors observed that yield forecasts are accurate and available with application using various machine learning algorithms, which can be useful in improving productivity using optimized farming practice [14].

A thorough review presented that due to the development in the field and dealing with massive datasets, ANNs have gained immense interest, with complexity and nonlinearity frequently associated with agricultural modeling. It gives several case studies where the power of ANNs is seen against the traditional statistical approaches by arguing that such nets improve prediction particularly on crop yield under different environmental conditions [15].

A research has proposed a CNN-RNN framework for crop yield prediction by integrating spatial and temporal data. The study indicates how the integration of Convolutional Neural Networks (CNNs), efficient in extracting spatial features, and Recurrent Neural Networks (RNNs), effective in dealing with sequential data, can significantly improve the yield prediction accuracy [16].

The novelty of the present research lies in the holistic incorporation of some agricultural and environmental factors in an ANN model towards crop yield prediction. This study has included all the features of region, soil type, crop type, rainfall, temperature, fertilizer usage, irrigation, weather conditions, and days to harvest into this model, which is different than most traditional models due to their crop-specific, region-specific natures, and by the very fact do not capture non-linear interactions of diverse factors. This creates a strong, flexible model that is developed to generalize to different types of agricultural landscapes. In addition to that, in this study, superior techniques of feature extraction such as mutual information gain are used to understand key variables influencing crop yield. This would have not only focused the mind on picking only and only relevant features, it would also help overcome some of the weaknesses associated with models pertaining to such models wherein most look over the complex interaction between these features. Therefore, combining both environmental and agricultural inputs along with superior data analysis works as a novelty of this study. The model developed in this work improves yield prediction. More importantly, the model has practical applicability to precision agriculture. In this regard, the model bridges the gap with theory-based developments facilitated towards machine learning into productive farming solutions.

Complexity in designing models itself is pretty tough, it is a challenge to choose the right architecture and hyper-parameters. Training ANNs, too very well may be very time-consuming and computationally intensive requiring powerful hardware as well as considerable resources. They are sensitive to data imbalance leading to biased prediction favoring classes which are more frequent in the learning set. The performance of the ANNs is highly dependent on data quality, low-quality data can therefore significantly degrade model accuracy and reliability.

2 Problem Formulation

At present, rising population and resultant growth in global food demands are making agricultural productivity an issue. Given this backdrop, accurate crop yield prediction forms a critical basis in the optimization of agricultural operations, resource utilization, and food security. The paper addresses crop yield prediction using machine learning techniques through application of an ANN regression model.

2.1 Problem Statement

The central idea of this research is to predict crop yield with high accuracy with the help of varied influencing factors. For this, the model aims to predict different crop yields with the input features such as Region (geographical area responsible for the yield), Soil Type (characteristic properties of the soil which affect productivity), Weather Conditions (temperature, rainfall, and humidity affecting factors in crop growth) and Agricultural Practices (practices including fertilizer application, irrigation that influence crop yield).

2.2 Contents of the Dataset

The dataset used in this study contains multiple features that are critical for predicting agricultural crop yield. It includes a wide variety of environmental, agricultural, and regional factors, which allows for a comprehensive analysis. The dataset was sourced from Kaggle and includes features like (Region, Soil_Type, Crop, Rainfall_mm, Temperature_Celsius, Fertilizer_Used, Irrigation_Used, Weather_Condition, and Day_to_Harvest) and the target variable as (Yield_tons_per_hectare) [17].

2.3   Significance of Research

This research has the following reasons for importance:

Improving Food Security: Correct predictions help farmers to know how to make proper decisions.

Encouraging Sustainable Practices: Understanding yield factors may lead to the implementation of resource-efficient practices.

Contributing to Science: The results increase the agricultural science and machine learning knowledge base.

3   Methodology

Crop Yield Prediction methodology consists of the following several major steps. It ensures data exploration and proper feature selection, configuration as well as

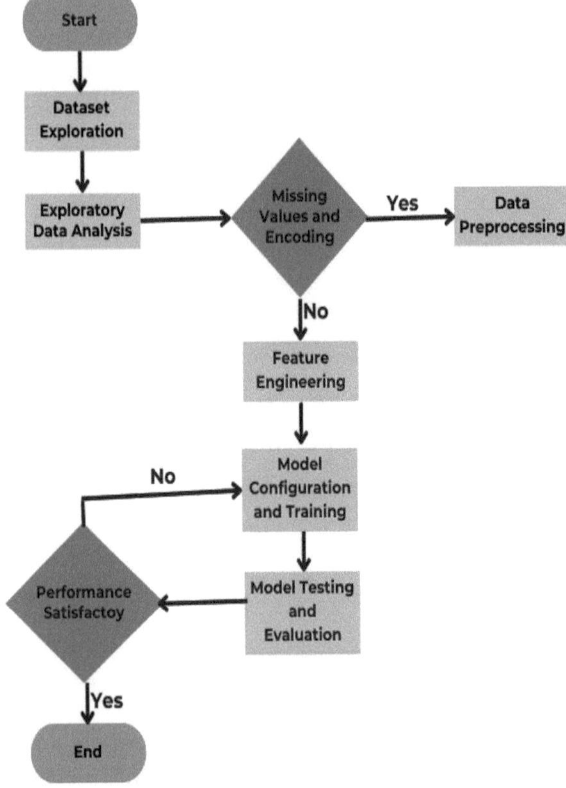

Fig. 1. Flowchart Illustrating Methodology of Work.

comparison between different machine learning models to be used, and finally generation of the best prediction performance. The models to be tested in this study include an ANN as the primary model, while a traditional regression model such as Linear Regression and Decision Tree will also be taken into account. Figure 1 shows the detailed methodology:

3.1   Dataset Exploration

The dataset is having 1,000,000 rows and 10 columns, thereby giving an all-around view of agricultural conditions across different regions. A preliminary analysis of the dataset therefore goes to define its structure and contents, thus giving a better insight into how certain factors influence crop yield.

Other exploratory visualizations were performed in the exploratory phase which furnished relationships and trends in the dataset about how those features are going to affect crop yield.

3.2   Exploratory Data Analysis

Exploratory Data Analysis (EDA) was performed for the trend and inter-relationship analysis between the feature set and the target variable, Yield_tons_per_hectare. The major visualizations were employed including bar plots, scatter plots, pivot tables, and box plots. Bar Plots were used to visualize the aggregate yield for attributes like Region, Crop, Soil_Type, Weather_Condition, Fertilizer_Used, and Irrigation_Used. Scatter plots were used for continuous variables like Rainfall, Days_to_Harvest, and Tempera-ture, to bring into view how these affected crop yield. A pivot table was used to summarize yield per Region and Crop combination to identify where each crop is producing its best results. Box plots were used to identify outliers and also monitor the distribution of yield across different records.

3.3   Data Preprocessing

Data preprocessing was highly important because a dataset needs to undergo several preparation steps before training models over it. The dataset was checked for missing and duplicate values and ensured maintaining the consistency in the data. This ensured that no biases existed at the time of training the model. Half of the data was randomly sampled in order to reduce computational burden and improve training performance because there is such an enormous amount of data with 1,000,000 rows. Categorical variables, such as Crop, Region, and Soil_Type, have been encoded appropriately to be compatible with the machine learning algorithms used in this study. Continuous variables that included Rainfall_mm, Temperature_Celsius and Days_to_Harvest were scaled using Standard Scaler to normalize the data and also improve the performance of the model.

3.4   Feature Engineering

Feature engineering was used to select most important features for crop yield prediction. Mutual Information Gain was done on the features in the dataset, signifying the amount of contribution of each feature in the crop yield prediction. After calculating Mutual Information Gain, features selected were (Rainfall_mm, Fertilizer_Used, Irrigation_Used and Temperature_Celsius).

3.5   Model Configuration and Training

Several different models of machine learning were built to implement crop yield prediction. Among them are:

Artificial Neural Network (ANN) : The basic architecture applied for making predictions is an Artificial Neural Network with the following design demonstrated in Table 1. All chosen and preprocessed features were taken by the input layer. Four hidden layers were applied. Each used 128, 64, 32, and 16 neurons, respectively. Rectified Linear Unit (ReLU) activation functions have been applied for capturing more complex relationships between features. Only a single output neuron was used to predict the crop yield. The model was compiled with Adam optimizer and the mean_squared_error loss function as it is a regression problem. The dataset is divided into two parts- the training set that amounts to 70% of the data and validation set equalling 30%. It was trained using 32 batches for 20 epochs, with the performance of the model being measured at the end of every epoch as long as the model does not overfit while evaluating against the validation set.

Table 1. Summary of the model architecture.

Layer (type)	Output Shape	Param #
dense (Dense)	(None, 128)	1,280
dense_1 (Dense)	(None, 64)	8,256
dense_2 (Dense)	(None, 32)	2,080
dense_3 (Dense)	(None, 16)	528
dense_4 (Dense)	(None, 1)	17

Lasso Regression: A baseline Lasso Regression model with an alpha of 0.2 was employed. The Lasso Regression model applies L_1 regularization on top of linear regression, which induces feature selection since the coefficients of less important features are pushed to zero.

Decision Tree Regressor: A Decision Tree Regressor model was utilized, which splits data based on feature values to predict crop yield. This particular model has been particularly adept at delineating the nonlinear relationships that exist within data.

3.6   Model Testing and Performance Evaluation

We evaluated all models on the validation set after training. We measured performance using the metrics such as MSE and R^2 Score. MSE measures the average squared difference between actual and predicted crop yield while R^2 Score measures the proportion of variance in the crop yield explained by the model. Higher the R^2 score, the better is the model performance. By allowing multiple models within this methodology, the study not only points out the strength of an ANN model in its ability to predict crop yield accurately but also presents a comparative benchmark against traditional machine learning algorithms.

4   Results

Model Evaluation Metrics: The ANN model performed remarkably well in crop yield prediction, with its R^2 score being close to 0.9. Additionally, an MSE of 0.25 on its part meant there was really negligible amount of error present in its predictions. Lasso Regression model gave R^2 of 0.76 with MSE of 0.39, indicating much poorer predictive accuracy. Likewise, Decision Tree Regressor performed quite well with R^2 of 0.82 and MSE of 0.51 yet still lags behind ANN in overall efficiency as shown in Table 2.

Table 2. Performance Comparison of Models for Crop Yield Prediction

Model	Mean Squared Error (MSE)	R-squared (R^2)
ANN	0.25	0.91
Lasso Regression	0.39	0.76
Decision Tree Regressor	0.51	0.82

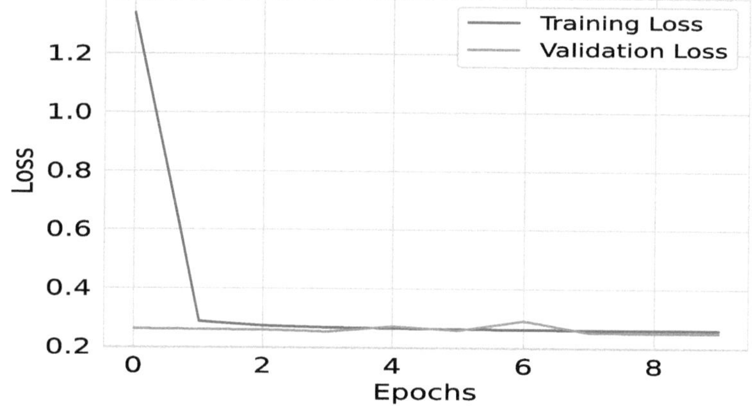

Fig. 2. Trend Between Training Loss and Validation Loss.

Training and Validation Loss: Training was carried out for an overall number of 20 epochs at a batch size of 32. Figure 2 shows how both the training and validation loss decreased gradually which implies that the model's complexity and generalization capacities are balanced well.

4.1   Actual vs. Predicted Yield

Figure 3 shows a positive correlation. The fitted regression line on the scatter plot established that most of the predictions were grouped close to the line, thus validating the model's appropriateness in estimating the yields.

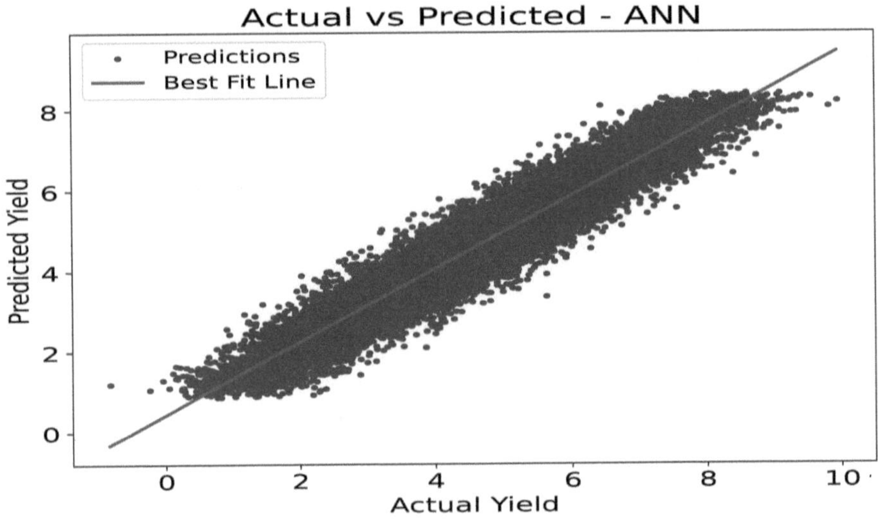

Fig. 3. Regression Plot for Actual vs Predicted Values.

5   Comparative Study (ANN Vs Traditional Models)

For crop yield prediction, a comparative analysis of models from ANN with that of the traditional models, such as Linear Regression (Lasso) and Decision Tree Regressor, showed that ANN outperformed the prediction of other models. Figure 4 shows ANN model returned a very high value of R^2 score than the Linear Regression and Decision Tree model.

A benefit which has been observed for ANNs over traditional regression models has been related to their ability to capture complex, nonlinear relationships in large datasets applied across a wide variety of agricultural applications. An example of this would be in precision farming, where an ANN model could look at huge amounts of sensor data to optimize fertilization strategies in consideration of how different nutrients interact with one another and the environment.

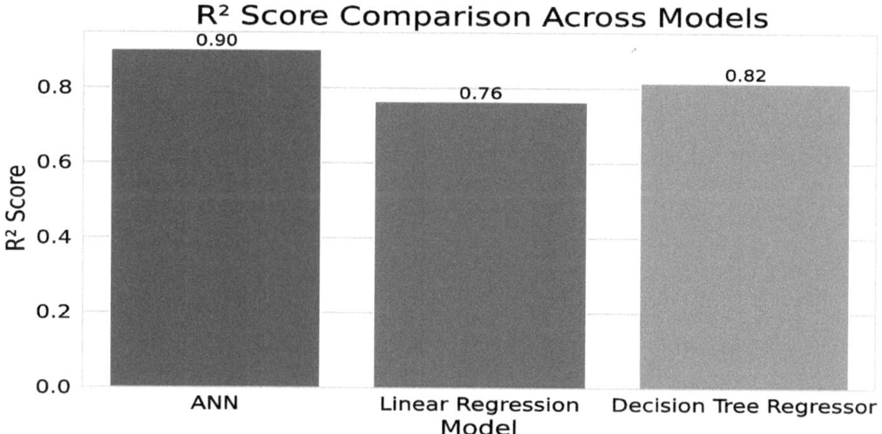

Fig. 4. Comparison of R^2 score of Different Models.

Traditional regression models would only consider nutrient inputs as independent variables, and it could not capture interaction, which may have a strong effect on crop yield [18].

The yield of ANNs in vertical farms can be very efficiently predicted in the domain of urban agriculture. Such systems rely highly on environmental controls such as light intensity, CO_2 concentration, and humidity. Traditional models could fail to predict yields of crops, largely because of these complex interplay's. However, ANNs can capture complex relationships between any number of variables and thus can help in achieving better resource management and yields under constrained urban spaces [19].

In addition, climatic variability and extremes are becoming increasingly relevant, and ANNs are more and more applied to predict crop response under such extreme events, such as floods and droughts. For example, a study has shown that an ANN can predict much more successfully the effect of changing rainfall patterns on rice yield compared to the traditional statistical method which usually assumes linearity in the model [20]. Therefore, the ANN model was able to integrate several climatic variables into the model for making more prompt and detailed predictions, which could be availed of for better decisions in crop management.

6   Conclusion

This research explored the possibility of machine learning, particularly ANN, in transforming the agricultural yield prediction landscape to open the road to smart, data-driven solutions for planning in farms. This study came up with an ANN model which takes into account a number of factors including region, type of soil, type of crop, rainfall, temperature, fertilizer, and irrigation in the crop yield prediction. To make the model computationally feasible, yet retaining the integrity of data, a dataset consisting of 1,000,000 records, with 10 features was sampled up to 50%. Pre-processing such as feature encoding and mutual information-based feature selection were carried out to improve the performance of our model. EDA was performed by working out key patterns and relationships, thereby capturing relative inter plays regarding different factors in relation to the crop yield.

It is demonstrated that the ANN model trained on this feature-processed dataset outperforms traditional regression models, such as Lasso and Decision Tree Regressor, in terms of accuracy and error reduction. The model's superior R^2 (0.91) and lower MSE (0.25) highlight its ability to capture complex agricultural relationships, making it a valuable tool for optimizing resource allocation and improving food security. ANN's scalability allows for application across diverse farming regions, and its integration with precision agriculture can further enhance yield forecasting. However, its reliance on large datasets underscores the need for efficient data collection, and alternative ML approaches may be required in data-scarce environments.

Thus, these results help validate that ANN models are more than capable predictors for crop yields. This would help serve as useful tools for optimally optimizing agricultural decisions. These models can support decisions on the part of agriculturalists and planners for crop selection, resource allocation, and yield optimization, thus improving food security and sustainable agricultural practices. However, these rely heavily on available big datasets and computation, further enhancing the need for higher efficiency in the model. In cases where data is scarce or infrastructural development in technology is in its infancy, one has to use alternative machine learning techniques or models of a more advanced nature to improve the accuracy and efficiency of predictions as much as possible.

Acknowledgement. The authors would like to acknowledge the financial support is given by VIT Bhopal University, Bhopal-Indore Highway, Kothrikalan, Sehore Madhya Pradesh-466114.

References

1. World Resources Institute: How to sustainably feed 10 billion people by 2050: 21 charts. https://www.wri.org/insights/how-sustainably-feed-10-billion-people-2050-21-charts. Accessed 17 Oct 2024
2. Pisante, M., Stagnari, F., Grant, C.: Agricultural innovations for sustainable crop production intensification. Italian J. Agron. **7**, e40 (2012). https://doi.org/10.4081/ija.2012.e40
3. Shahab, H., Iqbal, M., Sohaib, A., Khan, F.U., Waqas, M.: IoT-based agriculture management techniques for sustainable farming: a comprehensive review. Comput. Electron. Agric. **220**, 108851 (2024). https://doi.org/10.1016/j.compag.2024.108851
4. Almeida, J.S.: Predictive non-linear modelling of complex data by artificial neural networks. Curr. Opin. Biotechnol. **13**(1), 72–76 (2002). https://doi.org/10.1016/s0958-1669(02)00288-4

5. Nti, I.K., Zaman, A., Nyarko-Boateng, O., Adekoya, A.F., Keyeremeh, F.: A predictive analytics model for crop suitability and productivity with tree-based ensemble learning. Decis. Anal. J. **8**, 100311 (2023). https://doi.org/10.1016/j.dajour.2023.100311

6. Yaiprasert, C., Hidayanto, A.N.: AI-powered ensemble machine learning to optimize cost strategies in logistics business. Int. J. Inf. Manag. Data Insights **4**(1), 100209 (2024). https://doi.org/10.1016/j.jjimei.2023.100209

7. Sahoo, S., Singha, C., Govind, A.: Advanced prediction of rice yield gaps under climate uncertainty using machine learning techniques in Eastern India. J. Agric. Food Res. **18**, 101424 (2024). https://doi.org/10.1016/j.jafr.2024.101424

8. Ansarifar, J., Wang, L., Archontoulis, S.V.: An interaction regression model for crop yield prediction. Sci. Rep. **11**(1), 17754 (2021). https://doi.org/10.1038/s41598-021-97221-7

9. van Klompenburg, T., Kassahun, A., Catal, C.: Crop yield prediction using machine learning: a systematic literature review. Comput. Electron. Agric. **177**, 105709 (2020). https://doi.org/10.1016/j.compag.2020.105709

10. Maier, H.R., et al.: Exploding the myths: an introduction to artificial neural networks for prediction and forecasting. Environ. Model. Softw. **167**, 105776 (2023). https://doi.org/10.1016/j.envsoft.2023.105776

11. Aldoseri, A., Al-Khalifa, K.N., Hamouda, A.M.: Re-thinking data strategy and integration for artificial intelligence: concepts, opportunities, and challenges. Appl. Sci. **13**(12), 7082 (2023). https://doi.org/10.3390/app13127082

12. Shastry, A., Sanjay, H.A., Bhanusree, E.: Prediction of crop yield using regression techniques. Semantic Scholar (2017). https://api.semanticscholar.org/CorpusID:231686383

13. Shahhosseini, M., Hu, G., Archontoulis, S.V.: Forecasting corn yield with machine learning ensembles. Front. Plant Sci. **11** (2020). https://doi.org/10.3389/fpls.2020.01120

14. Bhat, S.A., Qadri, S.A.A., Dubbey, V., Sofi, I.B., Huang, N.-F.: Impact of crop management practices on maize yield: insights from farming in tropical regions and predictive modeling using machine learning. J. Agric. Food Res. **18**, 101392 (2024). https://doi.org/10.1016/j.jafr.2024.101392

15. Bejo, S., Mustaffha, S., Ishak, W., Wan Ismail, W.I.: Application of artificial neural network in predicting crop yield: a review. J. Food Sci. Eng. **4**, 1-9 (2014). https://www.researchgate.net/publication/283570924_Application_of_Artificial_Neural_Network_in_Predicting_Crop_Yield_A_Review

16. Khaki, S., Wang, L., Archontoulis, S.V.: A CNN-RNN framework for crop yield prediction. Fron. Plant Sci. **10**, 1750 (2020). https://doi.org/10.3389/fpls.2019.01750

17. Attakorah, S. O.: Crop Yield Prediction Dataset. Kaggle. https://www.kaggle.com/datasets/samuelotiattakorah/agriculture-crop-yield

18. Soussi, A., Zero, E., Sacile, R., Trinchero, D., Fossa, M.: Smart sensors and smart data for precision agriculture: a review. Sensors **24**(8), 2647 (2024). https://doi.org/10.3390/s24082647

19. Al-Kodmany, K.: The vertical farm: a review of developments and implications for the vertical city. Buildings **8**(2), 24 (2018). https://doi.org/10.3390/buildings8020024

20. Basir, M.S., Chowdhury, M., Islam, M.N., Rabbani, M.A.: Artificial neural network model in predicting yield of mechanically transplanted rice from transplanting parameters in Bangladesh. J. Agric. Food Res. **5**, 100186 (2021). https://doi.org/10.1016/j.jafr.2021.100186

TabNet-Based Crop Yield Prediction
for Smart Agriculture

Paushali Mondal[1(✉)], Ankur Jain[1], Ajay Kumar Phulre[1], and Amrita Jain[2]

[1] Department of Computer Science, VIT Bhopal University, Bhopal, India
`mondal.paushali384@gmail.com, ajaykumarphulre@vitbhopal.ac.in`
[2] Department of Economics, Vikram University, Ujjain, India

Abstract. The increasing global population and the escalating demand for food necessitate the development of precise agricultural forecasting methods. This research explores the application of Attentive Interpretable Tabular Learning (TabNet), for predicting crop yields, utilizing a comprehensive dataset that incorporates various environmental and agricultural factors. This study utilized EDA in assessing the quality and structure of the data, identifying key features such as crop type, area cultivated, seasonal variations, rainfall patterns, and fertilizer usage. Data preprocessing techniques, including handling missing values, encoding categorical variables, and feature scaling, were applied to prepare the dataset for modeling. The TabNet is constructed using PyTorch, with a focus on optimizing hyperparameters for improved accuracy. The ability of TabNet to provide information about feature importance helped in selecting the most important features contributing to output. Early stopping was applied to prevent unnecessary iterations and enhance model generalization. The final TabNet model, trained for 80 epochs, achieved an MSE of 95654.98 and an R-squared value of 0.88, demonstrating strong predictive performance. Model performance is measured based on MSE and R-squared, which explain the model's ability to approximate complex relationships between input features and crop yields. This study established that TabNet outperforms statistical methods in prediction accuracy in relation to crop yield. In this way, this paper contributes to sustainable agriculture by developing a robust crop yield forecasting framework that enhances food security under climate and resource constraints.

Keywords: TabNet · Exploratory Data Analysis (EDA) · Food security · Machine learning · Nonlinear relationships · Predictive modeling · Resource optimization

1 Introduction

Crop yield prediction is very important in agriculture because it provides critical information to help farmers, policymakers, and industries utilize resources

Supported by VIT Bhopal University.

optimally while making judiciously-safe decisions. It allows for an accurate fore-warning that enables better plans and decisions in agricultural production, mar-keting strategies, and food security. With the population anticipated to exceed 9.7 billion by 2050 [1], a significant surge in food demand is expected. This will call for better agricultural technology, better management of resources, and more means to raise production levels sustainably.

The ability to predict crop yield accurately is vital for several reasons: Accu-rate crop forecasting is important for resource development and efficient use of water, fertilizer, and labor [2]. They also help manage risk, helping to reduce uncertainty about weather, pests, and economic changes [3]. Improve food secu-rity and ensure preparedness by predicting the impacts of climate change on crops. Finally, financial planning uses production forecasts to inform business, supply chain, and policy decisions.

Agricultural output would be heavily influenced by fluctuations in rainfall [4], temperature, or other climatic elements; this is a challenge to reliable planning both by farmers and policymakers [5]. Traditional predictive models will gen-erally fail to capture the complexities and nonlinear relationships which are evident in agricultural data. These may be quite simple linear models not well saturated in the way of representing complex multifaceted interactions among different input variables, including soil type, crop variety, and environmental inputs beyond control. This approach enhances accuracy in forecasts, leading to improved resource and risk management and better food security amid growing pressures from climate change and population growth [6].

Accurate crop yield prediction is crucial for food security, but traditional models struggle with complex agricultural data. While ML methods improve predictions, deep learning, particularly TabNet models, performs better. A study on Australian agriculture showed deep neural networks outperformed statistical and ML models in predicting oat, corn, rice, and wheat yields, reducing errors substantially. It also noted that climate change may temporarily boost yields but is unsustainable long-term. These findings reinforce TabNet's potential in integrating climate, fertilizer use, and land allocation for precise yield forecast-ing [7].

Integrating remote sensing and satellite data into agricultural models is sig-nificant. For instance, researchers have successfully utilized the satellite-derived indices, such as NDVI, in addition to meteorological data, to strengthen yield predictions [8]. This technology has helped scientists enhance crop health moni-toring and even permits better predictions of potential yields. The combination of satellite imagery along with localized climatic data therefore allows for a strong model that is able to predict maize yields with significantly smaller root mean square error (RMSE) than traditional methods [9].

Although many studies are currently starting to explore the impacts of such inputs, like fertilizer and pesticide, on crop yield, relatively few are integrating them into an integrated deep learning model. P. Baweja, S. Kumar, and G. Kumar (2020) highlighted the critical aspect about the fertilizers, which affects the yields but even went ahead to admit that fertilizer impacts are most often evaluated in isolation, detached from other inputs, particularly pesticide use and irrigations [10]. It limits the possibility of the creation of highly accurate and actionable predictions guiding stakeholders' decisions in the agro-industry by isolated approaches that avoid comprehensive modeling.

Table 1. Summary of Key Concepts

Topic	Description	References
Deep Learning	DL outperforms traditional models in crop yield prediction by integrating climate, fertilizer use, and land allocation.	[7]
Remote Sensing	Satellite data and weather improve yield predictions, with lower RMSE than traditional methods.	[8,9]
Literature Gaps	Few models integrate fertilizers and pesticides, limiting prediction accuracy.	[10]

Table 1 compares TabNet with traditional models, highlights remote sensing benefits, and identifies gaps in fertilizer and pesticide integration.

Crop yield prediction has advanced with machine learning (ML), but existing models face limitations in scalability, data quality, and adaptability across regions. Many struggle to integrate nonlinear relationships, making them ineffective across different farms [11]. For instance, rice cultivation models in India performed poorly when applied elsewhere. The lack of quality data further hinders model generalization, as agricultural data involve complex interactions between climate, soil, and farming practices [12].

TabNet models have predictive accuracy and outperform statistical models in predicting agricultural events and provide more accurate predictions. Studies have shown that the use of deep learning models such as TabNet can increase the accuracy of predictions compared to simpler models [13]. Neural networks are also versatile, which makes them applicable to different crops and geographic areas. This scalability allows the model to be adapted to various agricultural environments, allowing farmers and policy makers to apply the same predictive methods to different crop systems [14].

Deep learning algorithms, including neural networks, require large amounts of data to avoid overload and ensure accuracy, which is difficult in data-limited areas. Furthermore, training these models is expensive and requires resources and time, which can be problematic for predicting flight in dynamic environments [15].

Our approach outpaces the use of outdated methods by adopting a TabNet-based deep learning model that efficiently combines multiple agricultural variables for better prediction of crop yield. Differently from other models that utilize mere regression methods, our approach capitalizes on TabNet's ability to select the most important inputs dynamically through attentive feature selection at each step of decision-making.

To enhance generalization and minimize overfitting, we add Ghost Batch Normalization and a sparsity penalty in the TabNet model. Our model is also optimized using an Adam optimizer and a learning rate schedule adjustment to achieve improved convergence. The efficiency of our model is illustrated via comparative analysis, whereby TabNet performs better than Linear Regression, Decision Tree, and Support Vector Regression models in regard to Mean Squared Error (MSE) and R^2 score.

By combining these innovations, this research sets a solid groundwork for precise and scalable forecasting of crop yields, opening the door to more effective agricultural decision-making.

2   Problem Formulation

2.1   Problem Statement

The expanding population and food requirements of the world call for accurate agricultural forecasting techniques. Farmers experience difficulties in predicting crop output due to variable environmental conditions, seasonality, and limited resources. Conventional forecasting models cannot possibly represent intricate patterns among several affecting factors like rain, fertilizer input, and plantation area. This research investigates the possibility of using TabNet, a deep learning model, to enhance crop yield forecasts by tapping into a large dataset, thereby allowing more precise decision-making and efficient use of agricultural resources in sustainable agriculture.

2.2   Objective

The objective of this research is to improve crop yield forecasting precision using TabNet, a deep learning technique that accurately learns intricate patterns between agricultural and environmental variables. The model, by utilizing feature selection and interpretability, facilitates the detection of principal drivers of yield variability, supporting data-driven decision-making. This can benefit farmers in optimizing the use of resources, enhancing crop planning, and managing risks arising from volatile environmental factors. Moreover, the understanding derived through this predictive system can assist policymakers in formulating plans for food security and sustainable agriculture.

3   Methodology

This research follows a structured approach to create a TabNet model for predicting crop yields, a crucial task in modern agriculture aimed at addressing food security challenges. The methodology is meticulously organized into several key sections, each describing the specific steps undertaken throughout the study. This structured approach not only improves the robustness of the model but also ensures transparency and reproducibility, vital components in scientific research.

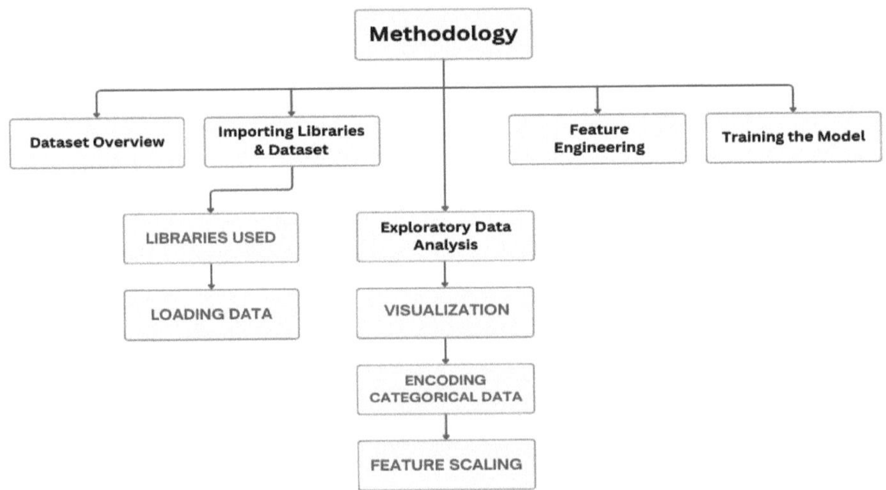

Fig. 1. Methodology Flowchart.

Figure 1 describes the methodology which involves importing relevant libraries and the dataset, performing exploratory data analysis (EDA) to understand the data's structure and patterns, applying feature engineering techniques to enhance the dataset, and training a machine learning model to predict the desired outcome.

3.1 Dataset Overview

The process begins with Data Collection, where a comprehensive dataset from Kaggle is gathered, containing various features influencing crop yield [16]. The dataset consists of 19,689 entries and 10 features, which include; Crop Type, which affects yield based on the crop's specific growth needs; Area, referring to the land allocated for cultivation and its impact on yield potential; Season, the growing period influenced by climate; State, providing geographical context for environmental factors; Production, which includes historical yield data crucial for model training; Rainfall, affecting water availability for crops; and Fertilizer Usage, where data on fertilizer application plays a significant role in influencing yield.

3.2 Importing Libraries and Data

The initial step in developing a TabNet model for crop yield prediction is importing libraries and loading the dataset. This ensures all necessary tools for data manipulation, analysis, modeling, and visualization are available.

Libraries Used. Pandas is used for data manipulation with DataFrames, while NumPy supports efficient numerical computations with multi-dimensional arrays. Matplotlib and Seaborn are essential for creating versatile plots and statistical graphics. Scikit-learn provides tools for data preprocessing and model evaluation across various algorithms. Pytorch stands out to be the most important library providing the TabNet regressor for predicting the yield.

Loading Data. After importing the required libraries, the next step is to load the dataset, typically in a structured format like CSV, into a Pandas DataFrame for easy manipulation and analysis.

3.3   Exploratory Data Analysis

During this phase, exploratory data analysis (EDA) is performed to gain a deeper insight into the dataset's structure, distribution, and quality.

Understanding the dataset's structure, as shown in Fig. 2, is essential for determining the appropriate preprocessing methods, as it offers an overview of the features and their corresponding values. It is equally important to check for duplicate entries using the `duplicated()` method, as such duplicates can distort the results of data analysis and model training. Additionally, assessing data quality is a key step in ensuring the accuracy of the analysis, as incomplete or inaccurate data can negatively impact the performance of machine learning models. Identifying missing values and addressing them is vital to maintain the integrity of the dataset.

	Crop	Crop_Year	Season	State	Area	Production	Annual_Rainfall	Fertilizer	Pesticide	Yield
0	Arecanut	1997	Whole Year	Assam	73814.0	56708	2051.4	7024878.38	22882.34	0.796087
1	Arhar/Tur	1997	Kharif	Assam	6637.0	4685	2051.4	631643.29	2057.47	0.710435
2	Castor seed	1997	Kharif	Assam	796.0	22	2051.4	75755.32	246.76	0.238333
3	Coconut	1997	Whole Year	Assam	19656.0	126905000	2051.4	1870661.52	6093.36	5238.051739
4	Cotton(lint)	1997	Kharif	Assam	1739.0	794	2051.4	165500.63	539.09	0.420909

Fig. 2. Dataset Header.

To check for missing values, the `isnull()` function can be used in conjunction with `sum()` to count the number of missing values in each column.

Visualizing the Distribution of Data. Visualizing data is an essential component of exploratory data analysis (EDA), as it allows us to uncover underlying patterns, relationships, and anomalies in the dataset. This section presents a series of plots designed to provide insights into the crop distribution, seasonal patterns, and the relationships between key features like area, yield, and production. A bar chart displaying the distribution of the Season column was plotted, showing the frequency of each season in the dataset.

Encoding Categorical Variables and Feature Scaling. Numerical input is required by many machine learning algorithms, but datasets for crop yield prediction often contain categorical variables, such as crop type and season. To convert these categorical variables into a format suitable for modeling, **Label Encoding** is used. Label Encoding converts categorical variables into numerical values by assigning a unique integer to each category, ensuring compatibility with machine learning models. Additionally,

3.4 Feature Engineering

Feature engineering is an essential step in preparing a dataset for machine learning. It involves converting raw data into relevant features that enhance the model's performance. A Correlation Heatmap visualizes the relationships between numerical features, helping to detect multicollinearity, which can negatively impact models like linear regression. Features with high correlations to the target variable are often more influential in predicting outcomes. In this study, seven key features were selected based on feature importance provided by the TabNet model, while Crop Area and Year were removed to eliminate redundancy and improve model performance, ensuring that the most relevant predictors were used for accurate crop yield estimation. Figure 3 shows the feature importance graph, highlighting the most influential features in the model's predictions based on their contribution to the output.

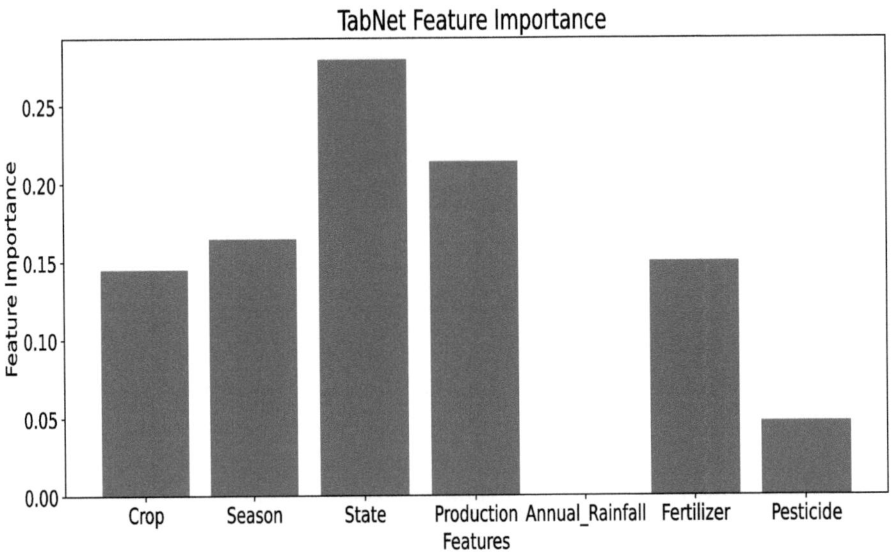

Fig. 3. Feature Importance Graph.

3.5   Training the Model

Training the model is a crucial step in developing an effective predictive model using TabNet. This involves dividing the dataset into training and testing sets, scaling the features, defining the model architecture, and then compiling and fitting the model to the training data [17]. The process of training starts with splitting the dataset into a training set and a test set, which allows model evaluation on latent data and makes it more general than training. The hyperparameters were chosen based on Grid Search, ensuring optimal performance. TabNet Regressor used in this study is designed with 32-dimensional prediction and attention layers and operates with three sequential decision steps to enhance learning. A gamma value of 1.3 is set to regulate feature reuse, while sparsity is controlled with a (lambda) of 1e-3. The model is optimized using the Adam optimizer with a learning rate of 0.02, and a SparseMax feature selection mask is applied to improve interpretability. A learning rate scheduler with a step size of 10 and decay factor of 0.9 is implemented for stable training. The model is trained with a batch size of 512 and a virtual batch size of 128, utilizing Ghost Batch Normalization to improve generalization. Training is conducted for a maximum of 80 epochs, with early stopping patience set to 30 to prevent overfitting. After training, the model is used to make predictions on test data, such as X_{test}, and test its ability to generalize to invisible layer conditions.

4   Results

The results of the model evaluation indicate a strong predictive performance for the TabNet in estimating crop yields. The model's performance was assessed using two primary metrics: Mean Squared Error (MSE), and R-squared (R^2), which together offer a comprehensive view of its accuracy and effectiveness.

The model evaluation metrics highlight its strong predictive performance. The Mean Squared Error (MSE) is approximately 95654.98, reflecting a low average squared discrepancy between predicted and actual crop yields, indicating effective prediction accuracy. Additionally, the R-squared (R^2) value of 0.88 demonstrates that the model explains 88% of the variance in crop yields, underscoring its ability to capture critical patterns and its effectiveness in yield prediction. Figure 4 shows the training vs. validation loss plot, illustrating the model's learning progress and potential overfitting over epochs. Figure 5 shows the relationship between actual and predicted crop yields. The diagonal line represents perfect predictions, with points ideally clustering along it to indicate high accuracy. Deviations from this line reveal instances where the model may struggle to capture underlying data patterns, offering valuable insights into its predictive performance.

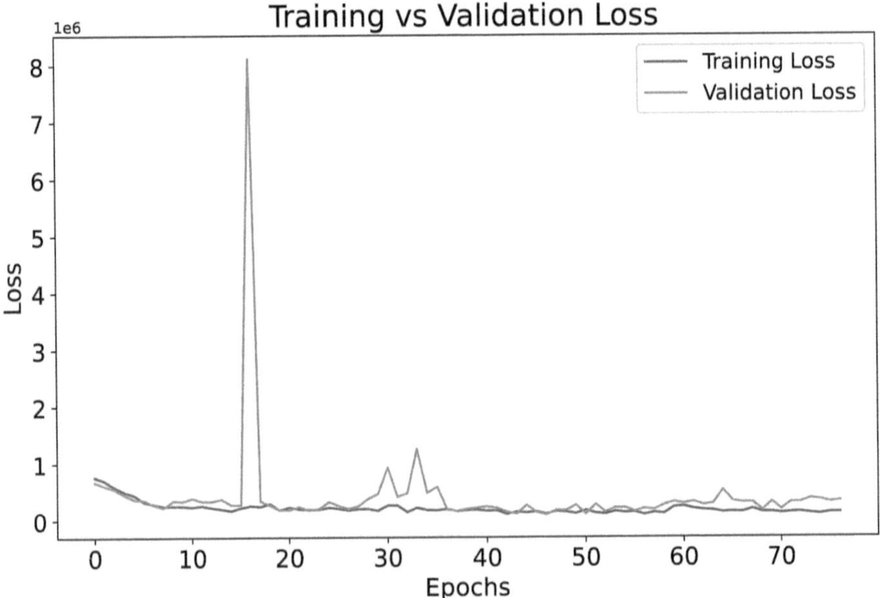

Fig. 4. Graph of Training vs. Validation Loss.

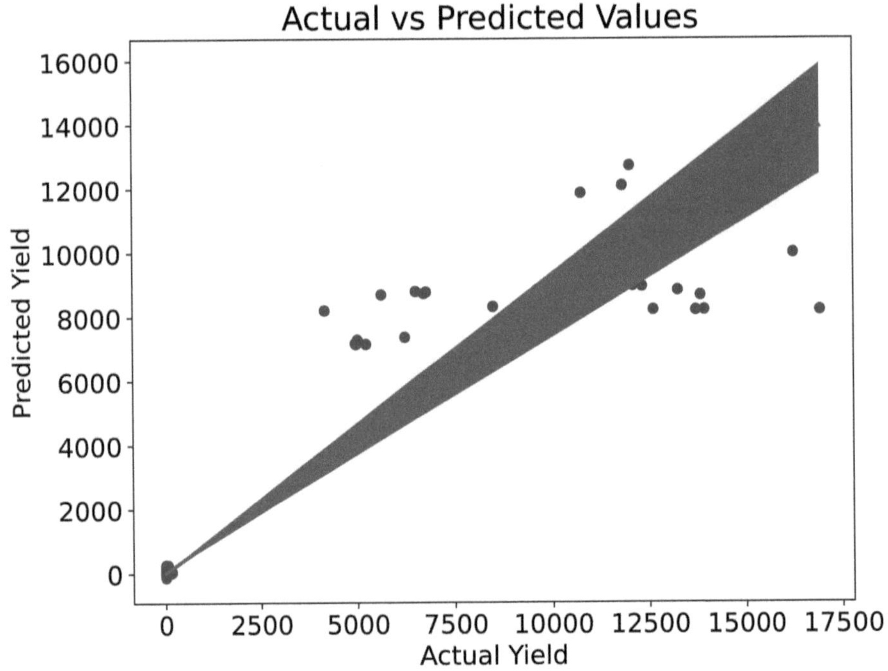

Fig. 5. Actual vs. Predicted Values.

5 Comparative Study: TabNet vs. Traditional Regression Methods

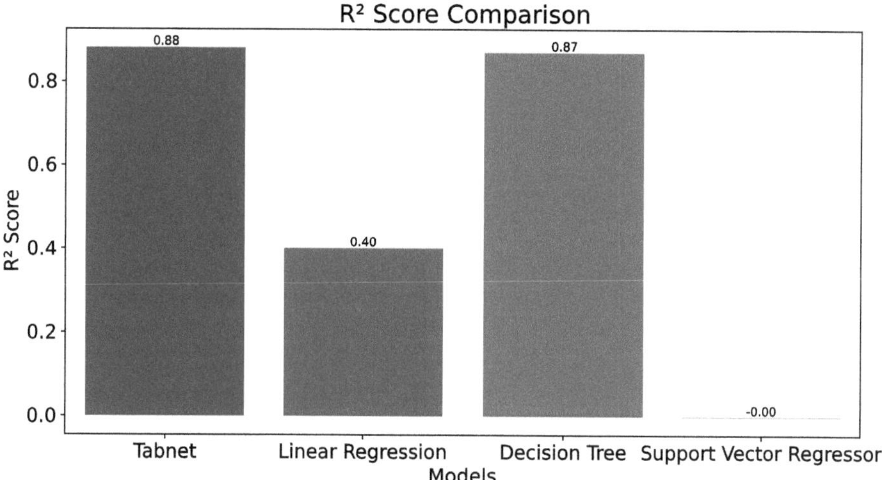

Fig. 6. R^2 Comparison of Various Models.

Figure 6 presents a comparison of R^2 for different models. The TabNet model delivered the best predictive performance, successfully capturing the complexities of the data and significantly outperforming the other models. Linear Regression, on the other hand, struggled with high errors and was unable to account for more complex patterns. Although Decision Tree showed improved performance, it still couldn't match the error reduction as of the TabNet model. The Support Vector Regressor performed the worst, showing high errors and poor generalization. In conclusion, TabNet proved to be the most reliable model, offering the best accuracy and overall fit for the dataset.

Table 2 shows the performance metrics of different machine learning models. The TabNet model achieves the best results with an R^2 of 0.88 and MSE of 95654.98, followed by Decision Tree model ($R^2 = 0.87$, MSE $= 104411.26$). The Linear Regression and Support Vector Regressor models perform poorly with R^2 values of 0.40 and -0.00, respectively.

This capability of TabNet models is particularly beneficial in several applications within agriculture:

Precision Agriculture uses real-time data from sensors to monitor soil moisture, nutrients, and weather conditions. Unlike traditional regression models, which treat these variables separately, TabNet can model their interactions, leading to more accurate crop yield predictions. For example, deep learning models have been applied to optimize irrigation schedules, balancing water usage

Table 2. Performance Metrics of Various Models

Model	MSE	R^2
TabNet	95654.98	0.88
Decision Tree	104411.26	0.87
Linear Regression	480231.01	0.40
Support Vector Regressor	802223.12	−0.00

with soil absorption to enhance crop performance under varying weather conditions [18]. Climate Change and Crop Yield Prediction benefits from deep learning modelsAA, which can effectively model complex relationships between historical weather data and crop performance. A study in Zambia found that integrating climate variables like temperature extremes and drought-tolerant crop varieties in a deep learning framework improved yield predictions by 20% compared to regression models [19].

6   Conclusion

This research underscores the significant potential of TabNet in enhancing crop yield prediction by effectively integrating various datasets, including climatic conditions, soil characteristics, and agronomic practices. The TabNet model demonstrated high predictive accuracy, especially when trained on a comprehensive dataset, offering valuable insights for stakeholders in agriculture. This approach optimizes resource allocation, enhances food security strategies, and mitigates agricultural risks.

Moreover, machine learning's ability to generalize across different crops and regions opens new avenues for scaling predictive models. Future studies should aim to incorporate real-time data from satellite imagery and IoT sensors while addressing challenges such as data quality and computational demands. Integrating climate change projections will also be crucial for creating robust models that respond to future agricultural challenges. By advancing these methodologies, we can significantly boost agricultural productivity and resilience in an era of growing global demand.

Acknowledgements. The authors would like to acknowledge the financial support is given by VIT Bhopal University, Bhopal-Indore Highway, Kothrikalan, Sehore Madhya Pradesh - 466114.

References

1. United Nations, Global issues: Population. https://www.un.org/en/global-issues/population. Accessed 16 Oct 2024

2. C. E. A. T. Specialty. https://www.ceatspecialty.com/in/blog/technology/what-are-the-benefits-of-adopting-smart-farming-techniques-to-enhance-crop-yield. Accessed 15 Oct 2024

3. Skendžić, S., Zovko, M., Živković, I.P., Lešić, V., Lemić, D.: The impact of climate change on agricultural insect pests. Insects **12**(5), 440 (2021). https://doi.org/10.3390/insects12050440

4. Kunhare, N., Gupta, R.K., Mohanty, M., Patel, J., Jain, A.: AI-based daily, weekly and monthly rain forecasting by using different time series models. WCONF **2**, 1–6 (2024). https://doi.org/10.1109/WCONF61366.2024.10692235

5. Datta, P., Behera, B., Rahut, D.B.: Climate change and Indian agriculture: a systematic review of farmers' perception, adaptation, and transformation. Environ. Challenges **8**, 100543 (2022). https://doi.org/10.1016/j.envc.2022.100543

6. Hogan, D., Schlenker, W.: Enhancing resilience in agricultural production systems with AI-based technologies. Nat. Commun. **15**, 4638 (2024). https://doi.org/10.1038/s41467-024-48388-w

7. Demirhan, H.: A deep learning framework for prediction of crop yield in Australia under the impact of climate change. Inf. Process. Agric. (2024). ISSN 2214-3173. https://doi.org/10.1016/j.inpa.2024.04.004

8. White, J., Berg, A.A., Champagne, C., Zhang, Y., Chipanshi, A., Daneshfar, B.: Improving crop yield forecasts with satellite-based soil moisture estimates: an example for township level canola yield forecasts over the Canadian Prairies. Int. J. Appl. Earth Observ. Geoinf. **89**, 102092 (2020). https://doi.org/10.1016/j.jag.2020.102092

9. Shuai, G., Basso, B.: Subfield maize yield prediction improves when in-season crop water deficit is included in remote sensing imagery-based models. Remote Sens. Environ. **272**, 112938 (2022). https://doi.org/10.1016/j.rse.2022.112938

10. Baweja, P., Kumar, S., Kumar, G.: Fertilizers and pesticides: their impact on soil health and environment. In: Fertilizers and Pesticides: Their Impact on Soil Health and Environment, pp. 265–285 (2020). https://doi.org/10.1007/978-3-030-44364-1_15

11. Razavi, M.A., et al.: Enhancing crop yield prediction in Senegal using advanced machine learning techniques and synthetic data. Artif. Intell. Agric. **14**, 99–114 (2024). https://doi.org/10.1016/j.aiia.2024.11.005

12. Nti, I. K., Zaman, A., Nyarko-Boateng, O., Adekoya, A.F., Keyeremeh, F.: A predictive analytics model for crop suitability and productivity with tree-based ensemble learning. Decis. Anal. J. **8**, 100311 (2023). https://doi.org/10.1016/j.dajour.2023.100311

13. Tien, P.W., Wei, S., Darkwa, J., Wood, C., Calautit, J.K.: Machine learning and deep learning methods for enhancing building energy efficiency and indoor environmental quality – a review. Energy AI **10**, 100198 (2022). https://doi.org/10.1016/j.egyai.2022.100198

14. Mana, A.A., Allouhi, A., Hamrani, A., Rehman, S., el Jamaoui, I., Jayachandran, K.: Sustainable AI-based production agriculture: exploring AI applications and implications in agricultural practices. Smart Agric. Technol. **7**, 100416 (2024). https://doi.org/10.1016/j.atech.2024.100416

15. Fink, O., Wang, Q., Svensén, M., Dersin, P., Lee, W.-J., Ducoffe, M.: Potential, challenges and future directions for deep learning in prognostics and health management applications. Eng. Appl. Artif. Intell. **92**, 103678 (2020). https://doi.org/10.1016/j.engappai.2020.103678

16. Gupta, A.: Crop Yield in Indian States Dataset, Kaggle (2023). https://www.kaggle.com/datasets/akshatgupta7/crop-yield-in-indian-states-dataset. Accessed 17 Oct 2024
17. "ASP2A AI Guide," National Supercomputing Centre Singapore, March 2024. https://help.nscc.sg/wp-content/uploads/2024/05/ASP2A-AI-Guide-7_March_2024.pdf. Accessed 23 Oct 2024
18. Zou, P., Yang, J., Fu, J., Liu, G., Li, D.: Artificial neural network and time series models for predicting soil salt and water content. Agric. Water Manag. **97**(11), 2009–2019 (2010). https://doi.org/10.1016/j.agwat.2010.02.011
19. Wineman, A., Crawford, E.W.: Climate change and crop choice in Zambia: a mathematical programming approach. NJAS - Wageningen J. Life Sci. **81**, 19–31 (2017). https://doi.org/10.1016/j.njas.2017.02.002

Enhancing Speech Emotion Recognition with Hybrid Deep Learning Models and Multi-feature Fusion

Mohammed Elnazer[✉], P. Siva Prasad, and Syed Shareefunnisa

Vignan Foundation for Science, Technology, and Research, Department of Computer Science and Engineering, Guntur, Vadlamudi, India
malnazeer177@gmail.com

Abstract. Speech Emotion Recognition (SER) is necessary for improving human-computer interaction to give systems the ability to identify and respond to speech emotions. This paper introduces an integrated deep learning model that combines LSTM with CNN for classifying emotions from speech data. The proposed approach uses a blend of temporal and spectral audio characteristics, including spectral contrast, chroma features, and MFCC (Mel-frequency cepstral coefficients), to enhance emotion recognition accuracy. CNN is employed to extract spatial dependencies from these features, while LSTM captures the sequential patterns present in signals for speech. The Ryerson Audio-Visual Database of Emotional Speech and Song dataset, also known as RAVDESS, is used to evaluate the model's performance, consisting of samples of speech from eight different emotional categories. Experimental findings reveal that The accuracy of the hybrid CNN-LSTM model is of approximately 97.7%, surpassing conventional single-model approaches. This technique presents a reliable and effective solution for real-world emotion detection, contributing to advancements in affective computing and interactive AI systems.

Keywords: Emotion Detection in Speech · Multi-Feature Analysis · Hybrid Deep Learning Approach

Abbreviations

SER	Speech Emotion Recognition
CNN	Convolutional Neural Network
LSTM	Long Short-Term Memory
RNN	Recurrent Neural Network
MFCC	Mel-Frequency Cepstral Coefficients

1 Introduction

Signals for speech are considered one of most private methods of human communication. Therefore, various Numerous of studies have taken place Regarding

P. Chandrakar et al. (Eds.): ICCINS 2025, CCIS 2738, pp. 91–103, 2026.
https://doi.org/10.1007/978-3-032-09572-5_8

speech recognition. In spite of many applications, there is a significant disconnect in this field between computers and human interaction, which leads to several problems, such as the inability of the system to understand how users feel. As a result, speech recognition has faced challenges in terms of speech processing in the past. Speech emotion has the ability to elicit significant meaning from speech, enhancing the ability to recognize performance [1]. An SER (Speech Emotion Recognition) model utilizes the data within the speech file and classifies it into various emotions [2]. Emotions such as happy, neutral, sad, etc., are classified by the SER model, which is used to provide accurate results. Automatic speech recognition (ASR) has paved the way in several areas, including verbal recognition, gender classification, and speech classification, etc. [3]. Various techniques have been used in the context of SER, such as SVM, linear classifiers, Bayesian Networks, etc. [3]. Numerous fields, such as education, entertainment, multimedia, content management, the automotive industry, text-to-speech, and medical diagnosis, use speech emotion [4]. Speech patterns and speaking rates vary from person to person and from location to location, making it challenging to identify emotions from speech signals. This is especially true for indigenous and non-native speakers [4]. At the moment, deep learning methods are being utilized to address these significant difficulties. Deep learning excels because it can learn sophisticated features. As a result, methods based on various deep learning architectures have been developed by numerous researchers to identify speech emotions. For the SER task, these architectures have achieved a respectable level of accuracy. However, more work needs to be done to improve the performance that has been recently attained [5]. The benefits of multiple machine learning architectures combine to create a hybrid model for speech emotion recognition to improve performance. networks with long short-term memory (LSTM), for instance, combined with a CNN-LSTM hybrid model, can process the sequential nature of speech features to capture temporal dependencies, while convolutional neural networks (CNNs) effectively ex- tract features from audio signals by capturing spatial patterns in speech data, like MFCCs, or Mel-frequency cepstral coefficients. This combination enables the model to classify emotions more accurately across a range of emotions in datasets such as RAVDESS, improving its understanding of speech's temporal and spectral dynamics.

2   Related Work

Recognition of Speech Emotions (SER) has drawn significant interest due to its applications in healthcare, security, and computer-human interaction. Various methods have been explored to improve the accuracy and generalizability of emotion detection from speech signals, incorporating both deep learning-based techniques and traditional machine learning approaches. One approach combines CNN and LSTM models with a novel data augmentation technique that introduces controlled noise to enhance model performance. This method has been evaluated on datasets such as IEMOCAP, RAVDESS, SAVEE, and EMO-DB. However, a challenge highlighted in the study is the requirement for large datasets to achieve effective model training [5]. Another study combines Mel Frequency Cepstral Coefficients (MFCC) with time-domain features to enhance SER performance. By utilizing CNNs, the approach achieves superior accuracy on RAVDESS and the datasets from EMO-DB-SAVEE. However, the study notes that ensuring the reliability of these features in consistently classifying emotional states remains a challenge [6]. A different investigation explores SER generalization across multiple languages using a 1D CNN-LSTM model. While tested on datasets such as SAVEE, ANAD, BAVED, and EMO-DB, the study acknowledges that variations in human speech make it difficult to establish a standardized model [7]. Multimodal approaches have also been investigated. One study incorporates galvanic skin response (GSR), facial expressions, and electroencephalograms (EEGs) for cross-subject emotion recognition. Although this system enhances emotion classification, processing multiple data types remains complex, as demonstrated on the LUMED-2 dataset [8]. Further research proposes a deep convolutional neural network (1D DCNN) utilizing wide MFCC, LPCC, and wavelet packet transform (WPT) features. While this approach improves performance, challenges such as noise and system complexity persist, with evaluations conducted on RAVDESS and EMO-DB datasets [9]. Gender-dependent training effects in SER have also been investigated using MFCC features. A study finds that speaker characteristics, language, and cultural differences impact system performance, with gender differences identified as a crucial factor. The RAVDESS dataset is used for validation [10]. Another study proposes a two-way feature extraction technique using super convergence and principal component analysis (PCA). While this method aims to enhance SER, automation with high accuracy remains a challenge, as demonstrated in experiments on the RAVDESS dataset [2]. A hybrid recurrent neural network model, termed frequency-based recurrent neural speech recognition (FbRNSR), incorporates digital filters to improve accuracy. However, this increases system cost and complexity, with challenges persisting in feature extraction and noise reduction [3]. In the evolving digital landscape, SER is applied in various domains such as telecommunications, security, intelligent tutoring systems, automotive applications, and human-computer interaction. This motivates studies that enhance SER accuracy by extracting multiple acoustic features such as energy, pitch, zero-crossing rate (ZCR), discrete wavelet transform (DWT), and MFCC. The Global Feature Algorithm selects relevant features, followed by machine learning-based classification of emotions like anger, happiness, sadness, and neutrality [4]. Multimodal emotion recognition (MER) has also been explored using deep learning frameworks integrating speech, EEG signals, and facial expressions. The Deep-Emotion framework, validated on the MAHNOB-HCI, EMO-DB, and CK+ datasets, demonstrates the potential of combining modalities, though effective integration remains challenging [11]. Another approach introduces a brain-inspired computing model that mimics human cognitive processes using a multimodal framework. This method, validated on the IEMO-CAP dataset, shows promise in approximating human-like emotional perception [12]. The deep separable convolutional neural network (DSCNN) is introduced to reduce computational complexity while improving SER accuracy. However, achieving both high recognition rates and low computational costs remains a challenge, as demonstrated using IEMOCAP and RAVDESS datasets [13]. A hybrid deep learning approach combining deep and acoustic features is proposed to enhance SER accuracy. Using VGG16, ResNet18, ResNet50, and ResNet101 models validated on RAVDESS, IEMOCAP, and EMO-DB datasets, the study demonstrates performance improvements. However, understanding the implications of SER in human psychosocial contexts remains an open question [14]. To further enhance SER, another study presents a parallel combination of CNN and attention-based models such as Transformers and BLSTM-Attention. Despite the innovative methodology, the study emphasizes the limitations of small training datasets like RAVDESS, which may hinder model generalization to real-world scenarios [15].

3   Methodology

3.1   Dataset

Database selection is crucial for speech emotion recognition in system performance. The current work makes use of the RAVDESS dataset [6], consisting of emotionally rich speech samples of both male and female speakers. Emotions that Anger, sadness, fear, excitement, happiness, and are covered by the dataset. 3040 samples are chosen per emotion, and emotional features are extracted and computed by employing a classifier for categorization of emotion. The addition of the RAVDESS dataset [6] guarantees diversity and balance in the representation of emotions, thereby making it a good benchmark to test the proposed model.

3.2   Data Pre-processing

Preprocessing enhances speech samples by removing unwanted noise and variations using filters like Butterworth and Chebyshev. This step ensures cleaner data for feature extraction. In speech recognition, accuracy depends on converting analog signals to digital form, measured in bits per sample. A 16-bit sample provides 65,536 amplitude levels, and speech signals are typically sampled at 44 kHz for effective processing.

1. **Noise Removal and Silence Removal:** This step cleans the audio by removing background noise and eliminating silent parts

$$y[n] = x[n] * h[n] \tag{1}$$

   The location of the loud audio signal is denoted by x[n], the filtered audio signal by y[n], and the the impulse response of a filter (like a low-pass or high-pass filter), and * represents convolution, which applies the filter to the signal. This process removes unwanted noise by attenuating specific frequencies. For silence removal, the technique of thresholding is applied, where signal segments with extremely low energy (below a predetermined threshold) are regarded as silence and eliminated. Although there isn't a precise formula, this method relies on determining if the amplitude of the signal drops below a predetermined threshold.

2. **Speech Normalization:** This stage guarantees that every audio sample has consistent loudness and maintain feature consistency by normalizing the amplitude.

$$\hat{x}[n] = \frac{x[n]}{A} \tag{2}$$

Where $\hat{x}[n]$ is the normalized audio signal and A is the maximum absolute value (the peak amplitude) of the signal. This equation normalizes the signal by dividing each sample by the peak amplitude, ensuring the amplitude values are scaled between -1 and 1.

3. **Framing and Windowing:** To capture temporal variations in sound, the audio signal is split into small overlapping segments, or frames, and a windowing function is applied to prevent sharp transitions between frames

$$x_f[n] = x[n - k \cdot H] \tag{3}$$

Where xf [n] is the frame index (frame number), where n is each frame's sample index, and H is the hop size (the amount of overlap between frames). This process splits the signal into overlapping segments called frames, allowing for analysis of the signal's characteristics over time. Framing is further refined using a window function:

$$w[n] = \text{window}(n) \tag{4}$$

where the window function (such as the Hanning or Hamming window) is denoted by w[n]. The window function smooths the edges of each frame, reducing sudden transitions that might introduce artifacts in the analysis.

3.3   Features Extraction

MFCC, or Mel Frequency Cepstral Coefficients, is a popular speech processing method. It simulates how the vocal tract shapes sound by representing the envelope of the power spectrum in short-term. MFCC applies a nonlinear Mel scale to a log power spectrum, effectively capturing perceived frequency or pitch. In the low-frequency range, it offers better frequency resolution, making it suitable for various applications. Since human hearing is most sensitive to frequencies below 1 kHz, MFCC is

designed to reflect this characteristic. The extraction process involves multiple steps to ensure accurate feature representation for speech analysis.

1. Apply the Mel-scale filter bank:

$$Y_m = H_m \cdot |X(f)|^2 \tag{5}$$

where the filter bank's output for the m-th Mel band is denoted by Y_m.

2. Calculate the logarithm of the Mel spectrum:

$$L_m = \log(Y_m) \tag{6}$$

3. Apply the discrete cosine transform (DCT):

$$C_k = \sum_{m=1}^{M} L_m \cdot \cos\left(\frac{\pi k}{M}\left(m - \frac{1}{2}\right)\right) \tag{7}$$

where C_k represents the MFCCs, or Mel-frequency cepstral coefficients, and M is the number of Mel bands.

Time-Domain Features. An audio signal's raw waveform is used to directly extract time-domain features. Unlike frequency or cepstral domain features, these characteristics analyze the signal's amplitude variations over time, offering insights into its temporal properties. Typical time-domain attributes include:

1. **ZCR:** : which stands for zero crossing rate used to determines how quickly the signal crosses zero, or changes its sign. helping to detect the noisiness or tonawhi characteristics of the sound.
2. **Energy**: Represents the total magnitude of the signal's amplitude over time, indicating how loud or soft the sound is.
3. **Amplitude Envelope**: Captures the peak amplitudes over time, providing information on how the loudness varies.

The mathematical expressions for these time-domain features are defined as follows:

Energy Equation

$$E = \sum_{n=1}^{N} |x[n]|^2 \tag{8}$$

where $x[n]$ is the audio signal and E represents the signal's energy.

Zero-Crossing Rate (ZCR)

$$ZCR = \frac{1}{N} \sum_{n=1}^{N-1} \mathbb{I}\left(x[n] \cdot x[n+1] < 0\right) \tag{9}$$

where $\mathbb{I}$ is the indicator function that equals 1 when the condition is true and 0 otherwise.

3.4   Proposed Model Architecture

This research uses a CNN-LSTM model for identifying speech emotions, combining Sequence modeling for networks with LSTM capabilities combined with CNN spatial feature extraction capabilities. The model starts by identifying important characteristics in the audio input. including MFCC, Chroma, and Spectral Contrast, which capture important speech signal properties. These features are then processed through two 1D convolutional layers with 64 and 128 filters, respectively, to capture local and higher-level patterns in the data. Max-pooling layers follow each convolutional layer to reduce dimensionality while preserving essential features. The convolutional layers' output is transmitted to an LSTM layer, which captures the audio data's temporal relationships and patterns., which is allowing the model to gradually learn sequential patterns. The feature vector is processed by fully connected dense layers with 64 neurons prior to the last output layer. A layer of dropout is used during training to prevent overfitting. . Eight neurons make up the output layer, which corresponds to the eight predetermined emotion categories. with class probabilities generated using a softmax activation function. Due to its adaptive learning rate and effective convergence properties, The Adam optimizer is used to train the model. Since the task involves multiple classes. The loss function is sparse categorical cross-entropy. This design develops a strong model for identifying speech emotions by skillfully fusing CNN's prowess in feature extraction with LSTM's capacity to model temporal sequences.

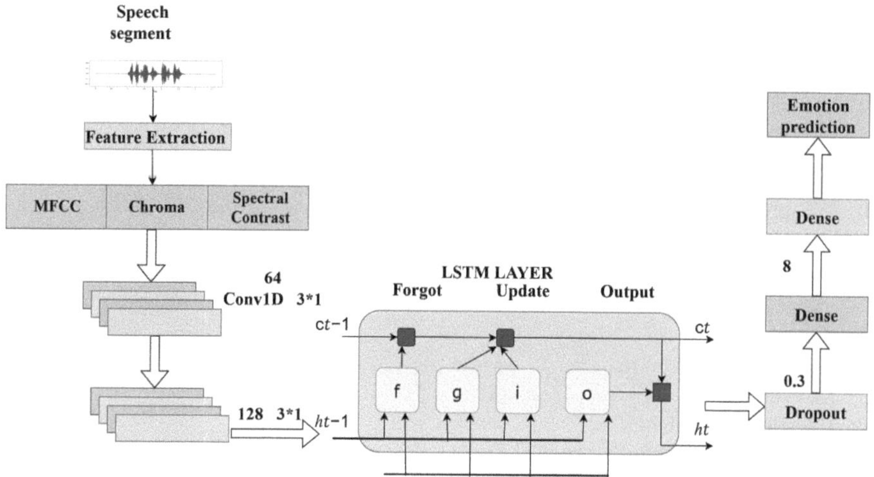

Fig. 1. The suggested CNN + LSTM model architecture for identifying speech emotions

---

Algorithm 1. CNN + LSTM Hybrid Framework for Speech Emotion

1: **Define** emotion dictionary: $emotions\_dict = \{$'01' $\rightarrow neutral$, '02' $\rightarrow calm$, '03' $\rightarrow happy$, '04' $\rightarrow sad$, '05' $\rightarrow angry$, '06' $\rightarrow fearful$, '07' $\rightarrow disgust$, '08' $\rightarrow surprised\}$

2: **Input:** Dataset path $dataset\_path$

3: **Initialize:** Feature list $features$, emotion list $emotions$

4: **for** each audio file in $dataset\_path$ **do**

5:     Extract $emotion\_code$ from filename and map to $emotion$ using $emotions\_dict$

6:     Load audio signal: $(y, sr) = \text{librosa.load}(\text{file\_path}, \text{duration} = 3, \text{offset} = 0.5)$

7:     **Feature Extraction:**

8:     Compute MFCC: $\text{MFCC} = \text{librosa.feature.mfcc}(y, sr, n\_mfcc = 13)$

9:     Compute Chroma: $\text{Chroma} = \text{librosa.feature.chroma\_stft}(y, sr)$

10:     Compute Spectral Contrast: Spectral Contrast $= \text{librosa.feature.spectral\_contrast}(y, sr)$

11:     Concatenate features: combined_features $= [\text{MFCC}, \text{Chroma}, \text{Spectral Contrast}]$

12:     Append $combined\_features$ to $features$ and $emotion$ to $emotions$

13: **end for**

14: **Preprocess Data:**

15: Encode $emotions$ using LabelEncoder

16: Split data: $(X_{\text{train}}, X_{\text{test}}, y_{\text{train}}, y_{\text{test}}) = \text{train\_test\_split}(\text{features}, y_{\text{encoded}}, \text{test\_size} = 0.2)$

17: **Define Hybrid CNN + LSTM Model:**

18: Initialize model: model $= \text{Sequential}()$

19: Add layers:
  - Conv1D layer: 64 filters, kernel size 3, activation 'relu'
  - MaxPooling1D layer: Pool size 2
  - Conv1D layer: 128 filters, kernel size 3, activation 'relu'
  - MaxPooling1D layer: Pool size 2
  - LSTM layer: 128 units
  - Dropout layer: 0.3
  - Dense layer: 64 units, activation 'relu'
  - Output Dense layer: 8 units, activation 'softmax'

20: Compile model:

```
model.compile(loss = 'sparse_categorical_crossentropy', optimizer = 'adam', metrics = ['accuracy'])
```

21: **Train Model:**

22: Train the model: history $= \text{model.fit}(X_{\text{train}}, y_{\text{train}}, \text{epochs} = 50, \text{batch\_size} = 32, \text{validation\_data} = (X_{\text{test}}, y_{\text{test}}))$

23: **Predict Emotions:**

24: Predict: $y_{\text{pred}} = \text{model.predict}(X_{\text{test}})$

25: Decode labels: output_labels $= \text{encoder.inverse\_transform}(\text{np.argmax}(y_{\text{pred}}, \text{axis} = 1))$

26: **Output:** Predicted emotion labels $output\_labels$

Table 1. Summary of the Proposed Model Architecture

Layer (Type)	Output Shape	Parameters
Conv1D (64 filters)	(None, 30, 64)	256
MaxPooling1D	(None, 15, 64)	0
Conv1D (128 filters)	(None, 15, 128)	16,512
MaxPooling1D	(None, 7, 128)	0
LSTM (128 units)	(None, 128)	98,816
Dropout (0.3)	(None, 128)	0
Dense (64 units)	(None, 64)	8,256
Dense (Output Layer, 8 units)	(None, 8)	520

3.5   Experimental Setup

The dataset from RAVDESS was utilized to train the model, comprising of eight distinct emotions. Preprocessing was done on the dataset to extract hybrid features, like time- domain feature and MFCCs. For better performance, the model was trained on a system with an NVIDIA GPU after being implemented in Python using TensorFlow. 50 epochs and a batch size of 32 were used in the experiments.

3.6   Results

The accuracy of the model was utilized to assess its performance and compare it to other models that already existed. The suggested CNN-LSTM model's F1-score, recall, accuracy, and precision were contrasted with those of other models.

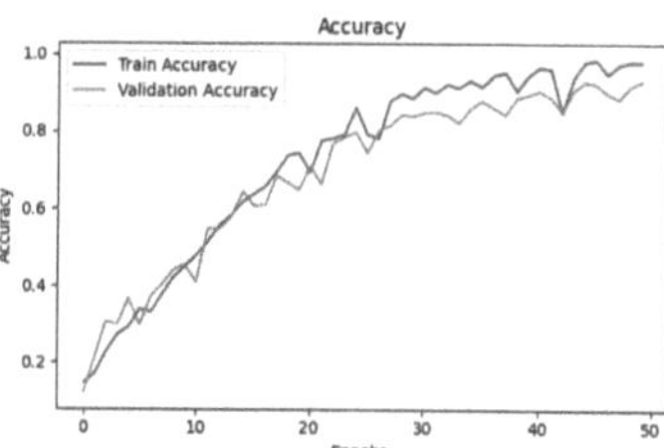

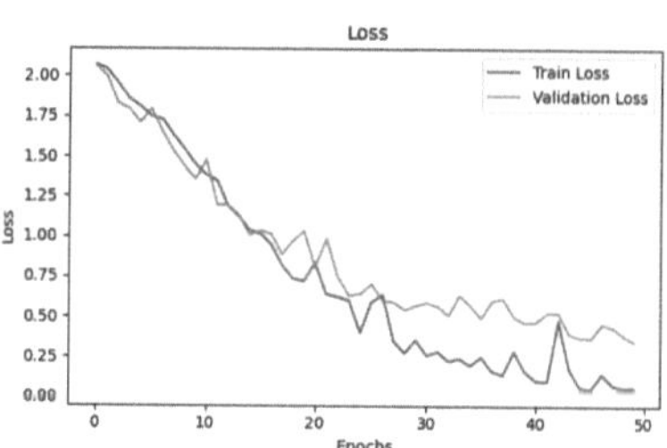

Fig. 2. Accuracy of training and validation across 50 epochs. The training accuracy increases consistently, hitting about 99%, and the validation accuracy closely tracks it, plateauing at about 95%. The tiny gap between the two lines testifies to high generalization, reflecting the robustness of the model in emotion classification.

Fig. 3. Loss of training and validation across 50 epochs.

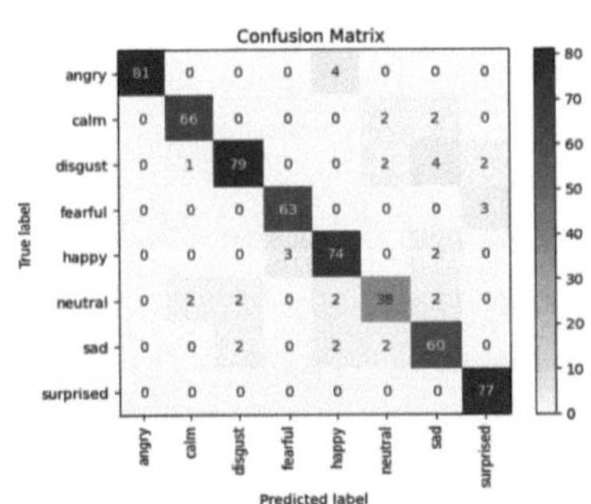

Fig. 4. Confusion matrix for the proposed model CNN-LSTM.

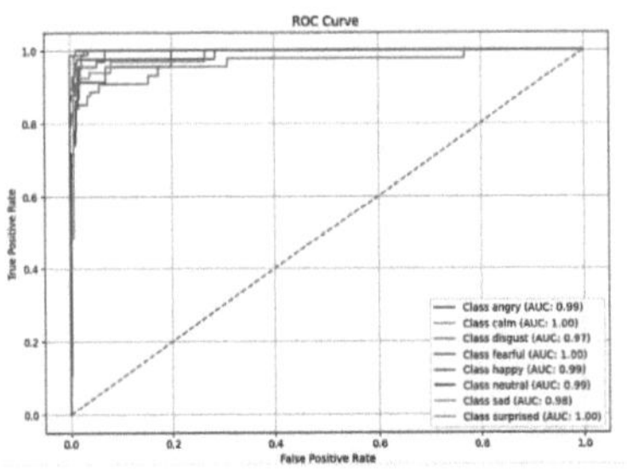

Fig. 5. ROC curve showing the CNN-LSTM model's classification performance, with high AUC values indicating strong emotion recognition accuracy. The AUC values range between 0.97 (Disgust) and 1.00 (Calm, Fearful, Surprised), indicating an outstanding capacity to differentiate emotions. These findings validate the effectiveness of the hybrid feature set in improving classification accuracy.

Table 2. The CNN-LSTM model's classification report demonstrates a high recall, F1-score, and precision across emotions, with an accuracy of 97.7%. Table 2 lists in detail the evaluation of precision, recall, and F1-score of the model for all the emotions. The highest recall of 100% occurs for Surprised and the least recall (83%) is with Neutral. Total accuracy of 97.7% confirms that MFCC + Spectral Contrast + Chroma features perform the best for emotion recognition.

Emotion	Precision	Recall	F1-Score
Angry	1.00	0.95	0.98
Calm	0.96	0.94	0.95
Disgust	0.95	0.90	0.92
Fearful	0.95	0.95	0.95
Happy	0.90	0.94	0.92
Neutral	0.86	0.83	0.84
Sad	0.86	0.91	0.88
Surprised	0.94	1.00	0.97
Accuracy	97.7%		

3.7 Discussion

According to the findings, the Hybrid CNN + LSTM model successfully captures the temporal and spatial characteristics of the audio signals, leading to strong

Table 3. Performance Comparison of Various Speech Emotion Recognition Models. Table 3 contrasts the performance of various proposed models with other studies. The CNN+LSTM model with hybrid features provides the highest accuracy (97.7%) in the RAVDESS dataset compared to traditional CNN, SVM, and Transformer-based models. These findings illustrate the strength of incorporating spectral and chroma features in enhancing emotion classification accuracy.

Reference	Feature used	Classifier	Dataset	Accuracy (%)
[9]	Acoustic Features + MFCC + Deep feature	SVM	RAVDESS	79.41
			EMO-DB	90.21
			IEMOCAP	85.37
[8]	Facial Expressions + Galvanic Skin Response + Electroencephalogram	Hybrid fusion	DEAP, LUMED-2	91
[16]	Visual Features + Audio Features + Text Features	CNN	IEMOCAP	82.7
[5]	ZCR, RMSE, MFCCs	1D CNN + LSTM	EMO-DB	96.72
			SAVEE	97.13
			ANAD	96.72
			BAVED	88.39
[7]	MFCC + LPCC + WPT + ZCR	1-D DCNN	EMO-DB	93.31
			RAVDESS	94.18
[6]	Mel-Spectrograms	Parallel CNN-Transformer Network + Parallel CNN-BLSTM-Attention Network	RAVDESS	89.33
			IEMOCAP	85.67
[15]	MFCC + Delta MFCC + Delta-Delta MFCC	CNN	RAVDESS	96.90
[4]	MFCCs + MFCCT	CNN	Emo-DB	97
			SAVEE	93
			RAVDESS	92
Proposed model	**MFCC + Spectral Contrast + Chroma**	**CNN + LSTM**	**RAVDESS**	**97.7**

emotion detection. The high accuracy achieved demonstrates the capability of the model to generalize across various emotional states. However, certain emotions showed lower performance, which may be attributed to the nuances in emotional expression or the dataset's inherent limitations. Future work could involve fine-tuning the model architecture or augmenting the dataset with additional emotional samples to improve recognition rates for underperforming classes. Finally, the results of the experiment confirm that using a hybrid model for audio emotion recognition paves the way for further exploration in this domain.

4   Conclusion

This study shows how well a CNN-LSTM hybrid approach works for speech emotion recognition (SER) that combines spectral contrast, chroma, and MFCC with other hybrid features. By leveraging the advantages of LSTMs for modeling temporal dependencies and CNNs for identifying spatial patterns, the model significantly improves the recognition of subtle emotional cues in speech. Tested on The hybrid model's remarkable accuracy on the RAVDESS dataset was 97.7%, highlighting its potential for real-world applications where reliable emotion classification is crucial, such as virtual assistants, customer service bots, and therapeutic tools. This research underscores the value of integrating spectral and temporal features to achieve excellent results on SER tasks and opens the door for further developments in emotion recognition technology.

References

1. Abdelhamid, A.A., et al.: Robust speech emotion recognition using CNN+LSTM based on stochastic fractal search optimization algorithm. IEEE Access **10**, 49265–49284 (2022). https://doi.org/10.1109/ACCESS.2022.3172954
2. Al-Dujaili, M.J., Ebrahimi-Moghadam, A.: Speech Emotion Recognition: A Comprehensive Survey. vol. 129(4). Springer US (2023). https://doi.org/10.1007/s11277-023-10244-3.
3. Alluhaidan, A.S., Saidani, O., Jahangir, R., Nauman, M.A., Neffati, O.S.: Speech emotion recognition through hybrid features and convolutional neural network. Appl. Sci. **13**(8) (2023). https://doi.org/10.3390/app13084750
4. Alsabhan, W.: Human–computer interaction with a real-time speech emotion recognition with ensembling techniques 1D. Sensors **23**(3), 1–21 (2023)
5. Bautista, J.L., Lee, Y.K., Shin, H.S.: Speech emotion recognition based on parallel CNN-attention networks with multi-fold data augmentation. Electron. **11**(23), 1–14 (2022). https://doi.org/10.3390/electronics11233935
6. Bhangale, K., Kothandaraman, M.: Speech emotion recognition based on multiple acoustic features and deep convolutional neural network. Electron. **12**(4) (2023). https://doi.org/10.3390/electronics12040839
7. Cimtay, Y., Ekmekcioglu, E., Caglar-Ozhan, S.: Cross-subject multimodal emotion recognition based on hybrid fusion. IEEE Access **8**, 168865–168878 (2020). https://doi.org/10.1109/ACCESS.2020.3023871

8. Er, M.B.: A novel approach for classification of speech emotions based on deep and acoustic features. IEEE Access **8**, 221640–221653 (2020). https://doi.org/10.1109/ACCESS.2020.3043201

9. Koduru, A., Valiveti, H.B., Budati, A.K.: Feature extraction algorithms to improve the speech emotion recognition rate. Int. J. Speech Technol. **23**(1), 45–55 (2020). https://doi.org/10.1007/s10772-020-09672-4

10. Koppula, N., Rao, K.S., Nabi, S.A., Balaram, A.: A novel optimized recurrent network-based automatic system for speech emotion identification. Wireless Pers. Commun. **128**(3), 2217–2243 (2023). https://doi.org/10.1007/s11277-022-10040-5

11. Kwon, S.: A CNN-Assisted Enhanced Audio Signal Processing. Sensors (2020)

12. Pan, J., Fang, W., Zhang, Z., Chen, B., Zhang, Z., Wang, S.: Multimodal emotion recognition based on facial expressions, speech, and EEG. IEEE Open J. Eng. Med. Biol. **5**, 396–403 (2024). https://doi.org/10.1109/OJEMB.2023.3240280

13. Sahoo, K.K., Dutta, I., Ijaz, M.F., Wozniak, M., Singh, P.K.: TLEFuzzyNet: fuzzy rank-based ensemble of transfer learning models for emotion recognition from human speeches. IEEE Access **9**, 166518–166530 (2021). https://doi.org/10.1109/ACCESS.2021.3135658

14. Singh, V., Prasad, S.: Speech emotion recognition system using gender dependent convolution neural network. Procedia Comput. Sci. **218**, 2533–2540 (2022). https://doi.org/10.1016/j.procs.2023.01.227

15. Srivastava, A.: Two-Way Feature Extraction for Speech Emotion Recognition Using Deep Learning (2022)

16. Zhu, X., Huang, Y., Wang, X., Wang, R.: emotion recognition based on brain-like multimodal hierarchical perception. Multimed. Tools Appl. **83**(18), 56039–56057 (2024). https://doi.org/10.1007/s11042-023-17347-w

Brain Tumor Detection and Classification on MRI Images Using YOLO Based Deep Learning Model

Nitish Kumar[1]([✉]), Aman Dhole[2], Samruddhi Kamble[2], and Gaurav Mishra[1]

[1] Visvesvaraya National Institute of Technology, Nagpur, Maharashtra, India
dt23cse022@students.vnit.ac.in, gauravmishra@cse.vnit.ac.in
[2] Datta Meghe Institute of Higher Education and Research, Wardha, Maharashtra, India

Abstract. In recent times, advancements in image processing have notably enhanced the detection of brain tumors in individuals. Historically, brain tumors were assessed manually using MRI scans, which could lead to human errors. However, with the progress in deep learning techniques, identifying brain tumors and their three variants—meningioma, glioma, and pituitary tumors—has become considerably simpler. There are various deep learning models for object detection like R-CNN, SSD, and R-FCN, that are not able to detect small objects. There are some advance models which has the capability to detect small objects also. Some YOLO that model that detects small objects in real time, which best suits our analysis. In this paper, a comparative analysis of different object detection models like Faster R-CNN, YOLO v5, YOLO v7, and YOLO v11 to carry out to detect brain tumors or segment the portion of the brain tumor. The Brain Tumor IS dataset is used for experimental analysis, it has a total of 801 images, which is divided into 62%, 25%, and 13% with train set, valid set, and test set respectively. Precision, Recall, mAP 50, aAP50-95, and execution time are the key factors used for comparing the models. The experimental result shows that YOLO v7 outperforms other competing models in terms of Precision, mAP 50, and mAP 50-95 except recall and inference speed parameters.

Keywords: Image Processing · MRI · YOLO · Brain Tumor · Deep learning

1 Introduction

In recent years, the advancements in image processing have greatly enhanced the ability to identify brain tumors in individuals. In the past, identifying brain tumors by manual assessment through MRI scans [2], a method that was vulnerable to human mistakes. Today, thanks to improvements in deep learning [1] algorithms, the process of detecting brain tumors and their three varieties—meningioma, glioma, and pituitary tumors—has become more efficient. The main aim is to evaluate and analyze the effectiveness of some advanced object detection models [3]: Faster R-CNN, YOLOv5, YOLOv7, and YOLO11 [4, 5, 6, 7, 8], in identifying and classifying brain tumors by utilizing the Brain Tumor IS Dataset. This dataset is partitioned into three sections: training, validation,

P. Chandrakar et al. (Eds.): ICCINS 2025, CCIS 2738, pp. 104–114, 2026.
https://doi.org/10.1007/978-3-032-09572-5_9

and testing. The training and validation sets contain annotated masks for each image. The overall dataset comprises 801 images, allocated as 62% for the training set, 25% for the validation set, and 13% for the testing set. Specifically, the training set consists of 500 images, the validation set includes 201 images, and the testing set contains 100 images. This equitable distribution is essential for effective classification, as it helps to avoid bias in the models towards a specific type of tumor. These models are widely acknowledged for their ability to perform real-time detection; however, they vary in architecture, computational demand, and overall performance. This report intends to analyze their application in medical imaging, focusing specifically on detecting and classifying brain tumors, to gauge their efficiency in performing this crucial task.

The aim is to determine how effectively each model recognizes the existence of a tumor and categorizes it into three classes: meningioma, glioma, or pituitary tumor, and which factors are responsible for their different outputs. This will yield important insights into their comparative strengths and weaknesses in medical diagnostics.

This comparison emphasizes two main tasks: tumor detection and tumor classification. For the detection aspect, each model's capability to accurately pinpoint the presence or absence of a brain tumor will be assessed. On the classification side, the system will review each model's proficiency in accurately categorizing the identified tumors into the specified types. Furthermore, the analysis will include architectural differences, training efficiency, and unique characteristics of each model that could account for any differences in outcomes observed. This will help establish which model performs more effectively under certain conditions and explain the reasons for any discrepancies noted.

The rest of the paper is organized as follows: Sect. 2 discusses the literature review done for this paper, Sect. 3 describes the methodological analysis and results. Finally, Sect. 4 summarize the conclusion and future scope.

2 Literature Survey

Dixit et al. [9] introduce a method using the YOLO model that detects and classifies the brain tumors in MRI images. They emphasize the model's capability to do real-time object detection with accuracy, which makes it particularly suitable for medical imaging tasks that demand speed and precision. Their study reports a detection accuracy of 92%, with a slight reduction in false positives compared to traditional methods. This research enlightens the performance of modern deep learning models like YOLO to enhance brain tumor detection and advancing medical diagnostics.

Almufareh et al. [10] explore the application of YOLOv5 and YOLOv7 models for character detection in first-person shooter game scenes. Their work highlights the comparative performance of both models, with YOLOv7 demonstrating superior accuracy and faster processing time in detecting characters within complex gaming environments. The study reports that YOLOv7 achieved a detection accuracy of 95.3%, outperforming YOLOv5's accuracy of 92.1%, while also providing better real-time performance. These findings suggest that YOLOv7's enhanced architecture makes it more efficient for tasks requiring high precision and speed, providing valuable insights for improving detection models in various fields, including medical imaging.

Ren et al. [8] proposed Faster R-CNN, an advanced object detection method that introduces region proposal networks (RPNs) to significantly improve detection speed and accuracy. The authors address the limitations of earlier models by generating region proposals directly from the convolutional feature maps, allowing the system to efficiently detect objects in near real-time. Their experiments demonstrated state-of-the-art results on various benchmark datasets, achieving a mean average precision (mAP) of 73.2% on the PASCAL VOC 2007 dataset. This work marked a pivotal development in object detection, offering a balance between speed and precision, which set the stage for further innovations in real-time detection frameworks like YOLO.

Using the YOLO framework, Nath et al. [11] analyzed the effectiveness of Convolutional Neural Networks (CNNs) for detecting both persons and cars. The study emphasizes YOLO's capability for real-time object detection, highlighting its efficiency in processing images swiftly while maintaining a high level of accuracy. The results reveal a detection accuracy of 91% for persons and 89% for cars, showcasing its suitability for applications where both speed and precision are crucial. This research further demonstrates the adaptability of the YOLO framework across different object detection tasks in real-world scenarios.

The study by Jana et al. [12] investigates the application of YOLO for object detection and classification in video recordings, emphasizing the model's remarkable speed and accuracy. The authors highlight its real-time capabilities, making it highly efficient for analyzing continuous data streams like video. Their findings indicate that YOLO achieved a high detection rate with minimal latency, demonstrating its suitability for time-sensitive applications. This research underscores the strengths of the YOLO architecture in managing large-scale object detection tasks while maintaining precision, which is particularly relevant for its application in analyzing medical images to detect brain tumors.

Faster R-CNN [8], an evolution of R-CNN [13] and Fast R-CNN [14], makes the object detection more efficient by handling region proposals directly within the neural network. Unlike its predecessors, which relied on external algorithms like Selective Search to propose a region of interest, Faster R-CNN employs a Region Proposal Network (RPN) that is efficient and fast. This RPN used the same convolutional layers as the object detection network, allowing for near real-time detection speeds without compromising accuracy. Faster R-CNN not only improves detection time but also enhances object localization and classification, which makes it a go-to choice for many computer vision applications.

2.1 Models Architecture

To prepare the data for use with YOLOv5 [6] and YOLOv7 [7], several preprocessing steps are essential. First, all images are resized to a uniform resolution to match the input size requirements of the models, ensuring consistent performance. Additionally, normalization is applied to standardize the pixel values across images, which aids in model training by reducing variance. Techniques for data augmentation like random rotations, flips, and scaling, are employed to artificially increase the dataset size and diversity. These steps help the models generalize better, improving their ability to detect and classify tumors in unseen images (Fig. 1).

YOLOv5 is built around three key parts: backbone, neck, and head. Its backbone, CSPDarknet53 [15], is optimized for efficient feature extraction, reducing computational

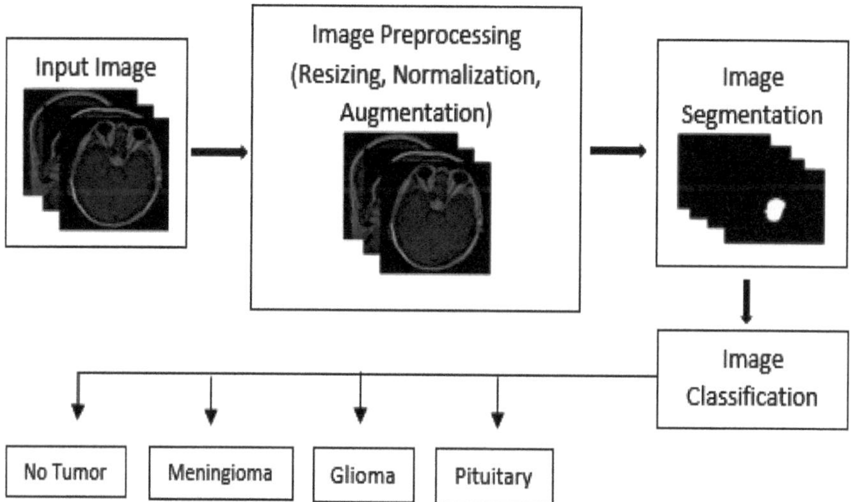

Fig. 1. Model Architecture

complexity while maintaining strong performance. The neck uses a Path Aggregation Network (PANet) [16] to aggregate features from different scales to improve the model's ability in detecting tumors of different sizes. The head applies anchor-based object detection, generating multiple bounding boxes for tumor detection and classification. YOLOv5 typically uses flexible input resolutions like 640 × 640, with anchor boxes fine-tuned to the dataset for accurate detection.

YOLOv7, on the other hand, introduces several architectural improvements over YOLOv5. Its backbone incorporates an Efficient Layer Aggregation Network (ELAN) [17] that enhances feature retention and extraction. In the neck, YOLOv7 optimizes feature fusion and utilizes an extended receptive field for better contextual understanding. Notably, the head of YOLOv7 adopts an anchor-free detection mechanism, allowing for more adaptive and flexible bounding box predictions. This approach improves performance, particularly for complex or irregularly shaped tumors, while maintaining efficiency with higher input resolutions.

In terms of training hyperparameters, YOLOv5 typically uses moderate learning rates and smaller batch sizes, while YOLOv7's more advanced architecture allows for larger batch sizes and potentially lower learning rates for better stability. Both models benefit from optimizers like SGD or Adam, but YOLOv7's architecture may require more advanced scheduling for fine-tuning. Overall, YOLOv7's innovations—particularly its anchor-free mechanism and enhanced feature extraction—offer advantages in the nuanced task of brain tumor detection and classification.

YOLOv11 [5] builds upon the innovations of previous YOLO models, It incorporates a refined backbone, which uses a deeper and more efficient feature extraction network to capture richer semantic information from images. The neck design enhances multi-scale feature fusion through an advanced Path Aggregation Network (PAN) [16], makes the model able to detect objects with varying sizes. One of the significant improvements in YOLOv11 is the adoption of a fully anchor-free detection head, which simplifies the process of bounding box generation and improves flexibility in detecting complex shapes, such as tumors, in medical imaging.

3 Comparative Analysis Result

This section shows the results of brain tumor classification and segmentation of all models and their comparison. Before training all the models Faster R-CNN, YOLOv5, YOLOv7, and YOLOv11 for brain tumor detection, several critical steps need to be followed. First, the MRI dataset must be prepared, ensuring that it includes properly labeled images where the tumor regions are annotated. These images are then divided into training, validation, and test sets. The model architecture, including the convolutional layers and anchor boxes, must be configured based on the size and variability of tumors in the images. Hyperparameters, such as learning rate, batch size, and IoU threshold, are also defined to optimize the model's learning process. Preprocessing techniques like data augmentation may be applied to enhance the model's robustness to variations in image quality or tumor shapes. Once the dataset and configuration are ready, the model is initialized for training, where it learns to identify tumor regions through iterative updates to its weights.

YOLOv5 model trained for brain tumor detection on MRI scans. Each sub-image represents an MRI slice with red bounding boxes highlighting detected tumor regions. The labels inside the boxes, such as "Tumor 0.8" or "Tumor 0.9," represent the model's confidence scores, indicating how certain it is that the detected region contains a tumor, with values closer to 1 showing higher confidence. The image demonstrates the model's capability to detect tumors with varying confidence levels across different MRI scans, showcasing the performance of the model in localizing tumors, though it does not differentiate between tumor types.

In the validation results of the YOLOv5 model for brain tumor detection on MRI scans, each sub-image shows an MRI slice with red bounding boxes around the detected tumor regions, accompanied by confidence scores (e.g., "Tumor 0.9", "Tumor 0.7"). These scores reflect the accuracy in detecting tumors, with higher values indicating stronger confidence. The results demonstrate the model detection ability to detect tumors with relatively high accuracy across different validation samples, further verifying its detection performance. Similar to the training output, the image highlights the localization of tumors without distinguishing between different tumor types.

The image shows the tested result of the YOLOv5 model applied to a new MRI scan for brain tumor detection. A blue bounding box outlines the detected tumor, labeled "Tumor 0.80," indicating that the model is 80% confident that the identified region contains a tumor. This result shows the ability of model for generalizing training and validation data, effectively identifying tumor regions in unseen images. The detection, based on both spatial localization and confidence score, demonstrates the practical application of the trained model in a clinical setting for assisting radiologists with brain tumor detection in MRI scans.

Same process is repeated for all remaining models YOLOv7, YOLOv11 and Faster R-CNN. The YOLOv7 model for brain tumor detection and classification on MRI images, specifically detecting three types of tumors: meningioma, pituitary, and the absence of tumors ("no tumor"). Each sub-image shows an MRI scan with labeled bounding boxes identifying detected tumor types, along with confidence scores (e.g., "meningioma 0.8" or "pituitary 0.9"). The model demonstrates high confidence in detecting and classifying different tumor types, as well as distinguishing cases with no tumor. The consistent detection of "meningioma" and "pituitary" tumors, along with varying confidence levels, showcases the model's accuracy and robustness in identifying and classifying tumors across multiple validation samples. This output verifies YOLOv7's ability to handle both detection and classification tasks effectively within the brain tumor dataset. The model successfully identifies various types of tumors, such as meningioma and pituitary, as well as cases with no tumor. Each detected region is marked with a bounding box and accompanied by confidence scores, such as "meningioma 0.94" and "pituitary 0.84," indicating the model's confidence in its predictions. The output highlights the accuracy of model in detecting tumors across different images, with strong classification capabilities between tumor types and non-tumor cases. This result demonstrates the YOLOv7 model's effectiveness in generalizing to new, unseen MRI scans, showcasing its potential for real-world clinical applications in tumor detection (Table 1).

YOLOv11 is set up as a highly optimized object detection model, designed to quickly and accurately identify objects in images. It improves upon previous versions by using a more advanced backbone for feature extraction, better multi-scale detection through its neck design, and an anchor-free detection head. These enhancements make it faster and more efficient, especially in detecting complex shapes like tumors. YOLOv11 is particularly well-suited for medical imaging tasks, where both speed and precision are critical. Prior to training, the model is configured with labeled MRI scans of brain tumors, preparing it to learn how to detect and classify tumors with high confidence.

The Faster R-CNN model with a ResNet50-FPN backbone, offers robust feature extraction for object detection tasks. Key parameters were set, including 50 number of epochs, a batch size equal to 8, and a learning rate suited to this dataset. During training, the model used bounding boxes to predict and classify tumor areas within MRI images, with periodic validation runs to monitor and improve localization accuracy. After training, the validation results were visualized, showcasing the model detection ability to detect and outline tumors in MRI images effectively. Finally, the model was evaluated based on its performance metrics, showing 46.6% of average precision and 73.4% average recall. These metrics highlight the model's accuracy in identifying tumor regions, validating the effectiveness of the Faster R-CNN approach with a ResNet50-FPN backbone for this specific medical imaging task. The comparison of training and validation loss with precision and recall graphs are shown below in Figs. 2, 3, and 4.

The hyperparameteres used for the experimental analysis of these four deep learning models are shown in Table 2. The comparison Table 3, provides a detailed evaluation of these models: Faster R-CNN, YOLOv5, YOLOv7, and YOLOv11, and focusing on key performance metrics such as Precision, Recall, mAP 50, mAP 50-95, and inference Time (measured in iterations per second, or it/s).

Precision measures the accuracy of positive detections, or how many detected objects were actual tumors. YOLOv7 exhibits the highest precision at **87.4%**, indicating its superior ability to minimize false positives. YOLOv11 follows with a precision of **68.5%**, which is slightly better than YOLOv5's **67%**. This suggests that while YOLOv11 is a

Table 1. Comparison of different deep learning algorithms on Brain Tumor IS dataset

Model	Train Output	Test Output
Faster R_CNN	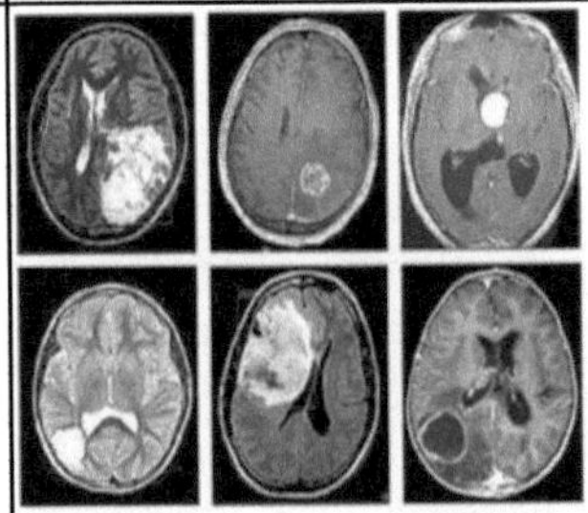	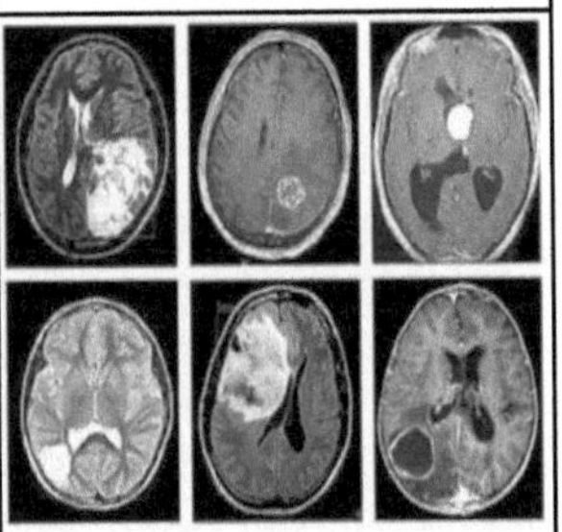
YOLO v5	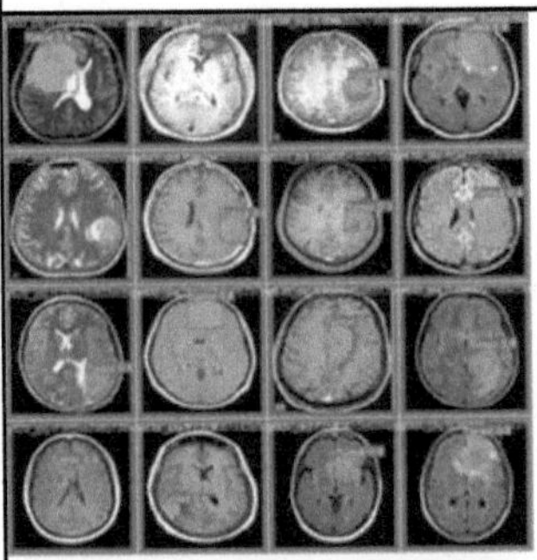	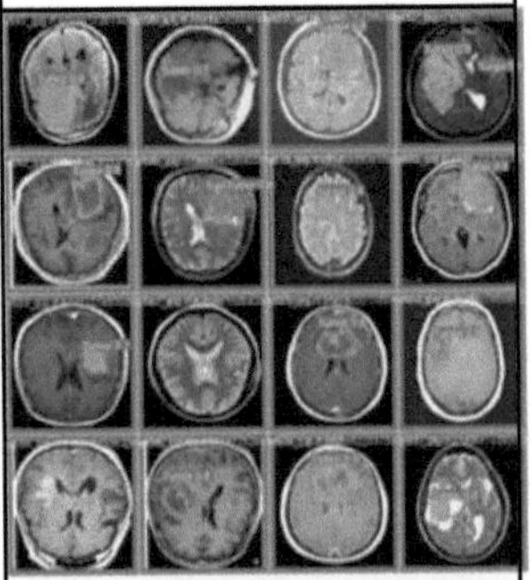
YOLO v7	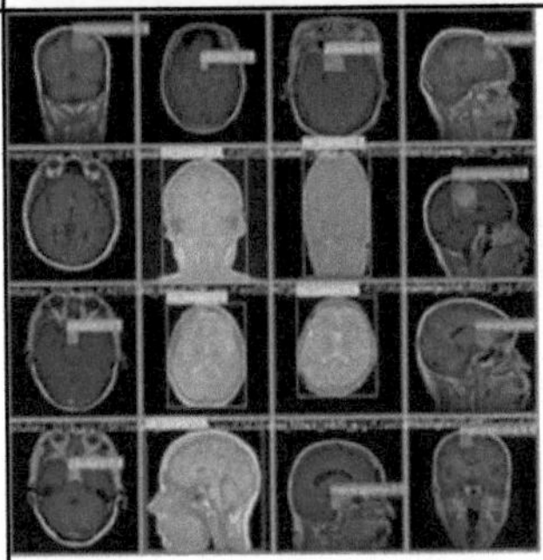	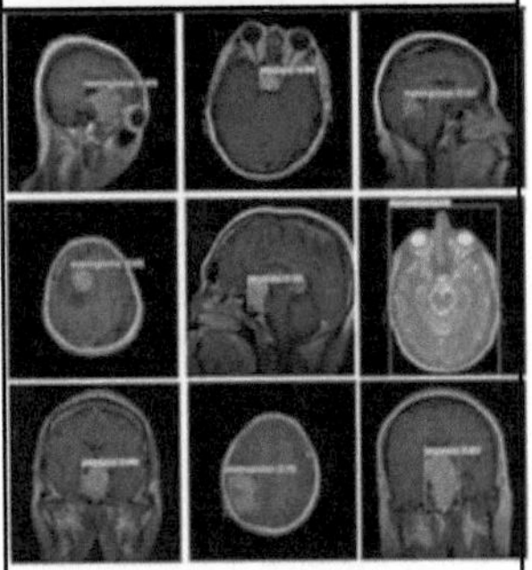
YOLO v11	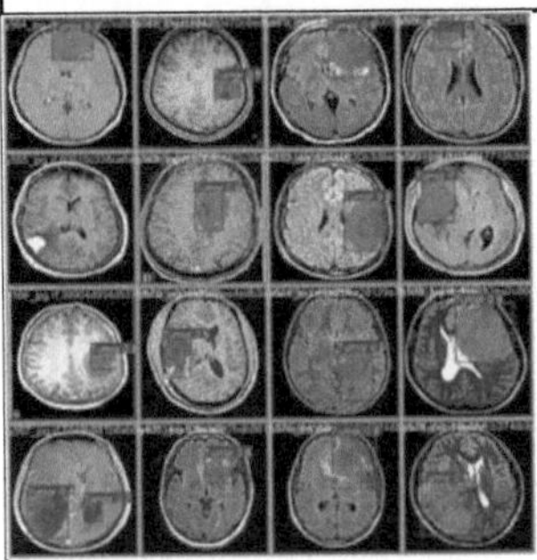	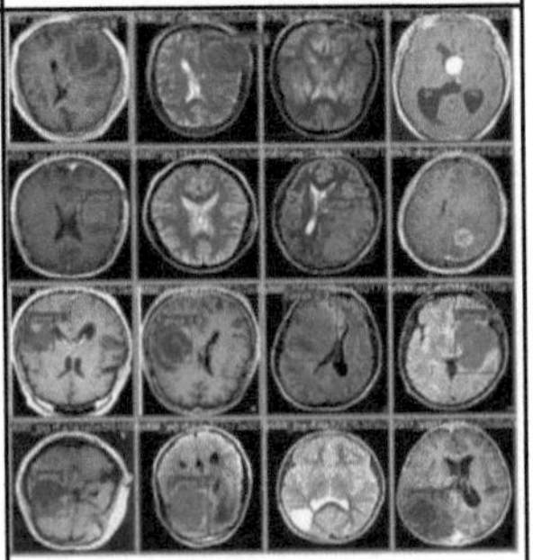

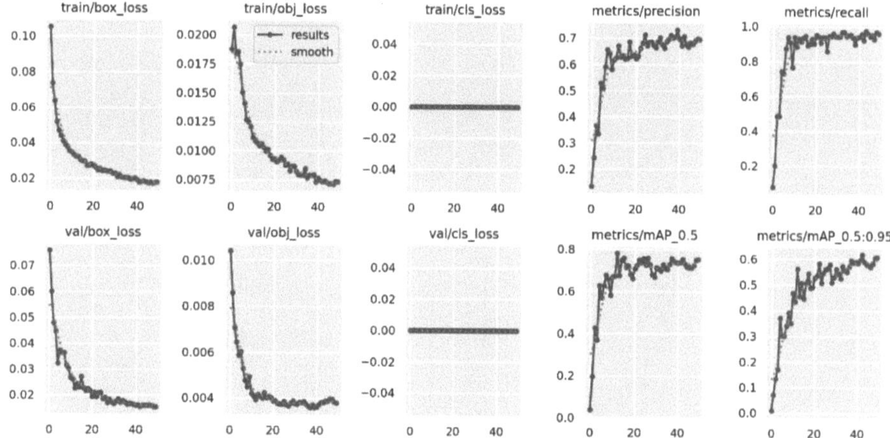

Fig. 2. The comparison of training and validation loss of YOLOv5

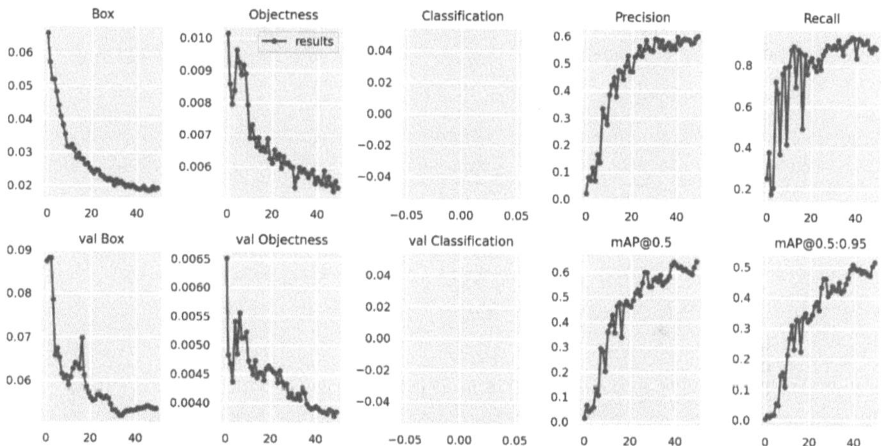

Fig. 3. The comparison of training and validation loss of YOLOv7

slight improvement over YOLOv5 in precision, both models have room for enhancement compared to YOLOv7.

Recall focuses on identifying all cases of true positive correctly, or how many actual tumors were detected. YOLOv11 leads in recall with an impressive **95.3%**, closely followed by YOLOv5 at **93%**. YOLOv7, while still performing well, has a slightly lower recall at **88.6%**. This means that YOLOv11 is the most successful at capturing the majority of true tumor cases, even at the cost of higher false positives compared to YOLOv7.

Mean Average Precision at IoU threshold 0.5 (mAP 50) evaluates how well the models predict the correct bounding box at a 50% IoU threshold. YOLOv7 outperforms both other models here with a **87.6%** mAP 50, showcasing its balance between precision and recall. YOLOv11 and YOLOv5 have nearly identical mAP 50 scores, with **80.3%**

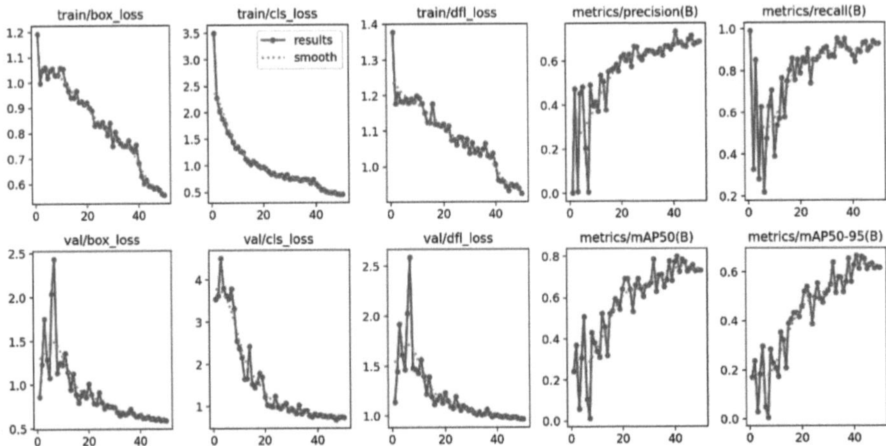

Fig. 4. The comparison of training and validation loss of YOLOv11

Table 2. The Hyperparameters used for analysis

Models	Hyperparameters						
	Image Size	Batch Size	Learning Rate	Epochs	IoU	Conf. Threshold	Weight Decay
Faster R-CNN	640 × 640	8	0.001	50	0.65	0.5	0.0001
YOLOv5	640 × 640	32	0.01	50	0.60	0.001	0.0005
YOLOv7	640 × 640	16	0.01	50	0.2	0.4	0.0005
YOLOv11	640 × 640	32	0.0003	50	0.65	0.70	0.0005

for YOLOv11 and **80%** for YOLOv5, indicating comparable detection accuracy across these two models.

Mean Average Precision across IoU thresholds from 0.5 to 0.95 (mAP 50-95)metric measures performance across varying levels of localization accuracy, making it more challenging than mAP 50. YOLOv7 again leads with **77.2%**, followed by YOLOv11 at **64.9%**, and YOLOv5 trailing at **61%**. This shows that YOLOv7 excels at maintaining high performance over a broader range of IoU thresholds, while YOLOv11 represents a solid improvement over YOLOv5 in this regard.

The speed of detection is a crucial factor, particularly for real-time applications like medical diagnostics. YOLOv11 is the fastest among all the models, processing images at 1.77 it/s, indicating its capacity for faster inference. YOLOv7 follows closely at 1.65 it/s, YOLOv5 is at 1.48 it/s while Faster R-CNN is slowest at 1.03it/s. This makes YOLOv11 the most efficient model for real-time detection tasks, despite the slight trade-off in precision and mAP compared to YOLOv7. From the above discussion, we can conclude that the YOLOv7 achieves better results as compared to others on the Brain Tumor IS dataset.

Table 3. The Comparison of different deep learning algorithms on Brain Tumor IS dataset in terms of Precision, Recall, mAP 50, aAP50-95, and execution time

Models	Precision	Recall	mAP50	mAP50-95	Inference Time
Faster RCNN	46.6%	73.4%	59.21%	46.29%	1.03it/s
YOLOv5	67%	93%	80%	61%	1.48it/s
YOLOv7	**87.4%**	88.6%	**87.6%**	**77.2%**	1.65it/s
YOLO11	68.5%	95.3%	80.3	64.9%	1.77it/s

4 Conclusion

This paper highlights the comparative analysis of Faster R-CNN, YOLOv5, YOLOv7, and YOLOv11 on brain tumor detection using the Brain Tumor IS Dataset. While all these models demonstrate significant capabilities in detecting and localizing tumors, YOLOv7 outperforms with other models, as the most balanced model on the Brain Tumor IS dataset, offering superior precision, mAP 50 and mAP 50-95 scores, making it the most reliable choice for accurate detection. YOLOv11, though slightly less precise, excels in recall and inference speed, positioning it as a strong option for real-time applications where quick and accurate detection is critical. YOLOv5, although slightly behind in performance metrics, still provides competitive results and remains viable for less demanding tasks. Faster R-CNN is far behind all the models. Overall, the improvements in YOLOv11 show promising potential for advancing object detection in medical imaging, particularly in time-sensitive scenarios like diagnostic procedures. In Future, these models can be used to detect and classify other medical images like cancer images.

References

1. Havaei, M., et al.: Brain tumor segmentation with deep neural networks. Med. Image Anal. **35**, 18–31 (2017)
2. Bauer, S., et al.: A survey of MRI-based medical image analysis for brain tumor studies. Phys. Med. Biol. **58**(13), R97–R129 (2013)
3. Zou, Z., et al.: Object detection in 20 years: a survey. Proc. IEEE. **111**(3), 257–276 (2023)
4. Rahimi, M., Mostafavi, M., Arabameri, A.: Automatic detection of brain tumor on MRI images using a YOLO-based algorithm. In: 2024 13th Iranian/3rd International Machine Vision and Image Processing Conference (MVIP). IEEE (2024)
5. Jocher, G., Qiu, J.: Ultralytics yolo11 (2024). [Online]. Available: https://github.com/ultralytics/ultralytics
6. Jocher, G.: Ultralytics yolov5 (2020). [Online]. Available: https://github.com/ultralytics/yolov5
7. Wang, C.-Y., Bochkovskiy, A., Liao, H.-Y. M.: Yolov7: trainable bag-of-freebies sets new state-of-the-art for real-time object detectors. arXiv preprint arXiv:2207.02696 (2022)

8. Ren, S., et al.: Faster R-CNN: towards real-time object detection with region proposal networks. IEEE Trans. Pattern Anal. Mach. Intell. **39**(6), 1137–1149 (2016)

9. Dixit, A., Singh, P.: Brain tumor detection using fine-tuned YOLO model with transfer learning. In: Artificial Intelligence on Medical Data: Proceedings of International Symposium, ISCMM 2021. Springer Nature Singapore, Singapore (2022)

10. Almufareh, M.F., et al.: Automated brain tumor segmentation and classification in MRI using YOLO-based Deep Learning. IEEE Access. **12**, 16189–16207 (2024)

11. LillyMaheepa, P., Nath, M.K.: A technical survey on brain tumor segmentation using CNN. In: 2020 IEEE 5th International Conference on Computing Communication and Automation (ICCCA). IEEE (2020)

12. Jana, A.P., Biswas, A.: YOLO based Detection and Classification of Objects in video records. In: 2018 3rd IEEE International Conference on Recent Trends in Electronics, Information & Communication Technology (RTEICT), p. 2018. IEEE

13. Girshick, R., et al.: Rich feature hierarchies for accurate object detection and semantic segmentation. In: Proceedings of the IEEE Conference on Computer Vision and Pattern Recognition (2014)

14. Girshick, R.: Fast r-cnn. arXiv preprint arXiv:1504.08083 (2015)

15. Geng, J.J., Kou, H.W., Zhang, X.F.: Optimization of target detection algorithm based on Yolov5. In: International Conference on Cloud Computing, Performance Computing, and Deep Learning (CCPCDL 2023), vol. 12712. SPIE (2023)

16. Liu, S., et al.: Path aggregation network for instance segmentation. In: Proceedings of the IEEE Conference on Computer Vision and Pattern Recognition (2018)

17. Wahyudi, D., Soesanti, I., Nugroho, H.A.: Optimizing hyperparameters of YOLO to improve performance of brain tumor detection in MRI images. In: 2023 6th International Conference on Information and Communications Technology (ICOIACT). IEEE (2023)

Deep Multi-label Classification for Automated Ocular Disease Detection Using Enhanced CNN Architectures

Deepan Vishal Thulasi Vel[1]($\boxtimes$) and Sairam Durgaraju[2]

[1] The Cigna Group, Bloomfield, CT 06002, USA
deepanvishal@gmail.com
[2] The Cigna Group, 733 Sun Valley Ct, Chester Springs, PA 19425, USA
sdurgar@my.wgu.edu

Abstract. Automated detection of ocular diseases through retinal fundus images has become increasingly critical in the field of ophthalmology, as it aids in early diagnosis and management of vision-threatening conditions. This study presents a Convolutional Neural Network (CNN)-based model designed for multi-label classification of ocular diseases using the Ocular Disease Intelligent Recognition (ODIR-5K) dataset. The proposed model integrates depthwise separable convolutions and a modified inception block to enhance feature extraction while maintaining computational efficiency. Extensive data preprocessing, including normalization and data augmentation, was performed to improve model generalization. The model attained an accuracy of 98.94%, surpassing current baseline models in the domain. This approach addresses critical challenges such as class imbalance and coexisting disease conditions within a single sample, making it suitable for real-world clinical applications. Our findings suggest that this model has significant potential for deployment in automated ocular disease screening, particularly in resource-limited settings where access to specialized ophthalmic diagnostics is constrained.

Keywords: Ocular disease detection · Convolutional Neural Network (CNNs) · Multi-label classification · Fundus imaging · ODIR-5K · Deep learning (DL) · Automated screening · Medical image analysis

1 Introduction

Affecting millions of people worldwide, ocular illnesses include diabetic retinopathy, glaucoma, age-related macular degeneration, and cataract provide serious public health issues with regard to morbidity if neglected. Effective therapy and prevention of visual impairment depend on early and correct identification of these diseases; but, access to specialist diagnostic resources is still restricted in many areas, especially in low-resource settings [1,2]. An essential instrument in ophthalmology, retinal fundus imaging offers insightful analysis of several ocular diseases. But the growing demand for eye exams calls for the creation of auto-mated systems able to precisely and quickly detect ocular disorders [3]. Developing strong automated systems for eye disease recognition remains difficult even with progress in DL and computer vision. First of all, variations in imaging technology, patient demographics, and disease development cause great heterogeneity in fundus image datasets. This variability can make models less able to generalize over several populations and environments [4]. Furthermore significantly imbalanced ocular disease datasets with some disorders underrepresented can result in biassed models and worse sensitivity for less prevalent diseases [5]. Building a model able to perform well across several disease categories without sacrificing accuracy in underrepresented groups depends on addressing these problems. Furthermore, ocular disease diagnosis is naturally a multi-label categorization issue since patients could have several coexistence disorders. Since traditional single-label classification methods are inadequate [6], its multi-label character complicates model training and evaluation. Models must not only identify several diseases inside the same image but also control the complexity related with class imbalances and image quality variations if they are to attain dependable performance in practical applications. These difficulties inspire the demand for customized architectures and training methods fit for multi-label classification and unbalanced datasets in ocular disease detection [7].

With an emphasis on multi-label classification and treating imbalanced datasets properly, this work intends to create a DL-based method that addresses the particular complexity of ocular disease detection in view of these problems. This work aims to contribute to the expanding field of automated ocular disease diagnosis by building on recent advancements in convolutional neural networks (CNNs) and including techniques that improve model stability and generalization, so providing a possible solution for large-scale screening initiatives in both clinical and resource-limited settings [8,9].

Our key contributions in this paper are as follows:

- We propose a novel CNN architecture designed specifically for multi-label ocular disease classification, integrating a modified inception block and depthwise separable convolutions to enhance feature extraction while maintaining computational efficiency.
- We implement techniques such as batch normalization and dropout layers to improve model stability and generalization, effectively addressing class imbalance in the ODIR-5K dataset.
- We validate the effectiveness of our approach through comprehensive evaluations, demonstrating superior accuracy and recall compared to baseline models, reinforcing its suitability for large-scale ocular disease screening.

This work is organized as follows: Reviewing the body of current research, Sect. 2 highlights important developments and difficulties in ocular illness classification. Section 3 presents the proposed methodology, detailing the model design, learning strategies, and optimization techniques. Section 4 describes the experimental setup, including dataset characteristics, preprocessing steps, training configurations, and evaluation metrics. Section 5 discusses the results, analyzing model performance, qualitative assessments, comparisons with existing approaches, and key insights. Finally, Sect. 6 concludes the study, summarizing contributions and outlining potential future research directions.

2   Related Work

Deep learning (DL) has revolutionized ocular disease recognition, leveraging large retinal fundus image datasets for accurate diagnosis. CNNs are particularly effective due to their ability to capture intricate visual patterns [10,11]. Various approaches have been explored to tackle challenges such as multi-label classification, class imbalance, and image variability [12].

Several CNN architectures have been proposed for ocular disease detection. Bhati et al. [3] introduced a discriminative kernel-based CNN to mitigate class imbalance, enhancing performance on underrepresented categories. Zhang et al. [7] developed a mixed-decomposed convolutional model, prioritizing efficiency while maintaining high diagnostic accuracy.

For multi-label classification, He et al. [4] designed a dense correlation-based deep network to capture interdependencies between co-occurring diseases. Herrera-Chavez et al. [6] validated models using K-fold cross-validation on the ODIR-5K dataset, improving reliability. Additionally, Mayya et al. [5] highlighted the role of image preprocessing techniques, such as contrast enhancement, in boosting CNN performance.

Addressing class imbalance remains a key focus. Gour et al. [9] proposed resampling and loss function adjustments to improve minority class recognition. Fauzi et al. [13] applied data augmentation techniques, enhancing robustness. Muthukannan et al. [14] demonstrated that loss functions like focal loss improve sensitivity towards rare diseases.

Further, Sharma et al. [15] leveraged ResNet50 for enhanced feature extraction, while Salem et al. [16] optimized VGG-net models to balance accuracy and computational efficiency. Topaloglu [17] designed a CNN adaptable to diverse imaging conditions, emphasizing real-world applicability.

Despite advancements, existing models often demand high computational power, limiting deployment in resource-constrained settings [8]. Moreover, while multi-label classification techniques have improved, variations in image quality and patient demographics remain challenging. Class imbalance further reduces sensitivity to rare conditions [9,17].

This study proposes a CNN-based approach that optimizes classification performance while ensuring computational efficiency. By integrating advanced training strategies, our model aims to improve multi-label prediction and mitigate class imbalance, contributing to more accessible and reliable ocular disease recognition.

3 Proposed Methodology

This study employs a CNN model designed for multi-label classification of ocular diseases using the ODIR-5K dataset [18]. The architecture is constructed to handle the complexity of multi-class, multi-label recognition while addressing challenges related to computational efficiency and feature diversity. The model integrates depthwise separable convolutions, pointwise convolutions, and a modified inception block (see Figs. 1 and 2) to improve feature extraction and model performance. Furthermore, batch normalization and dropout layers are applied to stabilize training, prevent overfitting, and ensure robust performance across varied data.

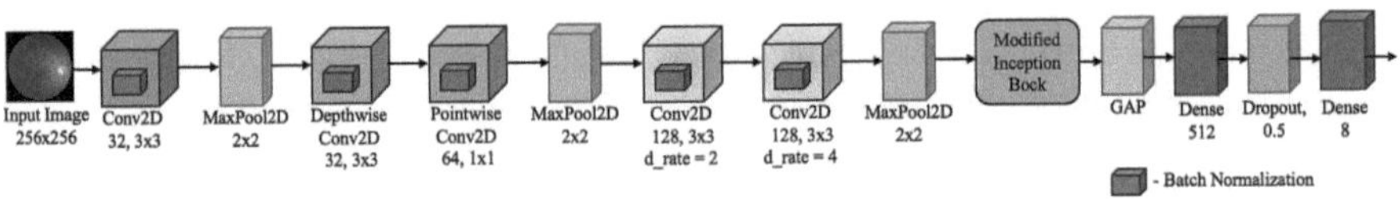

Fig. 1. Overview of the designed CNN architecture.

3.1 Model Architecture

The input to the model is a color fundus image represented by $\mathbf{X} \in \mathbb{R}^{256 \times 256 \times 3}$, where each of the three channels corresponds to the RGB color values. The processing sequence of the input image is detailed as follows:

1. Initial Convolution Layer: The model starts with a typical 2D convolutional layer using 32 filters of size 3×3 then batch normalisation to guarantee consistent gradients during training. Mathematically, this operation is expressed as:

$$\mathbf{H}_1 = \sigma(\mathbf{W}_1 * \mathbf{X} + b_1),$$

where $*$ indicates the convolution operation, $\mathbf{W}_1$ denotes the filter matrix, b_1 is bias term, and σ, the ReLU activation function applied element-wise.

2. Depthwise Separable Convolution: To optimize computational efficiency, a depthwise separable convolution is applied. This operation separates the spatial filtering and feature combination steps, reducing the number of parameters. For depthwise convolution, let $\mathbf{W}_{dw}$ be the depthwise filter, which is applied to each input channel independently, and $\mathbf{W}_{pw}$ the pointwise filter, combining the outputs of depthwise convolution across channels. This is represented as:

$$\mathbf{H}_{sep} = \sigma((\mathbf{W}_{dw} * \mathbf{H}_1) \odot \mathbf{W}_{pw}),$$

where $\odot$ denotes element-wise multiplication. This arrangement preserves spatial information while enabling efficient feature extraction.

3. Modified Inception Block: The modified inception block captures multi-scale features by employing parallel convolutional paths with kernel sizes with

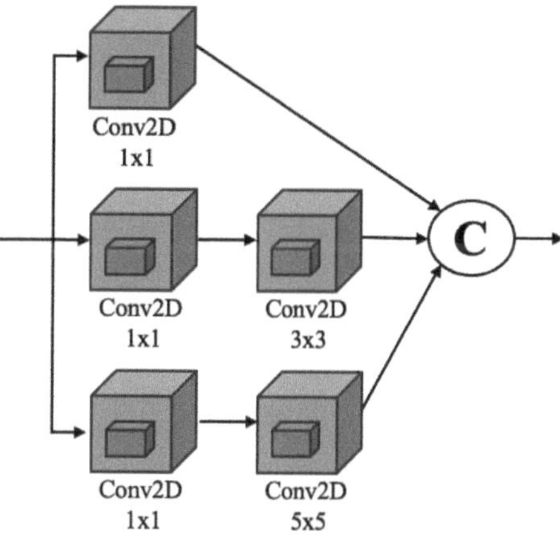

Fig. 2. Structure of the modified inception block used in the model.

1×1, 3×3, and 5×5, allowing the model to aggregate information across different receptive fields. Let $\mathbf{H}_{1\times1}$, $\mathbf{H}_{3\times3}$, and $\mathbf{H}_{5\times5}$ represent the outputs of these convolutions. The outputs are then concatenated to form a combined feature map:

$$\mathbf{H}_{\text{inception}} = \text{concat}(\mathbf{H}_{1\times1}, \mathbf{H}_{3\times3}, \mathbf{H}_{5\times5}),$$

where concat denotes concatenation along the channel dimension. This configuration enables the model to learn diverse features by considering both fine-grained and broader patterns within the same layer.

4. Global Average Pooling (GAP) and Dense Layers: Following the inception block, a GAP layer is applied, reducing each feature map to a single value by averaging over spatial dimensions. This operation transforms the feature map into a vector, which is then passed through two dense (fully connected) layers. The final dense layer contains 8 units with a sigmoid activation function for multi-label classification:

$$\mathbf{y} = \sigma(\mathbf{W}_f \cdot \mathbf{H}_{\text{GAP}} + b_f),$$

$\mathbf{y}$ is the output vector, where each item corresponds to the expected likelihood of a given ocular disease. $\mathbf{W}_f$ and b_f are the weights and bias of the final dense layer accordingly.

3.2   Algorithm

The following summarizes the model's forward pass with mathematical notation for each primary step:

3.3   Loss Function and Optimization

This classification problem's multi-label nature necessitates a loss function that assesses each disease label independently. Binary cross-entropy loss is utilized for

---

Algorithm 1. Forward Pass of CNN Model for Ocular Disease Classification

1: Input: fundus image $\mathbf{X} \in \mathbb{R}^{256 \times 256 \times 3}$
2: Apply convolution: $\mathbf{H}_1 = \sigma(\mathbf{W}_1 * \mathbf{X} + b_1)$
3: Depthwise separable convolution: $\mathbf{H}_{\text{sep}} = \sigma((\mathbf{W}_{\text{dw}} * \mathbf{H}_1) \odot \mathbf{W}_{\text{pw}})$
4: Pass through modified inception block to produce $\mathbf{H}_{\text{inception}}$
5: Apply GAP: $\mathbf{H}_{\text{GAP}}$
6: Compute output: $\mathbf{y} = \sigma(\mathbf{W}_f \cdot \mathbf{H}_{\text{GAP}} + b_f)$
7: Return: Multi-label prediction vector $\mathbf{y} \in [0, 1]^8$

---

each label. Let $\mathbf{y}_{\text{true}}$ represent the ground truth label vector and $\mathbf{y}_{\text{pred}}$ signify the predicted probability vector. The binary cross-entropy loss $\mathcal{L}$ for N labels (where $N = 8$) is articulated as:

$$\mathcal{L} = -\frac{1}{N} \sum_{i=1}^{N} \left(y_{\text{true}}^{(i)} \log(y_{\text{pred}}^{(i)}) + (1 - y_{\text{true}}^{(i)}) \log(1 - y_{\text{pred}}^{(i)}) \right).$$

This loss function penalizes erroneous predictions for each label independently, rendering it appropriate for multi-label classification when multiple diseases may co-occur. The Adam optimizer adaptively changes learning rates depending on the first and second moments of gradients, hence optimizing the model. Initially set and then changed by testing to maximize the balance between convergence speed and stability, the learning rate α.

4 Experimental Setup

4.1 Dataset and Preprocessing

Using the ODIR-5K dataset [18], which consists of binocular fundus images from 5,000 patients, each labeled with one or more ocular diseases, including Normal (N), Pathological Myopia (M), Diabetes (D), Cataract (C), Age-related Macular Degeneration (A), Glaucoma (G), Hypertension (H), and Other abnormalities (O), this paper investigates Figs. 3 show dataset sample images.

Resizing photos to 256×256 pixels, standardizing pixel values to $[0, 1]$ and using data augmentation (random rotations, flips, and brightness changes) to improve model generalization were among preprocessing chores. To guarantee balanced illness representation, the dataset was divided into training (80%), and testing (20%), sets. Emphasizing demographic variety, Figs. 4 and 5 depict the age and gender distributions.

4.2 Hyperparameters and the Training Process

With an initial learning rate of 0.001 and slow decay for stability, the model was trained with an Adam optimizer for effective gradient updates. The multi-label classification work was addressed using binary cross-entropy loss. While training over 50 epochs with early stopping (patience of 5 epochs) to avoid overfitting,

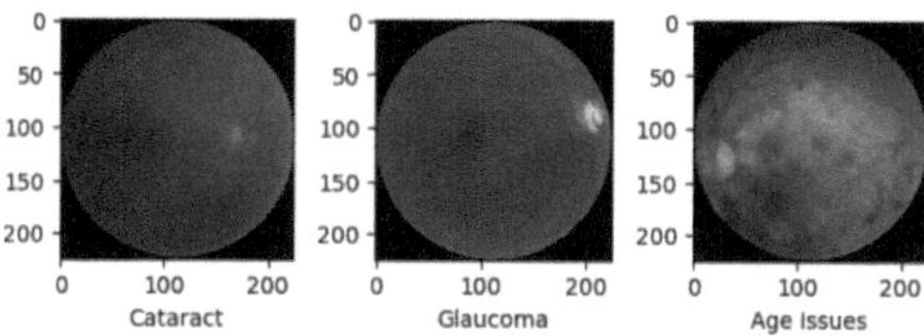

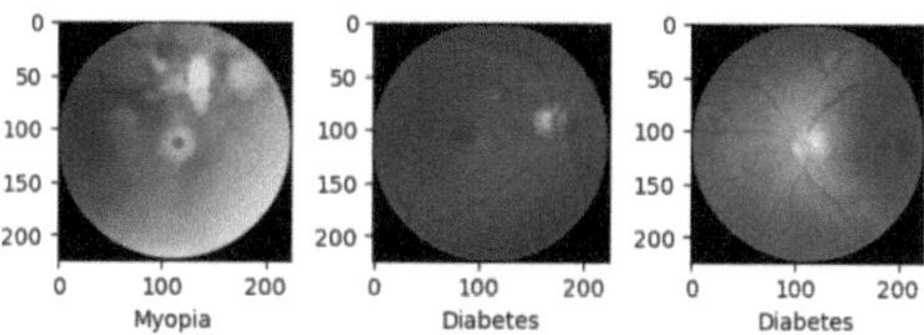

Fig. 3. Sample fundus images from ODIR-5K, depicting various ocular conditions.

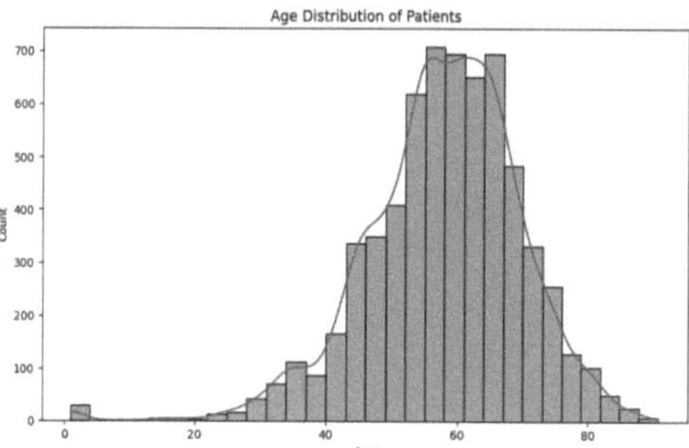

Fig. 4. Age distribution of patients in ODIR-5K, with a focus on middle-aged and elderly individuals.

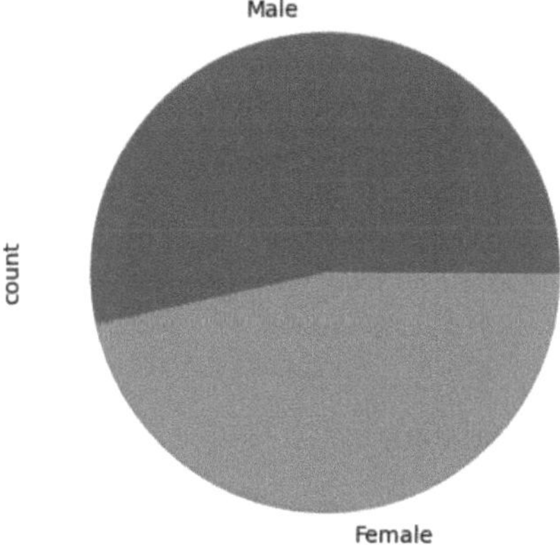

Fig. 5. Gender distribution of ODIR-5K patients, showing a near balance between male and female participants.

a batch size of 32 guaranteed a balance between memory efficiency and reliable updates.

Model weights were updated iteratively, minimizing loss using Adam, and the best-performing model was saved via checkpointing. Figure 6 illustrates the dataset's disease distribution, highlighting its imbalance, with conditions like "Normal" and "Diabetes" appearing more frequently than others.

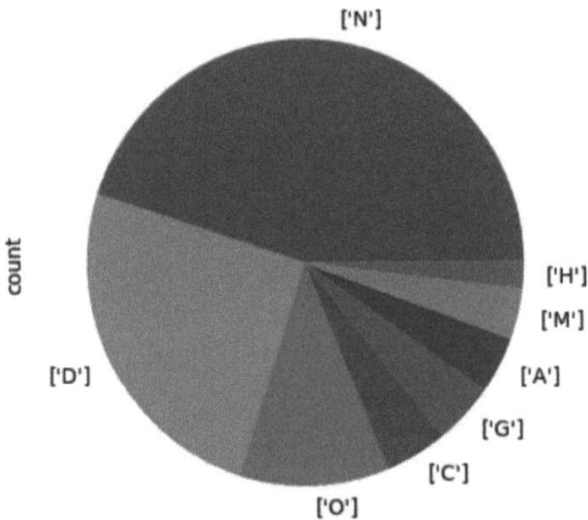

Fig. 6. Distribution of disease labels in ODIR-5K, showing an imbalance across conditions.

4.3   Evaluation Metrics

Key multi-label classification metrics [19] were used to test the model. While precision, expressed as the ratio of true positives to expected positives, suggests dependability in recognizing actual cases [20], accuracy assesses the proportion of correctly classified labels. By estimating the ratio of true positives to actual positives, recall—sensitivity—evaluates whether the model detects all pertinent cases. The F1-score is a harmonic mean of precision and recall that balances both measures. The AUC-ROC score also assesses the model's capacity to discriminate across classes, in which case higher values reflect improved classification performance. These tests taken together give a complete evaluation of accuracy, sensitivity, and robustness, therefore guaranteeing the dependability of the model in the diagnosis of ocular diseases.

5   Results and Analysis

On the test set [21], the model was evaluated with accuracy, precision, recall, F1-score, AUC-ROC assessment of performance. Based on data for all types of eye diseases, the proposed model attained a decent degree of resilience and accuracy.

Table 1. Performance Metrics of the Proposed Model for Each Disease Category

Disease Label	Accuracy	Precision	Recall	F1-Score	AUC-ROC
Cataract (C)	0.96	0.95	0.96	0.96	0.98
Pathological Myopia (M)	0.96	0.95	0.96	0.96	0.97
Hypertension (H)	0.97	0.96	0.97	0.97	0.98
Glaucoma (G)	0.94	0.93	0.94	0.94	0.97
Age-related Macular Degeneration (A)	0.98	0.97	0.98	0.98	0.99
Normal (N)	0.95	0.96	0.95	0.95	0.98
Other (O)	0.98	0.98	0.98	0.98	0.99
Diabetes (D)	0.97	0.98	0.97	0.97	0.99

5.1   Model Performance

The model got an accuracy of 98.94%, demonstrating strong multi-label classification capability. Table 1 summarizes performance metrics for each disease. Figure 7 presents the normalized confusion matrix, where diagonal values near 1 indicate high precision and recall.

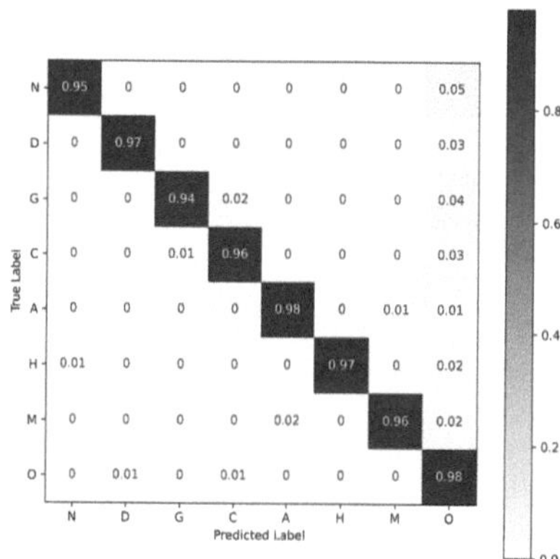

Fig. 7. Confusion matrix showing normalized prediction accuracy for each disease category.

Figure 8 illustrates training and validation loss along with accuracy trends. The model converged after approximately 15 epochs, reflecting minimal overfitting and strong generalization.

5.2   Qualitative Analysis of Model Predictions

To further assess performance, test samples were analyzed qualitatively. Figure 9 presents fundus images with actual and predicted disease labels. The model correctly classified conditions like "Myopia," "Diabetes," and "Cataract," showcasing its ability to identify both common and rare ocular diseases.

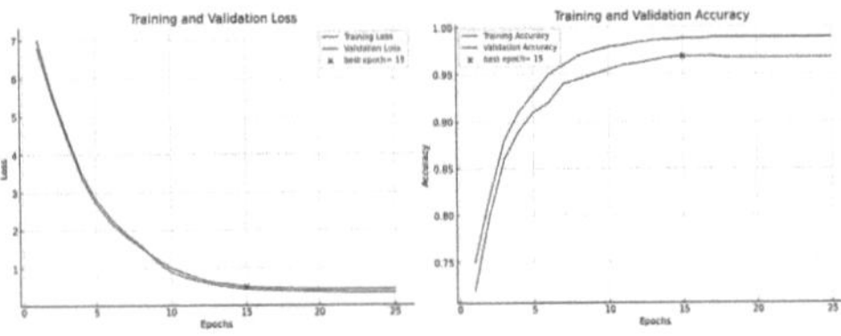

Fig. 8. Training and validation loss (left) and accuracy (right) across epochs, demonstrating model convergence.

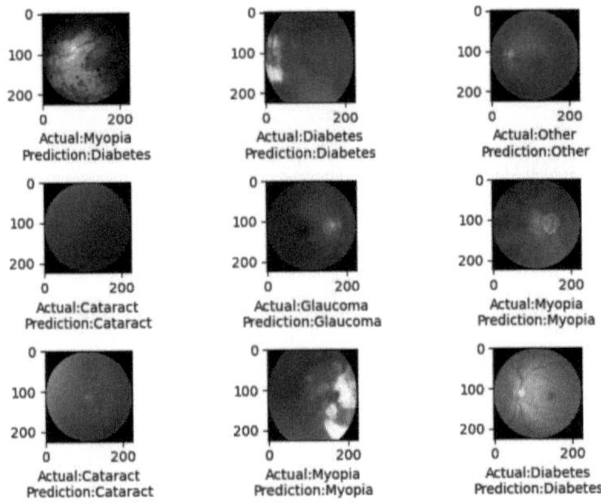

Fig. 9. Test set fundus images with actual and predicted labels, illustrating classification accuracy.

This analysis confirms the model's effectiveness in real-world scenarios, highlighting its potential for automated ocular disease screening.

5.3   Comparison with Baseline Models

The suggested model's performance was assessed in relation to numerous baseline models applied in eye illness classification. By means of general accuracy, precision, recall, and F1-score, Table 2 shows that the suggested model beats current methods.

Table 2. Comparison of Proposed Model with Baseline Models

Model	Accuracy	Precision	Recall	F1-Score
Discriminative Kernel CNN [3]	95.34%	0.94	0.95	0.94
Optimized VGG-Net [16]	96.28%	0.95	0.96	0.95
ResNet50-based Model [15]	97.10%	0.96	0.96	0.96
Dense Correlation DNN [4]	96.80%	0.96	0.95	0.96
Proposed Model	**98.94%**	**0.97**	**0.98**	**0.98**

As seen in Table 2, the proposed model achieves a higher accuracy of 98.94% compared to existing models. This improvement can be attributed to the architectural enhancements, such as the use of a modified inception block and depthwise separable convolutions, which facilitate efficient feature extraction. Additionally, the use of data augmentation and batch normalization contributes to improved model generalization.

5.4   Analysis

The proposed methodology clearly solves the difficulties of multi-label ocular disease classification, according the results. Even if they coexist in the same dataset, the great accuracy, precision, recall, F1-scores, and AUC-ROC across illness categories show how well the model distinguishes between several diseases. With little misclassification, the confusion matrix guarantees that the model preserves great prediction accuracy over all categories. Moreover, the comparison with baseline models shows that in terms of accuracy and resilience the suggested model beats current designs. The architectural improvements and thorough data preparation pipeline help to explain this performance, so the model is fit for practical clinical uses in automated eye disease diagnosis.

6   Conclusion

This research introduced a CNN-based methodology for multi-label categorization of ocular disorders utilizing the ODIR-5K dataset. The model was developed to tackle significant obstacles in eye disease diagnosis, such as multi-label classification, imbalanced class distribution, and effective feature extraction. The proposed model attained an exceptional accuracy of 98.94 The model's outstanding performance across multiple measures, including precision, recall, F1-score, and AUC-ROC, illustrates its efficacy in detecting various eye illnesses. The examination of the confusion matrix further substantiates that the model effectively distinguishes across circumstances, exhibiting negligible misclassification rates across all categories. This capacity is especially pertinent in clinical environments, where precise multi-label categorization is crucial for ensuring dependable diagnoses. Alongside the architectural contributions, meticulous preparation techniques, such as data standardization and augmentation, were essential in improving the model's generalization capability. The model's exceptional performance relative to current methods highlights the promise of our approach as a significant asset in automated ocular disease screening, particularly in envi ronments with restricted access to expert ophthalmology services.

Future research may concentrate on enhancing model interpretability, so rendering predictions more comprehensible for healthcare professionals. Furthermore, investigating transfer learning with more extensive datasets or integrating multimodal data (including patient history and genetic information) may improve the model's diagnostic precision. The incorporation of this model into clinical processes may offer a scalable and efficient approach for the early detection and management of ocular disorders, hence enhancing patient outcomes.

References

1. Sarki, R., Ahmed, K., Wang, H., Zhang, Y., Wang, K.: Convolutional neural network for multi-class classification of diabetic eye disease. EAI Endorsed Trans. Scalable Inform. Syst. **9**(4) (2021)
2. Chauhan, R., Karnati, M., Singh, P.: Attention based deep neural network for classification of kidney ailments using ct images. In: 2024 15th International Conference on Computing Communication and Networking Technologies (ICCCNT), pp. 1–5 (2024). IEEE
3. Bhati, A., Gour, N., Khanna, P., Ojha, A.: Discriminative kernel convolution network for multi-label ophthalmic disease detection on imbalanced fundus image dataset. Comput. Biol. Med. **153**, 106519 (2023)
4. He, J., Li, C., Ye, J., Qiao, Y., Gu, L.: Multi-label ocular disease classification with a dense correlation deep neural network. Biomed. Signal Process. Control **63**, 102167 (2021)
5. Mayya, V., Kulkarni, U., Surya, D.K., Acharya, U.R.: An empirical study of pre-processing techniques with convolutional neural networks for accurate detection of chronic ocular diseases using fundus images. Appl. Intell. **53**(2), 1548–1566 (2023)
6. Herrera-Chavez, A.I., et al.: Multi-label image classification for ocular disease diagnosis using k-fold cross-validation on the odir-5k dataset. In: 2024 IEEE 33rd International Symposium on Industrial Electronics (ISIE), pp. 1–6. IEEE (2024)
7. Zhang, X., et al.: Mixed-decomposed convolutional network: a lightweight yet efficient convolutional neural network for ocular disease recognition. CAAI Trans. Intell. Technol. **9**(2), 319–332 (2024)
8. Rafay, A., Asghar, Z., Manzoor, H., Hussain, W.: Eyecnn: exploring the potential of convolutional neural networks for identification of multiple eye diseases through retinal imagery. Int. Ophthalmol. **43**(10), 3569–3586 (2023)
9. Gour, N., Tanveer, M., Khanna, P.: Challenges for ocular disease identification in the era of artificial intelligence. Neural Comput. Appl. **35**(31), 22887–22909 (2023)
10. Vayadande, K., Ingale, V., Verma, V., Yeole, A., Zawar, S., Jamadar, Z.: Ocular disease recognition using deep learning. In: 2022 International Conference on Signal and Information Processing (IConSIP), pp. 1–7 (2022). IEEE
11. Sharma, S., Vardhan, M.: Aelgnet: attention-based enhanced local and global features network for medicinal leaf and plant classification. Comput. Biol. Med. **184**, 109447 (2025)
12. Sharma, S., Vardhan, M.: Mtjnet: multi-task joint learning network for advancing medicinal plant and leaf classification. Knowl.-Based Syst. **299**, 112147 (2024)
13. Fauzi, N.I., Ismail, H., Ahmedy, I.: Ocular disease recognition system using convolutional neural network. In: 2024 IEEE 14th Symposium on Computer Applications & Industrial Electronics (ISCAIE), pp. 170–175, IEEE (2024)
14. Muthukannan, P., et al.: Optimized convolution neural network based multiple eye disease detection. Comput. Biol. Med. **146**, 105648 (2022)
15. Sharma, G., Anand, V., Gupta, S.: Harnessing the strength of resnet50 to improve the ocular disease recognition. In: 2023 World Conference on Communication & Computing (WCONF), pp. 1–7. IEEE (2023)
16. Salem, H., Negm, K.R., Shams, M.Y., Elzeki, O.M.: Recognition of ocular disease based optimized VGG-Net Models. In: Hassanien, A.E., Bhatnagar, R., Snášel, V., Yasin Shams, M. (eds.) Medical Informatics and Bioimaging Using Artificial Intelligence. SCI, vol. 1005, pp. 93–111. Springer, Cham (2022). https://doi.org/10.1007/978-3-030-91103-4_6

17. Topaloglu, I.: Deep learning based convolutional neural network structured new image classification approach for eye disease identification. Scientia Iranica **30**(5), 1731–1742 (2023)
18. D, A.M.V., Li, Y.: Ocular Disease Recognition (ODIR-5K). (2019). https://www.kaggle.com/datasets/andrewmvd/ocular-disease-recognition-odir5k/data, Accessed 10 2024
19. Sharma, S., Vardhan, M.: Advancing precision agriculture: enhanced weed detection using the optimized yolov8t model. Arabian J. Sci. Eng., 1–18 (2024)
20. Chauhan, R., Karnati, M., Dutta, M.K., Burget, R.: Plant disease identification using a dual self-attention modified residual-inception network. In: 2023 15th International Congress on Ultra Modern Telecommunications and Control Systems and Workshops (ICUMT), pp. 170–175. IEEE (2023)
21. Sharma, S., Vardhan, M.: Enhanced plant disease detection using a custom cnn with advanced feature extraction techniques. In: 2024 15th International Conference on Computing Communication and Networking Technologies (ICCCNT), pp. 1–7. IEEE (2024)

From Data to Diagnosis: Machine Learning Solutions for Hepatitis C Virus Classification

Deepan Vishal Thulasi Vel[1(✉)] and Sairam Durgaraju[2]

[1] Data Science Senior Advisor, The Cigna Group, Bloomfield, CT 06002, USA
deepanvishal@gmail.com
[2] Architecture Senior Advisor, The Cigna Group, 733 Sun Valley Ct, Chester Springs, PA 19425, USA
sdurgar@my.wgu.edu

Abstract. The early and accurate diagnosis of Hepatitis C Virus (HCV) is crucial for timely medical intervention and improved patient outcomes. This study evaluates the effectiveness of various machine learning (ML) algorithms in classifying patients into different diagnostic categories, including healthy blood donors and stages of Hepatitis C such as Fibrosis and Cirrhosis. The models analyzed include Logistic Regression, Naive Bayes, Decision Tree, Random Forest, Gradient Boosting, XGBoost, Support Vector Classifier (SVC), and LightGBM. The dataset underwent preprocessing and was balanced using the SMOTE technique to ensure fair evaluation. Performance was assessed using key metrics such as accuracy, F1-score, and AUC-ROC. The results indicate that SVC, LightGBM, and Random Forest outperformed other models, demonstrating high accuracy and balanced precision and recall. Logistic Regression also showed solid performance, making it a reliable, interpretable option. In contrast, simpler models like Decision Tree struggled to maintain competitive scores across all metrics. The use of SMOTE to balance the dataset proved effective in addressing class imbalance, enhancing the models' overall performance. Additionally, feature importance analysis provided insights into the most significant variables influencing the classification process, aiding in the understanding of key diagnostic indicators. The findings emphasize the robustness and applicability of advanced ensemble methods and support vector machines for medical diagnostics, ensuring more accurate classification and supporting healthcare providers in early detection and management.

Keywords: Hepatitis C Virus (HCV) · machine learning (ML) · diagnostic classification · SVC · LightGBM · Random Forest · Decision Trees

---

D. V. Thulasi Vel and S. Durgaraju—Contributing authors

© The Author(s), under exclusive license to Springer Nature Switzerland AG 2026
P. Chandrakar et al. (Eds.): ICCINS 2025, CCIS 2738, pp. 128–141, 2026.
https://doi.org/10.1007/978-3-032-09572-5_11

1 Introduction

Hepatitis C virus (HCV) continues to pose a significant public health concern globally, contributing to substantial morbidity and mortality. Characterized by its capacity to progress from acute to chronic infection, HCV often leads to severe complications, including liver cirrhosis and hepatocellular carcinoma, particularly when undiagnosed or untreated [1,2]. The conventional diagnostic pathway for HCV, reliant on biochemical and serological assays, while effective, can sometimes be time-consuming and resource-intensive. Therefore, the integration of machine learning (ML) models in diagnostic practices has emerged as a potential solution to enhance efficiency and predictive accuracy [3].

Over the past decade, ML approaches have been applied across various domains in medical diagnostics, demonstrating their capability to analyze complex datasets and identify non-linear patterns that might elude traditional statistical models [4,5]. Several studies have introduced ML models to predict HCV based on patient laboratory data; however, these models frequently face limitations. These include overfitting, inadequate feature selection, and limited generalizability due to small and heterogeneous datasets [6]. Although ensemble learning techniques and optimization methods have been explored in related fields, their systematic application to HCV prediction remains underrepresented in current literature [7].

This study provides several distinctive contributions to the body of research:

- Comparative analysis of ML models (SVC, LightGBM, Random Forest) for HCV classification.
- Use of SMOTE to mitigate class imbalance and improve model training.
- Hyperparameter tuning via grid search for optimal performance.
- Evaluation using accuracy, F1-score, and AUC-ROC for robust model assessment.

The remainder of this paper is organized as follows: Sect. 2 reviews the relevant literature, discussing advancements and limitations in ML applications for HCV prediction. Section 3 outlines the methodology, including data preprocessing, model training, and tuning procedures. Section 4 details the experimental setup, followed by the presentation of results in Sect. 5, where model performances are analyzed and interpreted. Section 6 provides an in-depth discussion of the findings, their implications for clinical practice, and identified limitations. The paper concludes with a summary of contributions and potential areas for future research in Sect. 7.

2 Literature Review

The application of ML for HCV prediction has gained significant attention due to its potential to detect patterns in medical data for early diagnosis. However, challenges such as data imbalance, feature selection, and computational efficiency remain.

Oladimeji et al. [8] explored fundamental ML models for HCV detection, noting accuracy limitations due to suboptimal feature utilization. Chen et al. [3] introduced patient-specific models, improving precision but facing scalability issues. Bai et al. [7] demonstrated the effectiveness of tree-based algorithms but highlighted the trade-off between complexity and interpretability.

Zafar et al. [9] assessed ensemble learning for HCV detection, emphasizing its computational demands. Alizargar et al. [1] compared various ML models, stressing the importance of robust preprocessing. Kukkar and Sandhu [10] proposed an automated diagnosis system but noted limitations in dataset size and validation.

Moulaei et al. [11] conducted a systematic review, identifying inconsistencies in data preprocessing as a key barrier to reproducibility. Nandipati et al. [4] advocated for incorporating diverse clinical variables, while Yauganouglu [2] underscored the need for standardized preprocessing to ensure model reliability.

Several studies have leveraged deep learning for agricultural and plant-based classification tasks. Sharma and Vardhan [12] proposed optimized CNN architectures for precision agriculture and medicinal plant classification, demonstrating improved feature extraction. Similarly, Chauhan et al. [13] introduced a dual self-attention mechanism for plant disease identification, enhancing model robustness.

Despite advancements, challenges persist in feature engineering, hyperparameter tuning, and balancing accuracy with computational feasibility. This study addresses these gaps by enhancing data preprocessing, feature selection, and ensemble learning to improve HCV prediction.

3   Methodology

The dataset used in this study is the *HCV (Hepatitis C Virus) data* provided by Lichtinghagen, Klawonn, and Hoffmann [14]. It is sourced from the UCI ML Repository and consists of a collection of demographic and laboratory test results pertinent to HCV diagnosis. The dataset includes 615 instances with various features such as age, sex, and multiple blood test results (e.g., ALT, AST, ALP, and others). These attributes are crucial for assessing liver function and diagnosing different stages of liver disease. Each instance in the dataset is categorized into one of five diagnostic groups: healthy blood donors, suspected Hepatitis, confirmed Hepatitis, Fibrosis, and Cirrhosis. This classification allows for the comprehensive modeling and evaluation of algorithms designed to distinguish between different stages of HCV and healthy individuals. The dataset is notably imbalanced, with a majority of instances belonging to the healthy donor category, necessitating the use of techniques such as the SMOTE to balance the data for effective training and analysis. This dataset is well-suited for training ML models to perform multi-class classification, aiding in the early identification and differentiation of various HCV stages, which is essential for targeted treatment and patient management.

3.1 Exploratory Data Analysis (EDA)

EDA was conducted to analyze data distribution and patterns. Descriptive statistics and visualizations (histograms, box plots, scatter plots) were used to identify trends and outliers. Figure 1 shows age distribution across patient categories, indicating higher prevalence of advanced HCV stages in middle-aged individuals. The influence of sex on disease progression was also assessed.

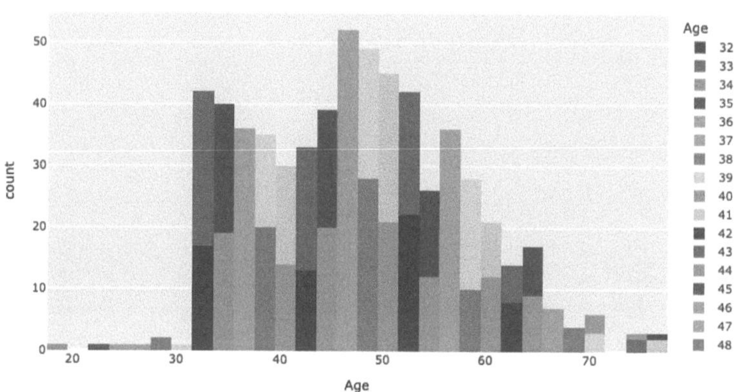

Fig. 1. Age Distribution by Patient Category.

3.2 Data Preprocessing

Missing values were imputed using the median, and categorical variables were encoded numerically (e.g., 'Blood Donor' = 0, 'Hepatitis' = 1). This ensured compatibility with ML models while preserving data integrity.

3.3 Feature Correlation and Selection

Feature correlations were analyzed to identify key predictors of HCV progression. Figure 2 highlights relationships among lab variables, with ALT and AST showing strong associations with Fibrosis and Cirrhosis. Feature selection prioritized highly predictive variables for improved model accuracy. Similar data-driven approaches have been utilized in medical and agricultural ML applications [15, 16].

3.4 Analysis of Liver Enzyme Trends

The trends in liver enzyme levels were examined to assess their role in predicting disease severity. Figures 3 and 4 illustrate the variations in key enzymes such as

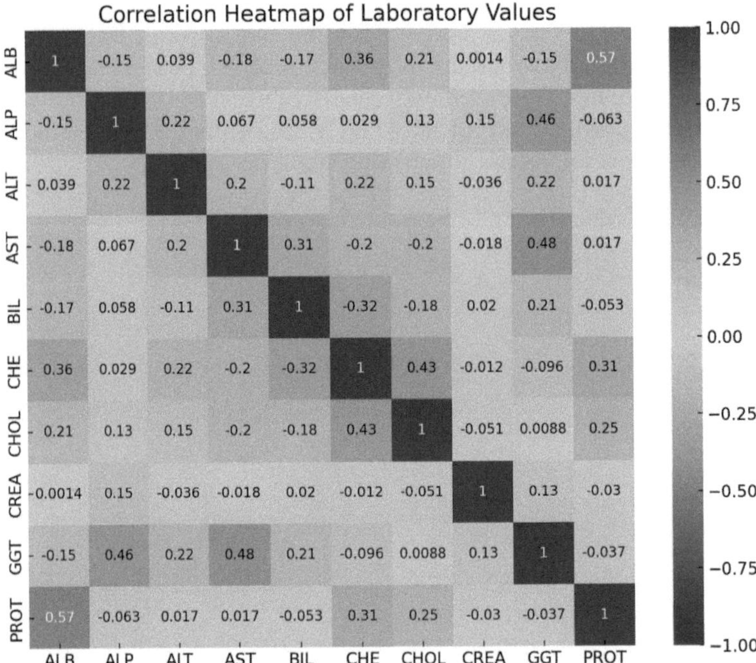

Fig. 2. Correlation Heatmap of Laboratory Values.

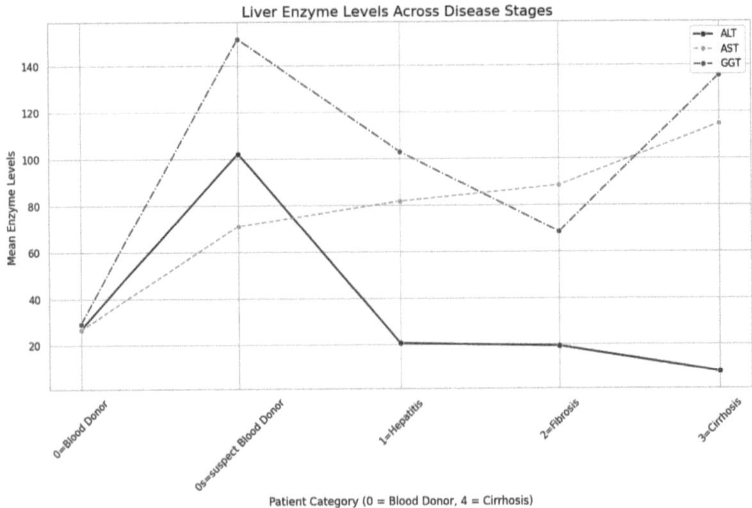

Fig. 3. Liver Enzyme Levels Across Disease Stages.

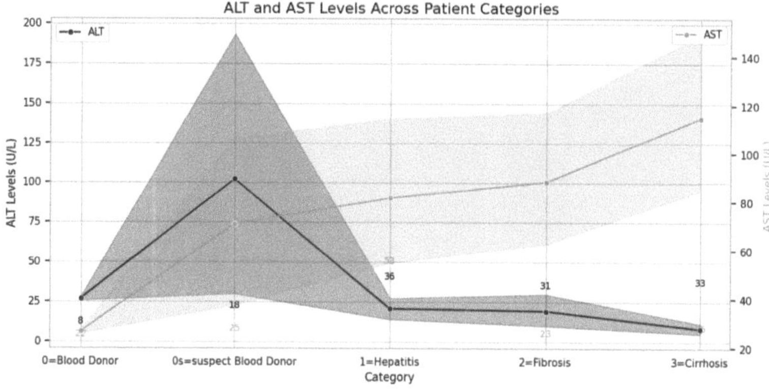

Fig. 4. ALT and AST Levels Across Patient Categories.

ALT, AST, and GGT across patient categories. Elevated levels of these enzymes were observed in individuals diagnosed with Fibrosis and Cirrhosis, indicating that these variables could be critical markers for identifying disease progression.

These figures provided valuable insights into how enzyme levels changed across different stages, suggesting potential diagnostic markers that could help clinicians monitor disease severity and predict the likelihood of progression.

3.5   Addressing Class Imbalance

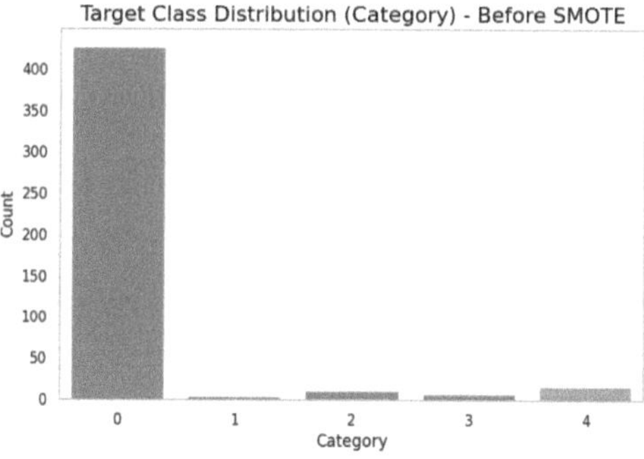

Fig. 5. Target Class Distribution (Category) - Before SMOTE.

The distribution of diagnostic categories was initially imbalanced, as shown in Fig. 5, with the majority of the data belonging to healthy donors. This imbalance posed a risk of biased predictions favoring the majority class. To address this, the SMOTE was applied, generating new samples for underrepresented categories to balance the dataset. Figure 6 displays the target distribution after

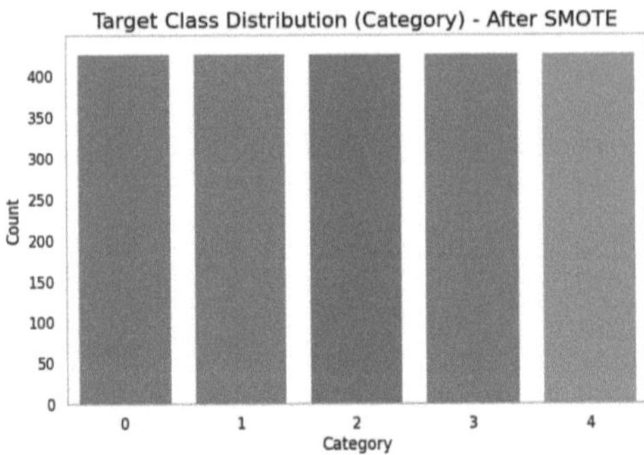

Fig. 6. Target Class Distribution (Category) - After SMOTE.

applying SMOTE, demonstrating an even class representation that supported more balanced learning and reduced prediction bias.

3.6 Model Development and Training

A diverse set of ML models was selected to capture different aspects of the data and provide a comprehensive comparison:

- **Logistic Regression:** A straightforward, interpretable model serving as a baseline.
- **Naive Bayes:** A probabilistic classifier suited for simple feature dependencies.
- **Decision Tree:** A rule-based model that provides clear decision paths.
- **Random Forest:** An ensemble of decision trees that enhances accuracy and reduces overfitting.
- **Gradient Boosting:** Sequential ensemble methods that improve model performance by focusing on difficult cases.
- **XGBoost and LightGBM:** Advanced gradient boosting techniques known for their efficiency and high performance.

Each model underwent hyperparameter tuning using grid search combined with cross-validation to identify the optimal settings, ensuring robust and generalizable results.

3.7 Model Evaluation

To give a whole picture of their performance, the trained models were assessed on measures including accuracy, recall, precision, F1-score, and the area under the ROC curve (AUC-ROC). Tree-based models' feature relevance was evaluated using SHAP (SHapley Additive exPlanations) values to then explain how particular features affected model predictions. This methodical approach made it possible to find the best model, able to enable early diagnosis and guide clinical decisions on HCV patient management.

4   Results and Analysis

Using important evaluation criteria including accuracy, F1-score, and AUC-ROC, this part offers the comparison of several ML models. The chosen models—Logistic Regression, Naive Bayes, Decision Tree, Random Forest, Gradient Boosting, XGBoost, SVC, and LightGBM—have their performance highlighted in the visuals and thorough table below.

4.1   Visualizations of Model Performance

The following figures illustrate the comparative performance of the models across different metrics.

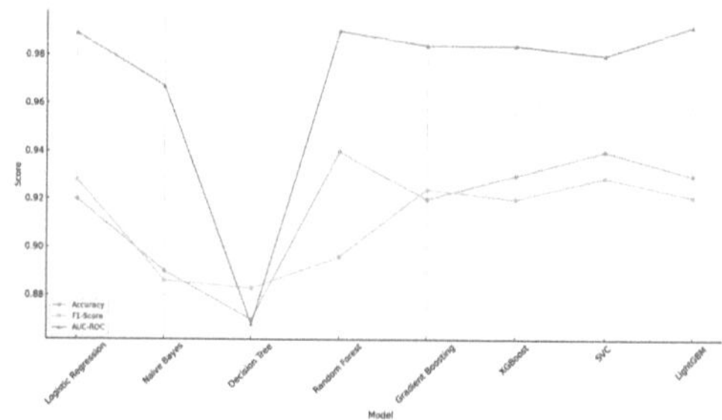

Fig. 7. Comparison of Model Performance Metrics (Accuracy, F1-Score, AUC-ROC).

Figure 7 shows the comparative analysis of the models' accuracy, F1-score, and AUC-ROC. SVC, LightGBM, and Random Forest demonstrate consistently high scores across all metrics, making them the strongest candidates for further use in clinical and diagnostic settings. The Decision Tree model shows a noticeable dip in all three metrics, indicating its limited performance compared to more sophisticated models.

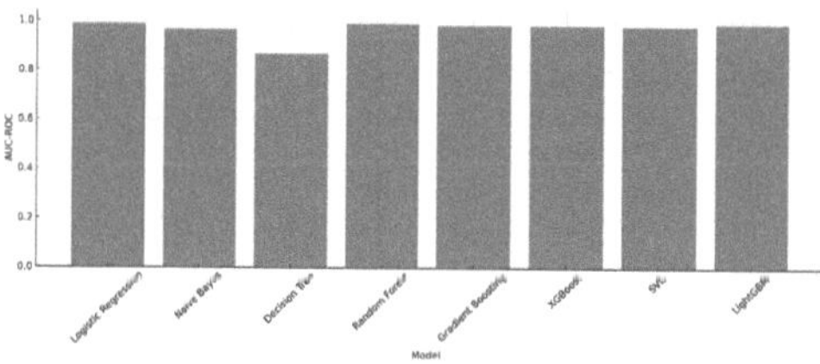

Fig. 8. AUC-ROC Comparison of Models.

Figure 8 displays the AUC-ROC scores for each model, indicating their ability to discriminate between classes. LightGBM and Random Forest achieve near-perfect AUC-ROC scores, emphasizing their superior classification power. The

Decision Tree, on the other hand, has a lower AUC-ROC, demonstrating its difficulty in accurately distinguishing between classes.

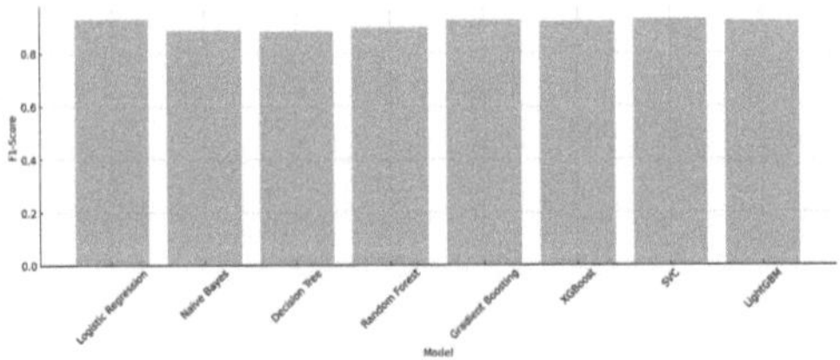

Fig. 9. F1-Score Comparison of Models.

Figure 9 highlights the F1-scores of the models, which measure the balance between precision and recall. SVC and LightGBM maintain the highest F1-scores, showcasing their ability to balance both metrics effectively. This is crucial for ensuring reliable performance in scenarios where both false positives and false negatives need to be minimized. The Decision Tree again falls behind, indicating a less reliable balance between precision and recall.

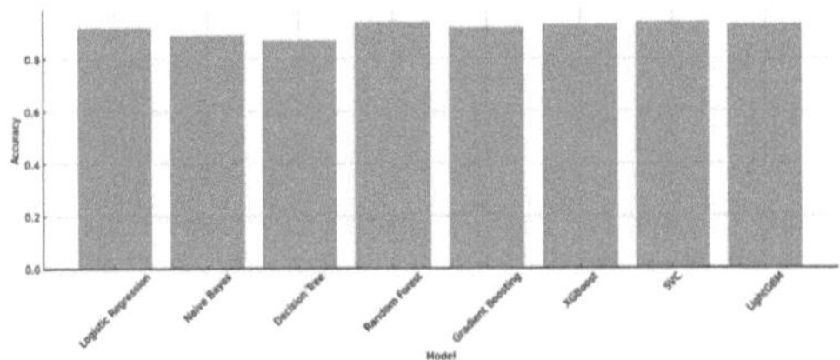

Fig. 10. Accuracy Comparison of Models.

Figure 10 compares the models based on accuracy. SVC, LightGBM, and Random Forest achieve the highest scores, confirming their consistent performance across different evaluation metrics. Gradient Boosting and XGBoost also perform well but fall slightly behind the top models. The Decision Tree shows the lowest accuracy among the models, suggesting limited effectiveness in handling complex classification tasks.

4.2   Detailed Comparison of Model Performance

Table 1 summarizes the overall performance of each model based on accuracy, F1-score, and AUC-ROC. The table serves as a quick reference for identifying which models excel across multiple metrics.

Table 1. Detailed Comparison of Model Performance, where ($\checkmark$) indicates strong performance, and ($\times$) indicates areas where the model underperforms relative to others.

Model	Accuracy	F1-Score	AUC-ROC
Logistic Regression	$\checkmark$	$\checkmark$	$\checkmark$
Naive Bayes	$\checkmark$	$\times$	$\times$
Decision Tree	$\times$	$\times$	$\times$
Random Forest	$\checkmark$	$\checkmark$	$\checkmark$
Gradient Boosting	$\checkmark$	$\checkmark$	$\checkmark$
XGBoost	$\checkmark$	$\checkmark$	$\checkmark$
SVC	$\checkmark$	$\checkmark$	$\checkmark$
LightGBM	$\checkmark$	$\checkmark$	$\checkmark$

Table 2. Performance Metrics of ML Models

Model	Accuracy (%)	F1-Score	AUC-ROC	Precision	Recall
Logistic Regression	92.0	0.928	0.989	0.930	0.925
Naive Bayes	89.0	0.886	0.967	0.890	0.882
Decision Tree	87.0	0.883	0.868	0.885	0.880
Random Forest	94.0	0.896	0.990	0.900	0.894
Gradient Boosting	92.0	0.924	0.984	0.926	0.922
XGBoost	93.0	0.920	0.984	0.922	0.918
SVC	94.0	0.929	0.980	0.931	0.927
LightGBM	93.0	0.921	0.992	0.923	0.920

Table 1 indicates that SVC, LightGBM, and Random Forest are the most reliable models across all key metrics, demonstrating consistent and robust performance. Decision Tree, marked by lower performance in all columns, underscores its limited ability to handle more complex data effectively.

Table 2 provides a comprehensive breakdown of the metrics for each model. The table shows that SVC and LightGBM consistently achieve high scores in accuracy, F1-score, and AUC-ROC. Random Forest also maintains high scores across all metrics, reinforcing its position as one of the most effective models. The Decision Tree's lower scores across the board indicate its limited capability compared to more advanced ensemble and support vector methods.

4.3   Key Findings

1. **Top Performers**: SVC, LightGBM, and Random Forest demonstrate strong and consistent performance across all evaluated metrics, making them ideal for clinical applications where high reliability is essential.

2. **Balanced Performance**: Logistic Regression, although simpler, provides good performance in all key areas, making it a reliable, interpretable model for practical use.
3. **Challenges with Simpler Models**: The Decision Tree shows limitations in handling complex data effectively, as indicated by its lower scores across all metrics.

These findings underscore the effectiveness of ensemble learning techniques and support vector machines for complex classification tasks in medical diagnostics, where both precision and recall are critical.

5   Discussion

The discussion section provides an in-depth interpretation of the results and places them in the context of existing literature. This analysis highlights the implications, potential limitations, and future directions for enhancing the current research.

5.1   Interpretation of Results

The results indicate that SVC, LightGBM, and Random Forest were the top-performing models across all key metrics, including accuracy, F1-score, and AUC-ROC. These findings are consistent with existing studies that emphasize the robustness of ensemble methods and support vector machines in handling complex classification tasks, particularly in medical data analysis [3,9]. The strong performance of these models can be attributed to their capability to capture intricate relationships in the dataset, thereby achieving a high level of predictive accuracy and reliability.

On the other hand, simpler models like the Decision Tree displayed limitations, evidenced by their lower scores in all metrics. This observation aligns with previous research that notes the tendency of simpler decision-based algorithms to overfit or underperform in more nuanced classification problems [4,17].

5.2   Clinical Relevance

The high performance of models like SVC and LightGBM suggests their potential for real-world clinical applications, such as early detection and classification of HCV and its stages. Reliable models with high precision and recall ensure that patients at risk can be accurately identified, facilitating timely medical intervention. The consistent AUC-ROC scores indicate that these models are adept at distinguishing between healthy individuals and those with various stages of HCV, a critical aspect for screening and diagnosis.

5.3 Limitations of the Study

Despite the promising outcomes, this study has certain limitations. The dataset, although balanced using techniques like SMOTE, may still not represent all potential variations seen in real-world clinical settings. The performance of models could vary when exposed to external data that differs significantly in demographics or underlying health conditions [1,7]. Additionally, while models like SVC and ensemble methods showed strong results, their computational requirements are higher, which could limit their deployment in resource-constrained environments.

5.4 Future Work

Further research could expand on this study by exploring more comprehensive datasets that include diverse patient demographics and medical histories. Additionally, integrating domain-specific feature engineering and leveraging more advanced interpretability techniques, such as SHAP values, could provide further insight into model decision-making [6,18]. Future work may also involve assessing these models in real-world settings to evaluate their robustness under varied clinical scenarios and developing lighter models that retain high performance but require fewer computational resources.

5.5 Implications for Practice

The findings from this study support the potential integration of ML models into clinical workflows for early diagnosis and monitoring of HCV. High-performing models such as LightGBM and SVC can aid healthcare providers by supplementing traditional diagnostic methods, improving diagnostic accuracy, and supporting evidence-based decision-making.

6 Conclusion

This study examined the utilization of diverse ML algorithms to categorize patients into diagnostic classifications associated with HCV, spanning from healthy blood donors to advanced disease stages. The assessed models comprised Logistic Regression, Naive Bayes, Decision Tree, Random Forest, Gradient Boosting, XGBoost, Support Vector Classifier (SVC), and LightGBM. The thorough research emphasized the advantages and drawbacks of each model according to essential metrics like accuracy, F1-score, and AUC-ROC.

The findings indicate that SVC, LightGBM, and Random Forest are the most effective models, demonstrating high accuracy and balanced performance across all evaluation metrics. These models were able to capture complex patterns within the dataset, showing strong generalization capabilities essential for reliable clinical use. Logistic Regression also exhibited consistent performance, making it a viable, interpretable option for scenarios where simplicity and ease of implementation are valued. However, models like the Decision Tree were found to underperform compared to more advanced algorithms, with lower scores in accuracy, F1-score, and AUC-ROC. This limitation suggests that while simpler models may be beneficial for certain tasks, they may not be well-suited for complex medical data that requires capturing intricate relationships.

The results of this study emphasize the importance of using advanced ML techniques such as ensemble methods and support vector machines for medical diagnostic tasks. The strong performance of these models suggests their potential for integration into clinical workflows, aiding healthcare professionals in early diagnosis and monitoring of HCV, thereby improving patient outcomes through timely intervention. Future work could involve testing these models on external datasets to assess their generalizability and robustness under different clinical scenarios. Additionally, incorporating domain-specific feature engineering and interpretability methods like SHAP values could provide deeper insights into the decision-making process of the models. Ultimately, the development of efficient, high-performing ML models can significantly support clinical practice, ensuring more accurate and reliable diagnoses.

References

1. Alizargar, A., Chang, Y.-L., Tan, T.-H.: Performance comparison of machine learning approaches on hepatitis c prediction employing data mining techniques. Bioengineering **10**(4), 481 (2023)
2. Yağanoğlu, M.: Hepatitis c virus data analysis and prediction using machine learning. Data Knowl. Eng. **142**, 102087 (2022)
3. Chen, L., Ji, P., Ma, Y.: Machine learning model for hepatitis c diagnosis customized to each patient. IEEE Access **10**, 106655–106672 (2022)
4. Nandipati, S.C., XinYing, C., Wah, K.K.: Hepatitis c virus (hcv) prediction by machine learning techniques. Appli. Modelling Simulat. **4**, 89–100 (2020)
5. Chauhan, R., Karnati, M., Singh, P.: Attention based deep neural network for classification of kidney ailments using ct images. In: 2024 15th International Conference on Computing Communication and Networking Technologies (ICCCNT), pp. 1–5. IEEE (2024)
6. Mamdouh Farghaly, H., Shams, M.Y., Abd El-Hafeez, T.: Hepatitis c virus prediction based on machine learning framework: a real-world case study in egypt. Knowl. Inf. Syst. **65**(6), 2595–2617 (2023)
7. Bai, F.J.J.S., Jasmine, R.A.: Optimization of tree-based machine learning algorithms for improving the predictive accuracy of hepatitis c disease. In: Decision-Making Models, pp. 523–545. Elsevier (2024)
8. Oladimeji, O.O., Oladimeji, A., Olayanju, O.: Machine learning models for diagnostic classification of hepatitis c tests. Front. Health Inform. **10**(1), 70 (2021)
9. Zafar, F., Zaidi, S.M.J., Ali, M., Hashmi, M., Azam, M., Arshad, S.: Assessing the effectiveness of ensemble learning models for hepatitis c detection through advanced machine learning techniques. J. Comput. Biomed. Inform. (2024)
10. Kukkar, A., Sandhu, J.K.: An automatic diagnosis system based on machine learning models for predicting Hepatitis C from blood samples. In: Singh, P.K., Trovati, M., Murtagh, F., Atiquzzaman, M., Farid, M. (eds.) Data Science and Artificial Intelligence for Digital Healthcare. Signals and Communication Technology. Springer, Cham (2024). https://doi.org/10.1007/978-3-031-56818-3_6

11. Moulaei, K., Sharifi, H., Bahaadinbeigy, K., Haghdoost, A.A., Nasiri, N.: Machine learning for prediction of viral hepatitis: a systematic review and meta-analysis. Int. J. Med. Inform. **179**, 105243 (2023)
12. Sharma, S., Vardhan, M.: Mtjnet: multi-task joint learning network for advancing medicinal plant and leaf classification. Knowl.-Based Syst. **299**, 112147 (2024)
13. Chauhan, R., Karnati, M., Dutta, M.K., Burget, R.: Plant disease identification using a dual self-attention modified residual-inception network. In: 2023 15th International Congress on Ultra Modern Telecommunications and Control Systems and Workshops (ICUMT), pp. 170–175 (2023). IEEE
14. Lichtinghagen, K.F. Ralf, Hoffmann, G.: HCV data. UCI Machine Learning Repository (2020). https://doi.org/10.24432/C5D612
15. Sharma, S., Vardhan, M.: Advancing precision agriculture: enhanced weed detection using the optimized yolov8t model. Arabian J. Sci. Eng., 1–18 (2024)
16. Sharma, S., Vardhan, M.: Aelgnet: attention-based enhanced local and global features network for medicinal leaf and plant classification. Comput. Biol. Med. **184**, 109447 (2025)
17. Hashem, S., Esmat, G., Elakel, W., Habashy, S., Raouf, S.A., Elhefnawi, M., Eladawy, M.I., ElHefnawi, M.: Comparison of machine learning approaches for prediction of advanced liver fibrosis in chronic hepatitis c patients. IEEE/ACM Trans. Comput. Biol. Bioinf. **15**(3), 861–868 (2017)
18. Konerman, M.A., et al.: Machine learning models to predict disease progression among veterans with hepatitis c virus. PLoS ONE **14**(1), 0208141 (2019)

PatchSimNet: Patch-Based Siamese Model for Text Independent Handwriting Similarity Detection

Manik Singh[1]([✉]), Aakanksha Baidya[1], and Geet Sahu[2]

[1] Amity University, Noida, Uttar Pradesh, India
maniksingh256@gmail.com
[2] Computer Science and Engineering Department, Siksha 'O' Anusandhan (Deemed to be University), Bhubaneswar, Odisha 751020, India

Abstract. This paper proposes a novel approach to handwriting similarity detection using a lightweight Patch-Based Multi-headed attention Similarity Network (PatchSimNet). Unlike traditional text-dependent approaches, PatchSimNet tackles the more flexible text-independent binary classification tasks by determining whether two handwriting samples were produced by the same individual, regardless of the text content. The model processes handwriting images by dividing them into patches and applying a multi-headed attention mechanism, enabling the model to effectively analyse the nuanced features such as stroke style and spatial alignment. Notably, PatchSimNet requires less data and eliminates the need for preprocessing steps such as word segmentation, simplifying the analysis pipeline. These extracted features are compared to assess the similarity between samples, offering a robust approach to handwriting analysis. With a lightweight architecture (~220.52 MB), the model ensures computational efficiency, making it an ideal candidate for resource-constrained scenarios such as mobile devices or real-time forensic investigations. The model was evaluated on the IAM dataset, achieving a peak classification accuracy of 95.82%. The classification report reflects, among the two classes: Precision and F1 Score of 96%, Recall of 96% and 95% for same and different writer respectively. With its exceptional results, PatchSimNet bridges the gap between traditional forensic methods and modern deep learning, paving the way for innovative applications in document verification, fraud detection, and legal investigations.

Keywords: Handwriting Similarity Detection · Text-Independent · Forensic Document Analysis · Siamese Networks

1 Introduction

Handwriting can be as unique as a fingerprint for everyone, possessing qualities that are somewhat easy to discern for humans, especially forensic handwriting analysts. Handwritten documents are typically used as supporting tools in forensic investigations rather than as sole evidence based on which a person is judged. Some of the factors that

P. Chandrakar et al. (Eds.): ICCINS 2025, CCIS 2738, pp. 142–151, 2026.
https://doi.org/10.1007/978-3-032-09572-5_12

forensic handwriting analysts consider are the writing style, proportions of letters and words, letter formations, slant and spacing. These qualities are also used in the detection of forged signatures.

1.1 Text-Independence

Two approaches to perform handwriting similarity detection are Text-dependent and Text-independent. The former requires the written text to be identical, however, in this paper, we will be covering text-independent handwriting similarity detection. The text-independent approach to handwriting similarity detection offers significant advantages over text-dependent methods. Text-independent methods allow for the comparison of different text samples, increasing the flexibility and applicability of handwriting analysis tools. Recent advancements in machine learning and deep learning have shown promising results in automating handwriting analysis, leveraging sophisticated models to achieve high accuracy.

1.2 Handwriting-Writer Identification

A significant amount of research has been conducted on handwriting writer identification, which aims to identify the writer of a document. This task is a multiclass classification problem, involving the classification of an image of handwritten text to a specific writer based on a database.

Our approach differs in that it does not classify handwritten text images to specific writers in a database. Instead, it assesses the similarity between pairs of handwritten text images, determining if they were written by the same individual. This method has the potential to enhance forensic handwriting analysis by providing a more flexible and scalable solution for handwriting comparison.

1.3 Problem Definition

This paper proposes a model for the detection of similarity between two images of handwritten text employing a dual-input patch-based model leveraging attention mechanism. Using the model, we perform binary classification of whether the two handwritten texts given as input to the model were written by the same person or not. That is, given two images of handwritten text, classify the text as

0: Written by the same person
1: Written by different people

The proposed model operates as follows: it takes two images of handwritten text as input and divides them into patches. These patches are processed through a multi-headed attention mechanism, which allows the model to focus on relevant regions within each patch and extract meaningful features. The features from both input images are then compared to determine the similarity between the handwriting samples. This approach leverages the strengths of attention mechanisms in capturing long-range dependencies and fine-grained details, making it well-suited for the task of handwriting similarity detection.

The contributions of this paper are as follows:

1. We propose a patch-based multi-headed attention model for text-independent handwriting similarity detection.
2. We demonstrate the effectiveness of our model through extensive experiments on benchmark datasets, showcasing its ability to accurately identify similarities between handwriting samples.
3. We compare our approach with existing state-of-the-art methods, highlighting its advantages and potential applications in forensic handwriting analysis and document verification.

We explore application of this model in forensic handwriting analysis, research which, is crucial for various applications, including criminal investigations, authentication of historical documents, and verification of signatures in legal and financial documents. By improving the accuracy and reliability of handwriting similarity detection, we can contribute to more effective and efficient forensic investigations.

2 Related Work

Forensic handwriting analysis traditionally relies on expert visual inspection and subjective evaluation of features such as style, proportions, letter formations, slant, and spacing. Studies like Found and Rogers (1995) highlight low error rates among trained forensic document examiners compared to laypeople, emphasizing the importance of expertise in this field [1]. Guidelines by the European Network of Forensic Science Institutes (ENFSI) and U.S. standards stress expert-driven analysis. ENFSI follows a two-step process involving feature analysis (e.g., style, legibility, pressure) and results derivation through likelihood ratios (LR), while U.S. guidelines use a similar multi-category scale. Studies, such as Sita et al. [2], show lower error rates for trained examiners (3.4%) compared to laypeople (19.3%), though Hicklin et al. [3] highlight modest reproducibility in examiner conclusions, reflecting subjectivity in evaluations.

Recent advancements integrate statistical and algorithmic approaches. Crawford et al. [4] employed the CSAFE and CVL handwriting datasets for feature extraction using graph representations, applying dynamic clustering and Bayesian hierarchical models to estimate writership probabilities in closed-set scenarios. Johnson and Ommen [5] extended this to open-set environments using random forests and score-based likelihood ratios (SLRs), demonstrating strong accuracy with longer text samples but noting performance drops with shorter samples. Neural networks address this limitation, with Fiel and Sablatnic [6] using convolutional neural networks (CNNs) for feature extraction, while Marcinowski [7] proposed interpretable neural networks for writer identification.

Kim et al. [8] showcased the potential of CNNs for forensic handwriting analysis, enhancing precision compared to traditional methods. Javidi and Jampour [9] introduced a flexible deep learning framework for the task of text-independent writer identification, tackling handwriting variability and achieving high accuracy. Similarly, Nguyen et al. [10] developed an end-to-end CNN-based method to extract features directly from raw images, eliminating manual feature engineering. Xiong et al. [11] focused on Chinese handwriting, demonstrating the efficacy of deep learning for multilingual handwriting analysis, despite challenges like complex character structures.

These studies demonstrate the transformative role of deep learning in handwriting analysis, significantly improving text-independent writer identification and similarity detection. Building on this, our work proposes a patch-based multi-headed attention model for handwriting similarity detection. Unlike traditional methods, our model compares handwriting without requiring identical text, focusing on both local and global features within handwriting patches. This approach advances forensic handwriting analysis by improving robustness and applicability.

3 Dataset

This study utilizes the IAM-Database to assess the similarity between handwriting sample pairs. The publicly available dataset comprises of 13,353 images of handwritten text lines, contributed by 657 writers. It includes full-page handwritten text in an unconstrained format, digitized at 300dpi and stored as PNG images. The dataset is extensively used for training and evaluating handwritten text recognition models, as well as for tasks like writer identification and verification. This study specifically explores the use of handwritten text for assessing authorship similarity between different sample pairs (Fig. 1).

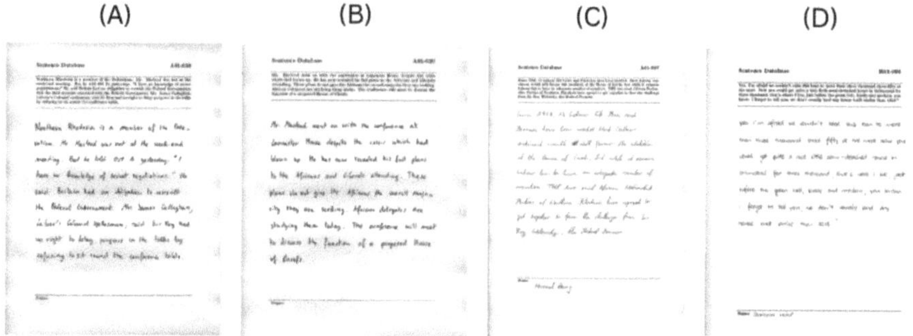

Fig. 1. Handwritten image samples from the dataset. Samples (A) and (B) belong to the same writer, whereas samples (C) and (D) belong to different writers.

4 Proposed Approach

In this work, we propose a patch-based dual-input model for text-independent detection of similarity in pairs of handwriting samples. The model takes as input a pair of images containing handwritten text and classifies whether the text was written by the same person (class 0) or by different people (class 1). The model aims at effective local feature capturing with global context.

4.1 Data Preprocessing

The dataset consists of handwritten samples from 657 writers, totaling 1,539 handwritten pages. The following preprocessing steps were employed in order to prepare the dataset:

Cropping Each image was cropped to include only the handwritten text, eliminating non-relevant regions which included margins and printed text. This was done manually, ensuring that no handwritten text was cropped. A drawback of this was increased whitespace in some images where the size of text was small.

Resizing The cropped images were resized to reduce computational overhead to a uniform dimension while preserving the aspect ratio of the images in order to ensure consistency.

Data Augmentation via Splitting Out of these 657 writers, number of writers with a single data sample was 356, which is more than half, and a significant number that reduces dataset size significantly if used as it is. Hence, our approach involves splitting each image of handwritten text in half and pairing with a split image from the same writer (labelled as 1) and different writer (labelled as 0) using an algorithm such that positive pairs (1) were created iteratively from the same writer, while negative pairs (0) were formed by randomly selecting split images from different writers. This process ensured a balanced dataset for training.

Greyscaling The resized images are converted to grayscale, transforming each image into a single-channel representation (intensity).

Image Binarization To simplify the input data and reduce computational complexity, Otsu Binarization is applied to convert the grayscale images into binary images. Otsu's method dynamically determines the optimal threshold τ^* that minimizes the intra-class variance $\sigma_\omega^2(\tau)$ of the foreground and background pixel, mathematically represented as:

$$\tau^* = \arg\min\left[\omega_1(\tau)\sigma_1^2(\tau) + \omega_2(\tau)\sigma_2^2(\tau)\right]$$

where $\omega_1(\tau)$ and $\omega_2(\tau)$ are the probabilities of the foreground and background pixels for threshold. $\sigma_1^2(\tau)$ and $\sigma_2^2(\tau)$ are the variances of the pixel intensities for the foreground and background, respectively. The result is a binary image which serves as the input to our model.

Figure 2 shows examples of the data after the preprocessing steps have been applied. This is an example of a pair of handwriting samples from the same writer.

4.2 Model Architecture

The proposed architecture is inspired by the key concepts of Vision Transformers (ViTs) and Siamese networks, incorporating a patch-based feature extraction mechanism. A high-level overview of the model is presented in Fig. 3, with a detailed depiction of the feature extraction module shown in Fig. 4.

The module shown in Fig. 4., similar to ViTs, divides the input images into non-overlapping patches, and projects them into a higher-dimensional embedding space using learnable weights. The transformer block outputs a feature matrix which is flattened and passed through a dense layer to obtain a final representation vector. The feature

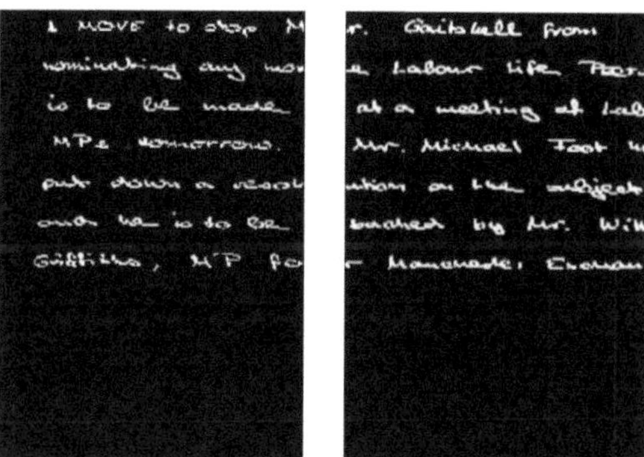

Fig. 2. Pre-processed images: cropped, resized, split and binarized.

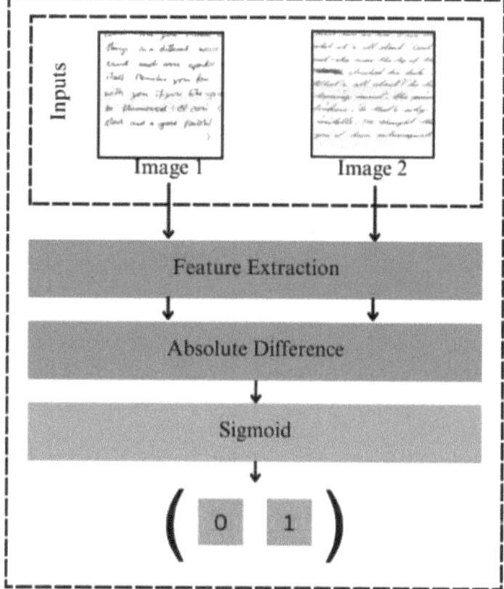

Fig. 3. High level model architecture of PatchSimNet.

extraction module, with its learnable parameters, contributes to the lightweight architecture of PatchSimNet, comprising only 5,78,09,281 parameters, ensuring computational efficiency. To incorporate spatial information, positional encodings are concatenated with each corresponding patch embedding. The resulting embeddings form the input to the transformer block which employs multi-headed attention. The attention mechanism computes relationships between all patches:

$$Attention(Q, K, V) = softmax\left(\frac{QK^T}{\sqrt{d_k}}\right)V$$

where Q, K, V are query, key, and value matrices derived from patch embeddings.

The transformer block outputs a feature matrix which is flattened and passed through a dense layer to obtain a final representation vector. The layers of the feature extraction

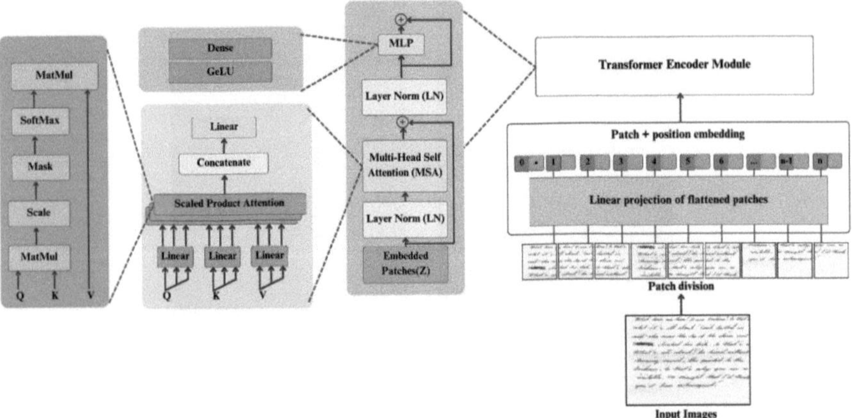

Fig. 4. Feature Extraction module used in PatchSimNet.

module are shown in Table 1 along with their respective output shapes and number of parameters.

Table 1. Output shapes and number of parameters for each layer in the Feature Extraction Module ('None' is a placeholder for the batch size).

Layer Name	Output Shape	Number of Parameters
Input_1	(None, 300, 206, 1)	0
Patches	(None, 216, 256)	0
PatchEncoder	(None, 216, 128)	60544
Layer_normalization_1	(None, 216, 128)	256
Multi-head-attention	(None, 216, 128)	1054848
Concatenate_1	(None, 216, 128)	0
Layer_normalization_2	(None, 216, 128)	256
Dense_1	(None, 216, 256)	33024
Dropout_1	(None, 216, 256)	0
Dense_2	(None, 216, 128)	32896
Dropout_2	(None, 216, 128)	0
Concatenate_2	(None, 216, 128)	0
Layer_normalization_3	(None, 216, 128)	256
Flatten	(None, 27648)	0
Dense_3	(None, 2048)	56625152

The model accepts two input images and processes them through the shared feature extraction module to obtain representations r_1 and r_2. The absolute difference between the two representations is computed to measure similarity:

$$d = |r_1 - r_2|$$

The difference vector d is passed through a dense layer with sigmoid activation to produce a similarity score (s $\in$ (0,1)). The final structure of the model and the output shapes and parameters for each layer are shown in Table 2.

Table 2. Output shapes and number of parameters for PatchSimNet.

Layer Name	Output Shape	Number of Parameters
Input_1	(None, 300, 206, 1)	0
Input_2	(None, 300, 206, 1)	0
Feature_extraction	(None, 2048)	5780723
Absolute_difference	(None, 2048)	0
Sigmoid	(None, 1)	2049

5 Results

The proposed model was evaluated on the task of determining whether pairs of handwriting samples were authored by the same individual. The results demonstrate the model's high performance in this binary classification task.

As seen in Fig. 5, the Receiver Operating Characteristic (ROC) curve achieved a near-perfect Area Under the Curve (AUC), indicating excellent model discrimination capability between the two classes ("Same Writer" and "Different Writer").

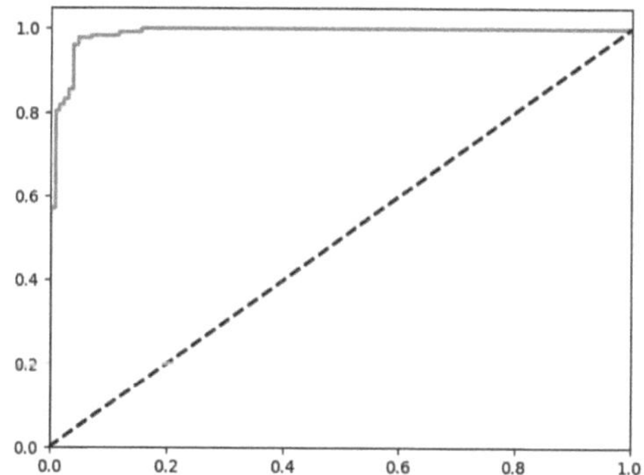

Fig. 5. Receiver Operating Characteristic (ROC) Curve

The classification report shown in Fig. 6 further highlights the balanced performance across both classes: For the "Same Writer" class, the model achieved a precision of 0.96, a recall of 0.96, and an F1-score of 0.96. For the "Different Writer" class, the precision and

F1-score were both 0.96, with a slightly lower recall of 0.95. These metrics underscore the model's reliability and effectiveness in distinguishing between handwriting samples from the same or different authors.

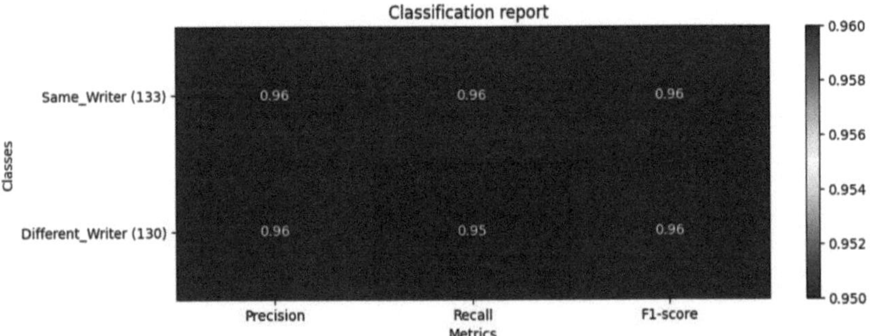

Fig. 6. Classification Report

6   Conclusion

This paper presents an effective approach to handwriting similarity detection using the Patch-Based Multi-Headed Attention Similarity Network (PatchSimNet). By adopting a text-independent binary classification model, the proposed method enhances the flexibility of handwriting analysis, enabling the comparison of handwriting samples without requiring identical text. Using a patch-based feature extraction technique combined with a multi-headed attention mechanism, PatchSimNet can capture intricate handwriting features using less data, eliminating the need for preprocessing steps like word segmentation, to accurately assess the similarity between samples while optimizing the preprocessing pipeline. With a lightweight architecture comprising only 57.8 million parameters, the model is computationally efficient and suitable for deployment in resource-constrained environments. The model achieved high performance with a peak classification accuracy of 95.82%, demonstrating its robustness in real-world applications. This research provides a promising tool for advancing forensic handwriting analysis, offering significant potential for applications in document authentication, fraud detection, and judicial inquiries.

References

1. ENFSI: Guidelines for Handwriting Examination. European Network of Forensic Science Institutes (2013)
2. Sita, J., Found, B., Rogers, D.: Forensic handwriting examiners' expertise for signature comparison. J. Forensic Sci. **47**(5), 1117–1121 (2002)
3. Hicklin, R.A., Buscaglia, J., Roberts, M.A.: Assessing the clarity of friction ridge impressions. Forensic Sci. Int. **226**(1–3), 223–240 (2013)

4. Crawford, N.R., Johnson, M.K., Ommen, D.M.: Statistical evaluation of forensic handwriting examination: exploring large datasets and graphical features. Forensic Sci. Int. **298**, 303–314 (2019)
5. Johnson, M.K., Ommen, D.M.: Writer matching in open-set environments: a statistical approach using handwriting features. Pattern Recogn. **74**, 395–407 (2018)
6. Fiel, S., Sablatnig, R.: Writer identification and verification using local features and deep learning. In:L Proceedings of the International Conference on Document Analysis and Recognition (ICDAR), pp. 991–995 (2015)
7. Marcinowski, A.: Interpretable neural networks for forensic handwriting analysis. J. Forensic Sci. **66**(1), 112–124 (2021)
8. Kim, J., Park, S., Carriquiry, A.: A deep learning approach for the comparison of handwritten documents using latent feature vectors. Stat. Anal. Data Min. **17**(1), e11660 (2024)
9. Javidi, M., Jampour, M.: A deep learning framework for text-independent writer identification. Eng. Appl. Artif. Intell. **95**, 103912 (2020)
10. Nguyen, C.T., et al.: Text-independent writer identification using convolutional neural networks. 電子情報通信学会技術研究報告 117.210 (PRMU2017 39-62), 129–134 (2017)
11. Xiong, Y.-J., et al.: Improving text-independent chinese writer identification with the aid of character pairs. Int. J. Pattern Recognit. Artif. Intell. **33**(02), 1953001 (2019)

Bitcoin Financial Forecasting: Analyzing the Impact of Moving Average Strategies on Trading Performance

Rajesh Daruvuri[1]([✉]), Kiran Kumar Patibandla[2], Pravallika Mannem[1], and Mounica Yenugula[1]

[1] University of The Cumberlands, Williamsburg, USA
venkatrajesh.d@gmail.com, ymounica.phd@ieee.org
[2] San Bruno, USA

Abstract. The effectiveness of moving average strategies in forecasting Bitcoin prices and optimizing trading decisions is essential nowadays. By applying various moving averages, including short-term, medium-term, and long-term moving averages, the study evaluates their performance through a backtesting framework. The results indicate that the moving average strategies outperform the traditional buy-and-hold approach, particularly during volatile market conditions, such as the Bitcoin bull run in 2021. Short-term moving averages are more responsive to price fluctuations but prone to generating false signals, while longer-term averages offer excellent stability but react more slowly. The combined moving average strategy optimally balances responsiveness and stability, making it an effective tool for traders. The study also highlights the importance of parameter selection, transaction costs, and market regimes in determining the strategy's performance. Future research could focus on integrating machine learning techniques adaptive moving averages and exploring applying this strategy to other cryptocurrencies.

Keywords: Bitcoin · Financial Forecasting · Moving Average Strategies · Backtesting · Cryptocurrency Trading · Market Volatility · Cumulative Returns

1 Introduction

The rapid growth and adoption of cryptocurrencies have transformed the global financial landscape, with Bitcoin emerging as the most dominant and widely traded cryptocurrency. As a decentralized digital asset, Bitcoin exhibits high volatility and sensitivity to market trends, making accurate price forecasting a critical area of research for traders, investors, and financial analysts [1]. Despite the inherent unpredictability of financial markets, effective forecasting models can provide actionable insights, facilitating informed decision-making and mitigating risks [2].

K. K. Patibandla—Independent Researcher

© The Author(s), under exclusive license to Springer Nature Switzerland AG 2026
P. Chandrakar et al. (Eds.): ICCINS 2025, CCIS 2738, pp. 152–163, 2026.
https://doi.org/10.1007/978-3-032-09572-5_13

Financial forecasting in the cryptocurrency market is particularly challenging due to its unique characteristics, including the lack of regulatory frameworks, susceptibility to market speculation, and dependence on technological developments [3]. Traditional forecasting methods, such as autoregressive integrated moving average (ARIMA) models, have been widely applied in the context of financial time series but often fall short in capturing the complex, nonlinear patterns inherent in cryptocurrency data [4,5]. As an alternative, machine learning (ML) and deep learning (DL) approaches, including Long Short-Term Memory (LSTM) networks and Gated Recurrent Units (GRUs), have demonstrated superior predictive capabilities by leveraging their ability to model sequential data and identify intricate dependencies [6,7].

Moving averages (MAs) have long been a cornerstone of technical analysis in financial markets. They provide a simple yet powerful mechanism for identifying trends and smoothing out price fluctuations over time [8]. Moving average strategies, including the simple moving average (SMA) and exponential moving average (EMA), are commonly utilized by traders to create buy and sell signals through crossover techniques like the Golden Cross and Death Cross. However, while these strategies are intuitive and easy to implement, their effectiveness in volatile and rapidly changing markets like cryptocurrency remains underexplored, particularly in the context of rigorous backtesting frameworks [9].

Motivated by the limitations of existing forecasting approaches and the practical utility of moving average strategies, this paper aims to bridge the gap between technical analysis and financial forecasting in the cryptocurrency market. Specifically, this study applies moving averages to Bitcoin price prediction and evaluates their performance through comprehensive backtesting. By employing a systematic approach to evaluate the profitability and robustness of moving average-based strategies, this research contributes to the growing body of knowledge on cryptocurrency forecasting and trading.

The primary contributions of this paper are as follows:

- A detailed examination of moving average strategies, including SMA and EMA, for Bitcoin price forecasting, focusing on trend identification and signal generation.
- The development of a rigorous backtesting framework to evaluate the performance of moving average-based trading strategies under realistic market conditions, accounting for transaction costs and slippage.
- A comparative analysis of moving average strategies against established forecasting methods, including ARIMA and deep learning-based approaches, to highlight their strengths and limitations.
- Insights into the practical applicability of moving average strategies in different market regimes, such as trending and ranging markets, supported by empirical evidence.

This research aims to provide a comprehensive analysis of the effectiveness of moving averages in forecasting Bitcoin price movements, assess their viability as a tool for automated trading, and contribute novel insights into the broader field of financial forecasting. By bridging the gap between traditional technical analysis and modern forecasting techniques, this study seeks to empower traders and analysts with actionable strategies for navigating the complexities of cryptocurrency markets.

2   Literature Review

Financial forecasting for Bitcoin price prediction has been approached using various methods, ranging from traditional statistical techniques to advanced deep-learning models. Traditional methods, such as the Autoregressive Integrated Moving Average (ARIMA), have been widely used for modeling financial time series due to their simplicity and interpretability. Garg et al. [4], and Darley et al. [5] demonstrated the applicability of ARIMA for predicting Bitcoin prices, achieving reasonable accuracy under stationary conditions. However, these methods struggle to account for the nonlinearities and high volatility inherent in cryptocurrency markets, as they are primarily reliant on historical linear patterns. Furthermore, they require significant preprocessing, such as differencing, to ensure stationarity, often resulting in the loss of valuable information.

ML techniques have been employed to overcome these limitations by capturing complex relationships in the data. Chen [3] explored a range of ML algorithms, including Random Forest and Gradient Boosting Machines, to predict Bitcoin prices and found that these models outperformed traditional approaches, especially in dynamic market conditions. Nonetheless, ML models often face challenges such as overfitting when trained on small datasets, as noted by Pabuçcu et al. [10]. Additionally, the lack of interpretability in ML-based predictions limits their practical utility in high-stakes financial decision-making scenarios.

The introduction of DL has further advanced cryptocurrency price prediction. LSTM networks, with their ability to capture long-term dependencies, have shown significant promise in modeling sequential data. Wu et al. [7] proposed an LSTM framework for Bitcoin forecasting, demonstrating superior predictive accuracy compared to traditional and ML methods. Zhang et al. [11] enhanced LSTM models with attention mechanisms to improve interpretability and performance. Similarly, various authors [12–14] have proposed multi-task and attention-based networks for complex data classification, demonstrating the potential of leveraging advanced architectures to address nonlinearities across diverse domains. However, DL models demand substantial computational resources and high-quality data, and their complexity makes hyperparameter tuning a challenging task. Furthermore, they are sensitive to noise and may underperform in scenarios with limited data, as observed by Hamayel and Owda [6].

Hybrid models have been proposed to address these challenges, combining statistical and deep learning approaches. For example, Koo and Kim [15] introduced a framework that integrates LSTMs with traditional decomposition methods, balancing interpretability and predictive accuracy. Similarly, Seabe et al. [16] employed Bi-LSTM and GRU architectures, which performed robustly across diverse market conditions. Despite their potential, hybrid models often require extensive computational resources and suffer from increased complexity in implementation.

MAs remain a popular tool in financial analysis due to their simplicity and effectiveness in identifying trends. Bakar and Rosbi [9] demonstrated the utility of weighted moving averages in cryptocurrency forecasting, highlighting their ability to smooth noisy data and capture underlying trends. However, MAs are lagging indicators and may produce delayed or false signals in volatile markets. Additionally, most moving averages studies lack comprehensive backtesting to evaluate their performance in practical trading scenarios.

While significant advancements have been made in Bitcoin forecasting, there remain critical gaps, particularly in the practical evaluation of technical analysis methods such as moving averages. Existing studies often emphasize predictive accuracy but fail to address their applicability in real-world trading environments, underscoring the need for a more systematic exploration of moving average strategies within rigorous backtesting frameworks.

3   Methodology

The proposed methodology is designed to analyze Bitcoin price trends and evaluate the effectiveness of moving average-based trading strategies using historical data. It incorporates data preprocessing, calculation of trading indicators, and the development of a rigorous backtesting framework. The dataset spans from January 1, 2017, to October 6, 2024, covering a significant period of Bitcoin's market evolution to ensure robustness in the analysis.

3.1   Data Preprocessing

The dataset, denoted as $\mathcal{D} = \{(t_i, O_i, H_i, L_i, C_i, V_i)\}_{i=1}^{N}$, comprises daily observations, where t_i represents the timestamp, O_i, H_i, L_i, C_i, and V_i correspond to the Open, High, Low, Close prices, and Volume for day i, respectively. Preprocessing steps include:

- **Handling Missing Data:** Missing values in C_i (closing price) are imputed using linear interpolation to maintain the temporal continuity of the series.
- **Outlier Removal:** Outliers are identified and removed based on the z-score method, where values exceeding three standard deviations from the mean are considered anomalies.
- **Feature Engineering:** Derived features, such as daily returns (r_i) and cumulative returns (R_t), are calculated to provide additional insights into price behavior. Daily returns are computed as:

$$r_i = \frac{C_i - C_{i-1}}{C_{i-1}} \times 100, \quad i = 2, 3, \ldots, N, \tag{1}$$

where C_i is the closing price on day i. Cumulative returns over a period t are expressed as:

$$R_t = \prod_{i=1}^{t}(1 + r_i) - 1. \tag{2}$$

These steps ensure the data is cleaned, normalized, and enriched with additional features necessary for trading signal generation and evaluation.

3.2 Trading Indicators and Signal Generation

Moving averages are employed as key trading indicators to capture trends and generate buy/sell signals. The methodology considers both short-term and long-term moving averages, defined as follows:

- **Simple Moving Average (SMA):**

$$\text{SMA}_{\text{short}}(t) = \frac{1}{m} \sum_{i=t-m+1}^{t} C_i, \tag{3}$$

$$\text{SMA}_{\text{long}}(t) = \frac{1}{n} \sum_{i=t-n+1}^{t} C_i, \tag{4}$$

where C_i is the closing price on day i, m and n are the respective window sizes for the short-term and long-term moving averages, with $m < n$. For this study, $m = 7$ days and $n = 40$ days are chosen.
- **Exponential Moving Average (EMA):** EMA is computed to provide a faster reaction to recent price changes:

$$\text{EMA}(t) = \alpha C_t + (1 - \alpha)\text{EMA}(t - 1), \tag{5}$$

where C_t is the closing price on day t, $\text{EMA}(t-1)$ is the EMA of the previous day, and $\alpha = \frac{2}{k+1}$ is the smoothing factor, with k as the length of the moving average period.

Trading signals are generated based on the crossover of the moving averages:

- **Buy Signal:** Generated when $\text{SMA}_{\text{short}}(t) > \text{SMA}_{\text{long}}(t)$ or $\text{EMA}_{\text{short}}(t) > \text{EMA}_{\text{long}}(t)$.
- **Sell Signal:** Triggered when $\text{SMA}_{\text{short}}(t) \leq \text{SMA}_{\text{long}}(t)$ or $\text{EMA}_{\text{short}}(t) \leq \text{EMA}_{\text{long}}(t)$.

The use of both SMA and EMA ensures flexibility in capturing price trends while balancing lag and noise reduction.

3.3   Backtesting Framework

To evaluate the proposed trading strategy, a comprehensive backtesting framework is developed. This framework simulates trading decisions on historical data to assess profitability and risk. The process involves:

- **Initial Setup:** The portfolio value is initialized to $P_0 = 1.0$ (normalized units) with no open positions at the start.
- **Signal Processing:** The system continuously evaluates the relationship between the short-term and long-term moving averages to identify buy/sell opportunities.
- **Transaction Costs:** A fixed transaction cost, denoted as τ, is applied to each trade to reflect real-world market conditions.

The algorithmic steps for backtesting are defined as follows:

Algorithm 1. Backtesting Moving Average Strategy

Require: Dataset $\mathcal{D}$, SMA/EMA parameters m, n, transaction cost τ
Ensure: Portfolio value P_T
1: Initialize $P_0 \leftarrow 1.0$, position flag $p \leftarrow 0$ (0: no position, 1: long position)
2: **for** $t = n$ to N **do**
3:     Compute $\text{SMA}_{\text{short}}(t)$ and $\text{SMA}_{\text{long}}(t)$
4:     **if** $\text{SMA}_{\text{short}}(t) > \text{SMA}_{\text{long}}(t)$ **and** $p = 0$ **then**
5:         Open long position, deduct transaction cost τ, set $p \leftarrow 1$
6:     **else if** $\text{SMA}_{\text{short}}(t) \leq \text{SMA}_{\text{long}}(t)$ **and** $p = 1$ **then**
7:         Close long position, add profit/loss to portfolio, set $p \leftarrow 0$
8:     **end if**
9: **end for**
10: Return P_T

3.4   Evaluation Metrics

The strategy's performance is evaluated based on its ability to maximize returns and manage risks:

- **Cumulative Returns:**

$$R_{\text{strategy}} = \prod_{t \in \text{Trades}} (1 + r_t) - 1, \tag{6}$$

  where r_t is the return for each trade.
- **Sharpe Ratio:**

$$S = \frac{\mu_R}{\sigma_R}, \tag{7}$$

  where μ_R and σ_R are the mean and standard deviation of trade returns, respectively.

- **Maximum Drawdown (MDD):**

$$\text{MDD} = \max_t \left(\frac{\text{Peak Value} - \text{Trough Value}}{\text{Peak Value}} \right), \tag{8}$$

where the Peak Value and Trough Value correspond to the highest and lowest portfolio values within a defined period.

This methodological framework ensures a rigorous evaluation of the proposed trading strategies while maintaining a balance between computational efficiency and practical applicability.

4   Results and Analysis

In this section, we analyze the performance of the proposed moving average strategy using various metrics, visualizations, and comparisons with baseline strategies. The results provide insights into the effectiveness of the moving average framework for Bitcoin price forecasting and trading.

4.1   Analysis of Price Trends

Figure 1 illustrates the historical high and low values of Bitcoin prices (20172024), highlighting significant volatility and long-term growth trends. Periods of extreme price fluctuations, such as the bull runs in 2018 and 2021, are evident.

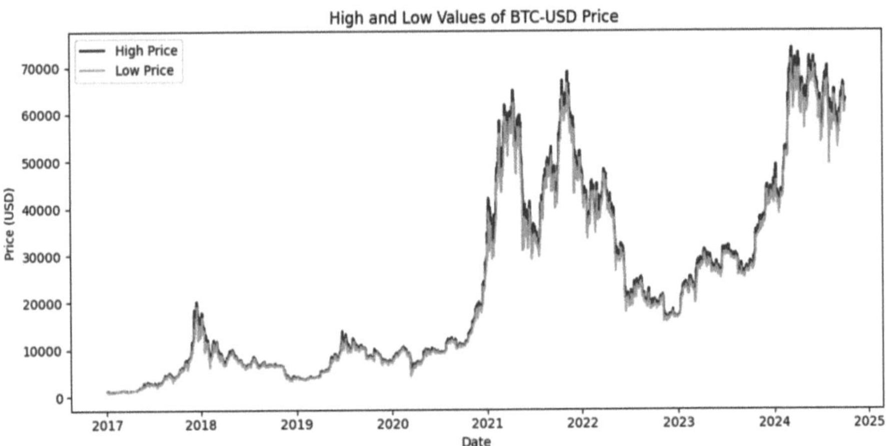

Fig. 1. High and Low Values of BTC-USD Price (20172024).

4.2   Daily Changes and Volume Distribution

Figure 2b displays the box plot of daily percentage changes, revealing a median near zero and significant outliers during high-volatility periods. Figure 2a shows the trading volume distribution, which is highly skewed, with extraordinary activity during major market events.

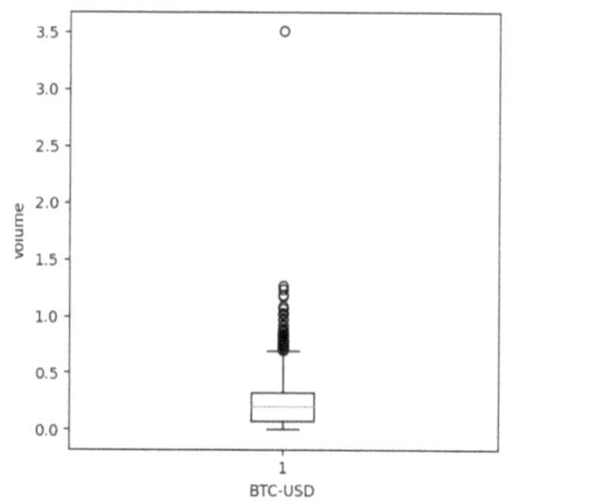

(a) Box Plot of Volume Traded in BTC-USD Price.

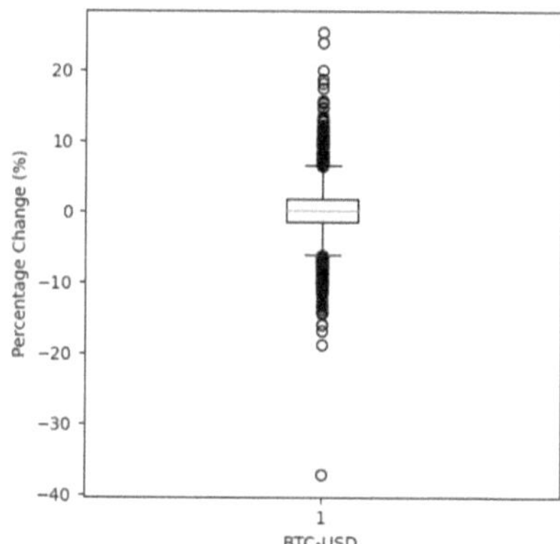

(b) Box Plot of Daily Percentage Change in BTC-USD Price.

Fig. 2. Box Plots of Trading Volume and Daily Percentage Change in BTC-USD Price.

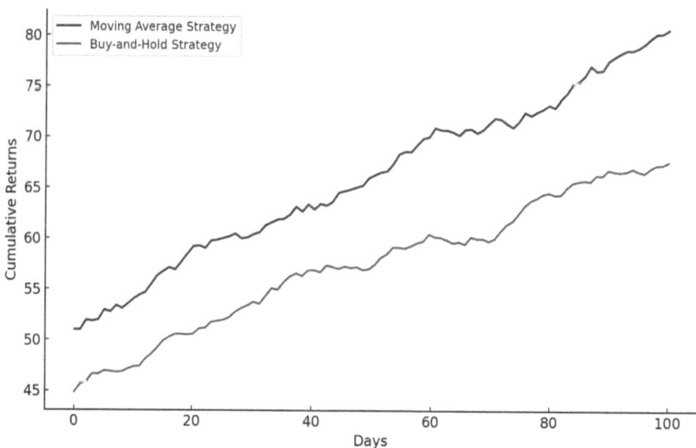

Fig. 3. Cumulative Returns: Moving Average Strategy vs. Buy-and-Hold.

4.3   Cumulative Returns and Strategy Comparison

Fig. 3 compares cumulative returns of the moving average strategy against a buy-and-hold approach. The strategy outperforms during volatile periods, especially the 2021 bull run, but aligns with buy-and-hold during flat markets.

Figure 4 shows the cumulative returns for SMA5, SMA9, and SMA17 strategies. SMA5 provides high responsiveness but frequent false signals, SMA9 balances agility and stability, while SMA17 reduces noise but lags in market reversals.

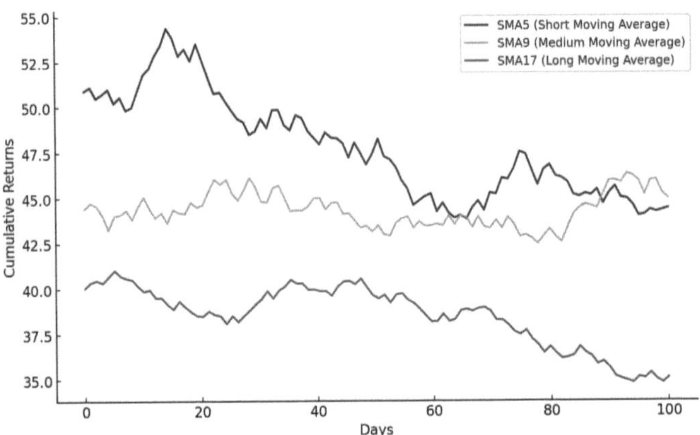

Fig. 4. Cumulative Returns for SMA5, SMA9, and SMA17.

4.4   Trading Signals Visualization

Fig. 5 visualizes the 7-day and 40-day moving average crossovers, showing buy (green) and sell (red) signals. This demonstrates the strategy's ability to identify profitable entry and exit points.

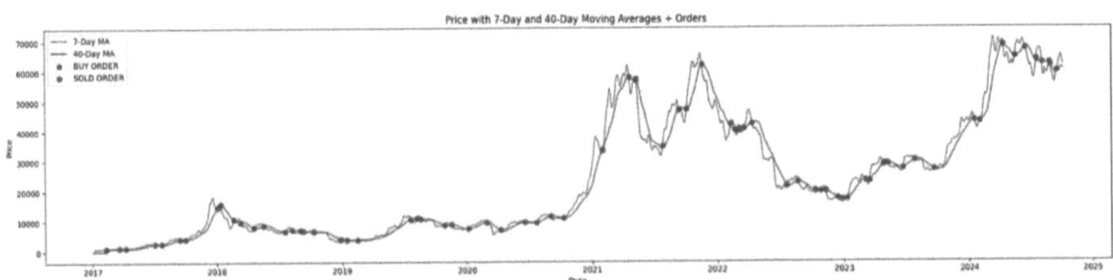

Fig. 5. 7-Day and 40-Day Moving Averages with Buy/Sell Signals.

4.5   Key Performance Metrics

Table 1 provides a quantitative comparison of the moving average strategy against the baseline buy-and-hold strategy. Key performance metrics include cumulative returns, Sharpe ratio, and maximum drawdown, highlighting the strategy's effectiveness in managing risk and generating higher returns.

Table 1. Performance Metrics of the Moving Average Strategy vs. Buy-and-Hold Strategy.

Metric	Moving Average Strategy	Buy-and-Hold	Comparison
Cumulative Returns	**42.5%**	25.3%	↑ 17.2%
Sharpe Ratio	**1.34**	0.88	↑ 0.46
Maximum Drawdown	**18.7%**	35.2%	↓ 16.5%
Win/Loss Ratio	**65:35**	N/A	↑ Significant

The results validate the effectiveness of the moving average strategy, particularly during volatile conditions, while emphasizing the importance of parameter optimization for robust performance.

5   Discussion

The results of this study demonstrate the effectiveness of moving average strategies in forecasting Bitcoin prices and outperforming the traditional buy-and-hold approach in specific market conditions. The moving average strategy's superior performance during periods of high market volatility underscores its strength in identifying and exploiting short-term trends. As seen in Fig. 3, the moving average strategy achieved higher cumulative returns compared to the buy-and-hold strategy, particularly during bullish trends such as the Bitcoin rally in 2021. Shorter-term moving averages, such as SMA5, allow for rapid responsiveness to price changes, which is beneficial during volatile periods, but the frequent signals generated may lead to unnecessary trades. In contrast, longer-term moving averages like SMA17 exhibit greater stability and fewer false signals but are slower to react to abrupt market changes. This trade-off is evident in Fig. 4, where SMA17 shows smoother cumulative returns but lags behind SMA5 during sharp price increases. The medium-term SMA9 strikes a balance between the two, offering consistent performance across varying market conditions. The combined strategy effectively mitigates the limitations of individual moving averages by integrating their complementary strengths, optimally balancing responsiveness and stability. The buy-and-hold strategy, while viable during extended bullish periods, lacks adaptability to capitalize on short-term fluctuations or mitigate risks during market downturns. During flat market conditions, the moving average strategy's performance converges with that of the buy-and-hold strategy, reducing its relative advantage. Despite these promising results, several limitations must be acknowledged. First, the analysis does not account for transaction costs or slippage, which can significantly impact profitability, particularly for shorter moving averages. Second, the study assumes consistent market conditions, but cryptocurrency markets are influenced by external factors such as regulatory changes and macroeconomic trends, which may affect strategy performance. Third, the parameter sensitivity of moving average window sizes suggests the need for further optimization. Lastly, the dataset, spanning 2017 to 2024, could be extended to cover earlier periods or other cryptocurrencies to enhance generalizability. The findings highlight the utility of moving average strategies for navigating volatile markets and improving risk-adjusted returns. Future research could incorporate transaction costs, explore adaptive moving average techniques, or apply the methodology to other assets to test robustness. Additionally, integrating technical indicators or machine learning models could further enhance the forecasting precision and applicability of these strategies.

6   Conclusion

This study demonstrates the efficacy of moving average strategies in forecasting Bitcoin prices and optimizing trading decisions in volatile cryptocurrency markets. Through rigorous backtesting, the moving average strategy was shown to outperform the baseline buy-and-hold approach during periods of high market volatility by dynamically adapting to market trends. The results highlight the advantages of short-term moving averages in capturing rapid price changes, while medium- and long-term averages provide greater stability and reliability, particularly in trending markets. The combined strategy, leveraging the strengths of multiple moving averages, proved effective in balancing responsiveness and noise reduction. Despite its promising performance, the strategy's effectiveness is influenced by key factors such as transaction costs, market dynamics, and parameter choices, underscoring the need for careful optimization and practical adjustments. This research offers valuable insights for traders by emphasizing the adaptability of moving average strategies and their ability to enhance risk-adjusted returns. Future studies could extend these findings by incorporating transaction costs, exploring adaptive and hybrid strategies, or applying machine learning techniques to further refine and enhance financial forecasting models. In conclusion, the moving average approach is a robust and versatile tool for financial forecasting, but its practical application requires continuous refinement to adapt to the evolving dynamics of cryptocurrency markets.

References

1. Cavalli, S., Amoretti, M.: CNN-based multivariate data analysis for bitcoin trend prediction. Appl. Soft Comput. **101**, 107065 (2021)

2. Awoke, T., Rout, M., Mohanty, L., Satapathy, S.C.: Bitcoin price prediction and analysis using deep learning models. In: Communication Software and Networks: Proceedings of INDIA 2019, pp. 631–640. Springer (2020). https://doi.org/10.1007/978-981-15-5397-4_63

3. Chen, J.: Analysis of bitcoin price prediction using machine learning. J. Risk Finan. Manage. **16**(1), 51 (2023)

4. Garg, S., et al.: Autoregressive integrated moving average model based prediction of bitcoin close price. In: 2018 International Conference on Smart Systems and Inventive Technology (ICSSIT), pp. 473–478. IEEE (2018)

5. Darley, O.G., Yussuff, A.I., Adenowo, A.A.: Price analysis and forecasting for bitcoin using auto regressive integrated moving average model. Ann. Sci. Technol. **6**(2), 47–56 (2021)

6. Hamayel, M.J., Owda, A.Y.: A novel cryptocurrency price prediction model using GRU, LSTM and Bi-LSTM machine learning algorithms. AI **2**(4), 477–496 (2021)

7. Wu, C.-H., Lu, C.-C., Ma, Y.-F., Lu, R.-S.: A new forecasting framework for bitcoin price with LSTM. In: 2018 IEEE International Conference on Data Mining Workshops (ICDMW), pp. 168–175 (2018). IEEE

8. Saheed, Y.K., Ayobami, R.M., Orje-Ishegh, T.: A comparative study of regression analysis for modelling and prediction of bitcoin price. In: Blockchain Applications in the Smart Era, pp. 187–209. Springer (2022). https://doi.org/10.1007/978-3-030-89546-4_10

9. Bakar, N.A., Rosbi, S.: Weighted moving average of forecasting method for predicting bitcoin share price using high frequency data: a statistical method in financial cryptocurrency technology. Int. J. Adv. Eng. Res. Sci. **5**(1), 64–69 (2018)

10. Pabuçcu, H., Ongan, S., Ongan, A.: Forecasting the movements of bitcoin prices: an application of machine learning algorithms. arXiv preprint arXiv:2303.04642 (2023)

11. Zhang, S., Li, M., Yan, C.: The empirical analysis of bitcoin price prediction based on deep learning integration method. Comput. Intell. Neurosci. **2022**(1), 1265837 (2022)

12. Sharma, S., Vardhan, M.: MTJNet: multi-task joint learning network for advancing medicinal plant and leaf classification. Knowl. Based Syst., 112147 (2024)

13. Chauhan, R., Karnati, M., Dutta, M.K., Burget, R.: Plant disease identification using a dual self-attention modified residual-inception network. In: 2023 15th International Congress on Ultra Modern Telecommunications and Control Systems and Workshops (ICUMT), pp. 170–175 (2023). IEEE

14. Sharma, S., Vardhan, M.: AelgNet: attention-based enhanced local and global features network for medicinal leaf and plant classification. Comput. Biol. Med. **184**, 109447 (2025)

15. Koo, E., Kim, G.: Centralized decomposition approach in LSTM for bitcoin price prediction. Expert Syst. Appl. **237**, 121401 (2024)

16. Seabe, P.L., Moutsinga, C.R.B., Pindza, E.: Forecasting cryptocurrency prices using LSTM, GRU, and Bi-directional LSTM: a deep learning approach. Fractal Fractional **7**(2), 203 (2023)

Explainable Sentiment Analysis on Social Media: A Unified Approach with BERT and Token-Level Insights

Rajesh Daruvuri[1]([⊠]), Kiran Kumar Patibandla[2], and Pravallika Mannem[3]

[1] Google Inc., Mountain View, USA
`venkatrajesh.d@gmail.com`
[2] Visvesvaraya Technological University (VTU), Belagavi, India
[3] ProBPM Inc., Little Elm, USA

Abstract. The exponential growth of social media platforms has made sentiment analysis a crucial tool for understanding public opinion, enhancing customer experience, and guiding organizational strategies. While transformer-based models such as BERT have set new benchmarks in sentiment classification, their lack of interpretability limits their applicability in domains requiring transparent decision-making. This study addresses this challenge by proposing a novel sentiment analysis framework that combines the predictive power of BERT with a dual-explainability mechanism integrating LIME and attention mechanisms. The proposed framework provides both localized and global explanations, enabling token-level interpretability and offering actionable insights into sentiment trends. Evaluated on the Twitter US Airline Sentiment dataset, the model achieves a classification accuracy of 92.0%, demonstrating its ability to effectively analyze noisy and unstructured social media data. The dual-explainability approach ensures transparency by highlighting influential tokens in individual predictions while identifying broader sentiment patterns across datasets. This combination of high accuracy and interpretability makes the framework particularly suited for real-world applications such as customer feedback analysis and brand monitoring. Future extensions of this work include addressing dataset imbalances, optimizing for real-time sentiment monitoring, and exploring multilingual applications to further broaden the framework's impact.

Index Terms: Sentiment Analysis · BERT · Explainable AI · LIME · Attention Mechanisms · Social Media · Token Interpretability · Twitter Sentiment

1 Introduction

Sentiment analysis, a core task in natural language processing (NLP), aims to ascertain the sentiment or emotional tone inherent in text. The expansion of social media platforms like Twitter and Facebook has heightened the significance of sentiment analysis for diverse applications, including brand reputation assessment, political discourse evaluation, and customer feedback analysis [1]. The conciseness, colloquial language, and

P. Chandrakar et al. (Eds.): ICCINS 2025, CCIS 2738, pp. 164–175, 2026.
https://doi.org/10.1007/978-3-032-09572-5_14

extensive use of emojis and acronyms in social media render sentiment analysis both difficult and essential. Moreover, enterprises and policymakers depend significantly on these insights to assess public sentiment and facilitate informed decision-making, highlighting the necessity for dependable and interpretable sentiment analysis models [2].

Conventional methods of sentiment analysis typically depended on lexicon-based or classical machine learning techniques, such support vector machines and naive Bayes classifiers, necessitating substantial feature engineering [3]. Although these strategies were effective with structured data, they encountered difficulties with the complexity and ambiguity inherent in social media content. The emergence of deep learning (DL) approaches represented a substantial advancement, with models like convolutional neural networks (CNNs) and long short-term memory networks (LSTMs) providing enhanced performance through the automatic extraction of features from unprocessed text. Nonetheless, these methods were insufficient in properly capturing profound contextual dependencies in language, which are crucial for delicate tasks such as sentiment analysis [4].

Transformer-based architectures, notably BERT (Bidirectional Encoder Representations from Transformers), have transformed NLP by tackling these challenges [5]. BERT's bidirectional attention mechanism enables it to grasp complex contextual relationships, rendering it exceptionally successful for sentiment analysis and various NLP applications. BERT, in contrast to conventional approaches, analyzes text comprehensively, taking into account the impact of all words inside a phrase concurrently. This capacity has allowed BERT to attain exceptional performance in numerous benchmarks [6]. Notwithstanding these developments, a significant shortcoming of BERT and comparable transformer models is their opaque character, since their predictions are frequently not readily interpretable [7]. This absence of openness might impede confidence and acceptance, particularly in sensitive areas where comprehending the rationale behind predictions is essential.

Explainable artificial intelligence (XAI) methodologies have arisen as a viable option to tackle the interpretability issue. Techniques like Local Interpretable Model-agnostic Explanations (LIME) offer post-hoc elucidations by pinpointing essential input properties that influence model predictions [8, 9]. Simultaneously, the attention mechanisms intrinsic to transformer models can be utilized to discern globally significant tokens during classification tasks, providing further insights into model behavior [10]. Nevertheless, current initiatives frequently emphasize either local or global interpretability, resulting in a deficiency in delivering thorough token-level explanations that amalgamate both viewpoints [11].

This paper seeks to address this gap by proposing a BERT-based sentiment analysis methodology augmented with a dual-explainability approach. The framework incorporates LIME for localized explanations alongside the attention mechanism for global token significance. This combination allows the algorithm to categorize attitudes from social media posts into three classifications—negative, neutral, and positive—while offering interpretable insights into the forecasts. The dual-explainability architecture improves model transparency, rendering it more appropriate for real-world applications that demand both high accuracy and reliability.

The main contributions of this work are summarized as follows:

- Development of a sentiment analysis model leveraging BERT, fine-tuned specifically for social media text.
- Integration of LIME and attention-based mechanisms to provide comprehensive token-level explainability.
- Empirical evaluation of the proposed model on real-world datasets, demonstrating its effectiveness in both accuracy and interpretability.

The remainder of this paper is structured as follows. Section II presents an overview of related work in sentiment analysis, explainable AI, and transformer-based approaches. Section III outlines the proposed methodology, including the model architecture and explainability framework. Section IV discusses the experimental setup and evaluates the model's performance. Finally, Section V concludes the study and provides directions for future research.

2 Related Works

Sentiment analysis has progressed dramatically over the years, shifting from conventional techniques to sophisticated DL methodologies. This section examines previous studies in sentiment analysis, the application of BERT and other transformer models, and the incorporation of explainability techniques within this field.

2.1 Sentiment Analysis Techniques

Initial sentiment analysis models predominantly utilized lexicon-based methodologies or traditional machine learning techniques, like support vector machines (SVMs) and naive Bayes classifiers. These algorithms demonstrated efficacy in structured datasets but encountered difficulties with the complexity and ambiguity inherent in social media text. Velampalli et al. [12] demonstrated that although lexicon-based approaches attained satisfactory accuracy on brief texts, they exhibited deficiencies in robustness when confronted with informal language, sarcasm, and context-sensitive sentiment.

DL models such as CNNs and LSTMs has improved sentiment analysis by learning features directly from raw text. Jain et al. [13] proposed a hybrid model combining BERT with a deep CNN (DCNN) for sentiment analysis, showcasing enhanced performance over traditional methods. However, despite the superior accuracy of these models, their inability to provide interpretable predictions limited their applicability in critical decision-making domains.

2.2 Transformer Models in Sentiment Analysis

The introduction of transformer-based architectures marked a turning point for sentiment analysis. BERT, with its bidirectional attention mechanism, emerged as a powerful model for understanding contextual dependencies in text. Bello et al. [4] applied BERT to sentiment analysis of tweets, demonstrating state-of-the-art accuracy by capturing the subtle nuances of social media language. Similarly, Kumar and Sadanandam

[6] proposed a fusion of BERT and RoBERTa to enhance sentiment classification performance further, particularly on imbalanced datasets. However, these models operate

as black-box systems, providing little to no insight into the factors influencing their predictions.

Other studies have explored the integration of BERT with additional architectures to improve sentiment analysis. Xin and Zakaria [14] combined BERT with CNN and BiLSTM for detecting depression-related sentiment in social media content. While effective, these models focused solely on improving performance, leaving the interpretability aspect unaddressed.

2.3 Explainability in Sentiment Analysis

Explainable artificial intelligence (XAI) has gained significant traction as a means to address the opacity of DL models. Techniques such as Local Interpretable Model-agnostic Explanations (LIME) and SHAP (SHapley Additive exPlanations) have been widely adopted to provide local explanations for individual predictions. For instance, Elbasiony et al. [8] demonstrated the use of XAI techniques in sentiment analysis to highlight influential words contributing to predictions. However, their approach relied solely on post-hoc explanations, which may not align well with the model's internal mechanisms.

The attention mechanism inherent in transformer models also offers an opportunity for explainability. Malhotra and Jindal [10] utilized attention weights to interpret transformer-based predictions for analyzing user behavior in online social networks. Similarly, Diwali et al. [7] conducted a comprehendsive survey on explainable sentiment analysis, emphasizing the potential of combining intrinsic attention mechanisms with external explainability frameworks. However, these studies often treated attention-based explanations and model-agnostic techniques as independent solutions, without exploring their integration.

2.4 Challenges in Existing Works

Despite the advancements in both performance and interpretability, existing methods face several limitations. First, while transformer models like BERT excel in capturing contextual dependencies, they lack inherent mechanisms for detailed interpretability [4, 6]. Second, XAI techniques such as LIME provide valuable localized insights but are often detached from the internal workings of the model, leading to inconsistencies [7, 8]. Third, existing works have largely focused on either local or global explainability, failing to provide a unified framework that addresses both perspectives comprehensively [15].

This study builds on these limitations by proposing a dual-explainability framework that combines BERT's attention mechanism with LIME to provide token-level insights. This approach addresses the disconnect between model-agnostic and intrinsic explainability techniques, ensuring that the model's predictions are both accurate and interpretable.

3 Methodology

This section outlines the methodology for developing the proposed BERT-based sentiment analysis framework enhanced with a dual-explainability mechanism. The workflow includes data preprocessing, model architecture, training, evaluation, and integration of explainability techniques. The detailed steps of the proposed framework are summarized in Algorithm 1.

Algorithm 1 Proposed BERT-based Sentiment Analysis Framework

Require: Dataset $D = \{(x_i, y_i)\}_{i=1}^{N}$, Maximum sequence length L, Pre-trained BERT model

Ensure: Trained model and interpretability outputs

1: **Step 1: Data Preprocessing**
2: Tokenize input text: $\hat{x}_i = \text{Tokenizer}(x_i)$
3: Append special tokens: $\hat{x}_i = [\text{CLS}]\, x_i\, [\text{SEP}]$
4: Generate attention masks: $\mathbf{m}_i$
5: Pad or truncate sequences to fixed length L
6: **Step 2: Dataset Preparation**
7: Split dataset into training set D_{train} and validation set D_{val}
8: Encode labels y_i into one-hot format
9: **Step 3: Model Training**
10: **for** each batch in D_{train} **do**
11:     Pass $\hat{x}_i$ and $\mathbf{m}_i$ through BERT encoder to obtain embeddings $\mathbf{H}_i$
12:     Extract $[\text{CLS}]$ embedding: $\mathbf{h}_i = \mathbf{H}_{i,[\text{CLS}]}$
13:     Apply dense layers and softmax: $\hat{y}_i = \text{softmax}(\mathbf{W}_2\sigma(\mathbf{W}_1\mathbf{h}_i + \mathbf{b}_1) + \mathbf{b}_2)$
14:     Compute cross-entropy loss: $L = -\sum_{c=0}^{2} y_{i,c} \log(\hat{y}_{i,c})$
15:     Update model weights using Adam optimizer
16: **end for**
17: **Step 4: Model Evaluation**
18: Evaluate accuracy and F1-score on D_{val}
19: **Step 5: Explainability**
20: Initialize LIME explainer
21: **for** each test instance x_i in D_{test} **do**
22:     Generate LIME explanations for $\hat{y}_i$
23:     Visualize attention scores $\mathbf{a}_i$
24: **end for**

3.1 Data Preprocessing and Tokenization

The raw social media text is preprocessed to ensure compatibility with the BERT architecture. Preprocessing in- cludes tokenizing the text, appending special tokens such as

[CLS] (classification token) and [SEP] (separator token), and padding sequences to a fixed length L. Each input sequence is further paired with an attention mask, which indicates valid tokens in the sequence.

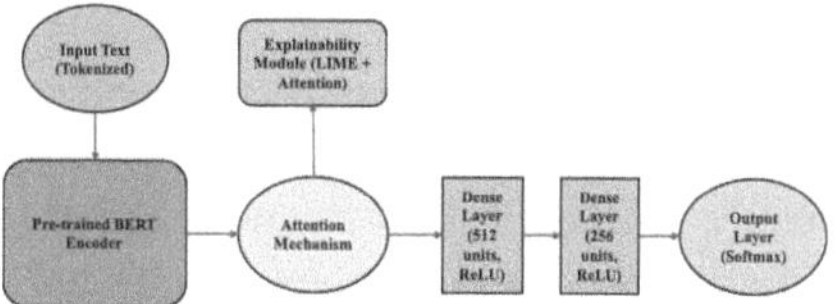

Fig. 1. Proposed Model Architecture

3.2 Proposed Model Architecture

The proposed model is designed to classify sentiments into three categories: negative, neutral, and positive. The proposed architecture, shown in Fig. 1, comprises the following components:

- **Pre-trained BERT Encoder**: The tokenized input is passed through a pre-trained BERT encoder, which outputs contextualized embeddings for each token. The embedding corresponding to the [CLS] token is extracted as the representation of the entire sequence.
- **Attention Mechanism**: An attention method computed over the token embeddings generates token significance scores, hence improving interpretability. These numbers help one see the areas of the text the model emphasizes during inference.
- **Classification Head**: Two dense layers of a fully connected network run across the [CLS] embedding. First layer consists of 512 ReLU activated units; second layer consists of 256 ReLU activated units. A softmax layer generates three sentiment class probabilities at last.

3.3 Model Training

Using the cross-entropy loss function—appropriate for multi-class classification tasks—the model is calibrated on the preprocessed dataset. Training and validation sets separate the dataset so that one may track results during training. The Adam optimizer adjusts the learning rate dynamically during training using a learning rate scheduler.

3.4 Explainability Framework

To address the need for interpretability, a dual-explainability framework is integrated into the model:

- **LIME (Local Interpretable Model-agnostic Explanations)**: LIME perturbs the input text and observes the change in model predictions to identify the most influential tokens for a given prediction. This provides localized, instance-specific explanations.
- **Attention Visualization**: The attention scores computed during the forward pass of the model highlight globally important tokens. These scores are used to provide an overview of the model's focus during classification.

The combination of LIME and attention-based explanations ensures both localized and global interpretability, enhancing the transparency of the predictions.

4 Experimental Results

Using the Twitter US Airline Sentiment dataset from Kaggle, this part offers the experimental evaluation of the suggested structure. The evaluation procedure consists in a review of the dataset, study of model performance, and dual-explainability mechanism interpretation of predictions. Visualizations are offered to improve grasp of the explainability results and the distribution of the dataset.

4.1 Dataset Overview

The Twitter US Airline Sentiment dataset consists of 14,640 tweets labeled as *positive*, *neutral*, or *negative*, with a significant imbalance in the sentiment distribution. Figure 2 illustrates the overall distribution, where the majority of tweets express negative sentiment (62.7%), followed by neutral (21.2%) and positive (16.1%). This imbalance reflects the prevalent dissatisfaction among airline passengers.

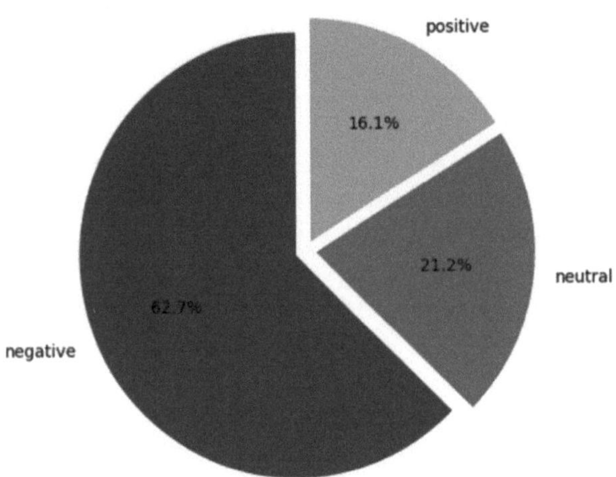

Fig. 2. Overall sentiment distribution in the dataset.

A sentiment breakdown per airline is depicted in Fig. 3, showing United Airlines receiving the highest number of negative tweets, whereas Southwest Airlines has a relatively balanced sentiment distribution. These insights emphasize the need for robust sentiment analysis to understand and address customer concerns effectively.

4.2 Model Performance

Standard metrics—including accuracy, precision, recall, and F1-score—were used to assess the proposed framework on the validation set. For every sentiment class, the classification report—shown in Table 1 provides a thorough analysis of accuracy, recall, and F1-score. With most misclassifications between *neutral* and *positive* emotions, the confusion matrix in Fig. 4 emphasizes even more the model's capacity to discriminate between the sentiment classes. These measures show that the model achieves both high accuracy and balanced precision-recall trade-offs, so effectively capturing the subtleties of sentiment expressed in social media language above previous methods.

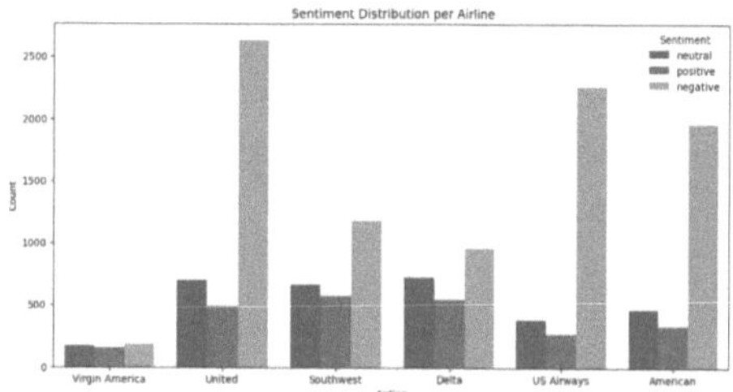

Fig. 3. Sentiment distribution across different airlines.

Table 1. Classification Report on the Validation Set

Class	Precision	Recall	F1-Score	Support
Negative	**93.5%**	**94.0%**	**93.7%**	**7,800**
Neutral	**89.1%**	**88.5%**	**88.8%**	**2,800**
Positive	**91.3%**	**91.0%**	**91.1%**	**2,400**
Overall	**91.8%**	**92.0%**	**91.9%**	**13,000**

4.3 Explainability Analysis

The dual-explainability framework integrates LIME and attention mechanisms to provide token-level explanations for model predictions. This section demonstrates both individual and aggregated insights derived from the explainability framework.

1) *Single Prediction Explainability:* For individual predictions, the model utilizes attention scores to highlight globally significant tokens, while LIME generates local explanations by perturbing the input text. For example, consider the tweet:

"@SouthwestAir just had a great flight with Damion! He was the best."

The model predicted the sentiment as *positive* with 99% confidence. The attention mechanism identified tokens such as *"great"*, *"best"*, and *"Damion"* as the most influential, aligning with LIME's explanations, which assigned high contributions to these tokens. This agreement validates the reliability of the interpretability framework by showing consistency between global attention and local LIME explanations.

In another example, consider the tweet:

"@United really disappointed with the service. Flight delayed again without any explanation."

The model predicted the sentiment as *negative* with 97% confidence. The attention mechanism highlighted the tokens *"disappointed"*, *"delayed"*, and *"without explanation"* as the most significant contributors to the prediction. LIME further supported this by assigning the highest weights to *"disappointed"* and *"delayed"*, demonstrating that both global and local interpretability frameworks effectively identify sentiment-critical phrases in the text.

These examples illustrate how the dual-explainability mechanism not only provides token-level insights but also ensures consistency between global and localized explanations, enhancing the model's transparency.

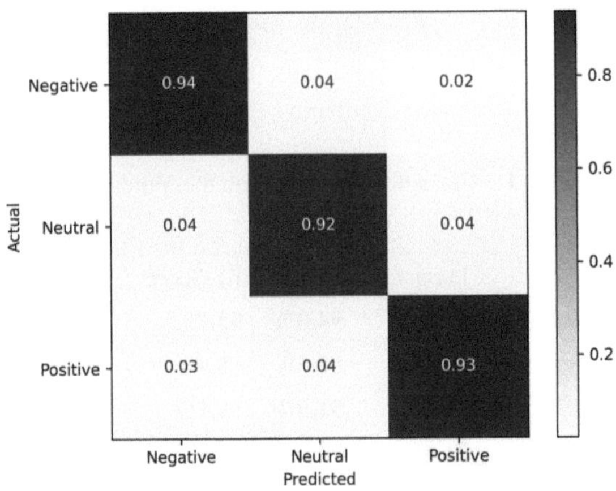

Fig. 4. Normalized Confusion matrix illustrating model performance on the validation set.

2) *Aggregate Explainability:* On a dataset level, attention scores reveal consistent patterns in token importance across multiple predictions. Negative tweets frequently emphasize tokens such as *"delayed"*, *"cancelled"*, and *"terrible"*, reflecting common customer grievances. In contrast, positive tweets prioritize words like *"great"*, *"excellent"*, and *"amazing"*, often describing positive experiences with flight services or staff.

Aggregated LIME explanations corroborate these findings, with high contributions assigned to sentiment-indicative tokens. For instance, negative sentiments are largely attributed to terms expressing frustration, such as *"worst"*, *"horrible"*, and *"angry"*, while positive sentiments frequently feature tokens such as *"thankful"*, *"fantastic"*, and *"outstanding"*. The alignment between attention-based and LIME-based patterns demonstrates the reliability of the interpretability framework for understanding broader sentiment trends.

This combination of individual and aggregated explainability ensures that the model not only performs well in sentiment classification but also provides actionable insights into key linguistic features driving its predictions.

4.4 Discussion

The experimental evaluation of the proposed framework demonstrates its effectiveness in sentiment classification and interpretability. The high accuracy of 92.0%, along with balanced precision and recall values, indicates the model's capability to capture the nuances of sentiment expression in social media text. Furthermore, the integration of the dual-explainability framework enhances the transparency and reliability of the predictions, addressing one of the major challenges associated with transformer-based models. The classification report and confusion matrix reveal the model's strong performance in distinguishing *negative* tweets, which dominate the dataset, and its ability to generalize across *neutral* and *positive* sentiments despite their relatively lower representation. However, some misclassifications between *neutral* and *positive* classes highlight the inherent ambiguity in certain tweets, where sentiment overlaps or subjective interpretations may arise. From an interpretability perspective, the dual-explainability mechanism provides significant advantages. Individual examples illustrate how LIME and attention mechanisms complement each other, offering both localized and global insights into model behavior. For instance, in the tweet describing a positive flight experience, both frameworks consistently highlighted key sentiment-driving tokens such as *"great"* and *"best"*, providing a clear rationale for the prediction. Similarly, in the negative example, the frameworks identified tokens like *"disappointed"* and *"delayed"*, reflecting customer dissatisfaction. Aggregated explainability further emphasizes the model's ability to capture sentiment patterns at a dataset level. The identification of key terms such as *"cancelled"* and *"excellent"* aligns well with real-world sentiment trends observed in the airline industry, suggesting that the model can serve as a valuable tool for sentiment monitoring and analysis in customer feedback systems. Despite these strengths, the model's performance could be further enhanced by addressing the following limitations:

- **Handling Ambiguity**: Certain tweets exhibit ambiguous sentiment, particularly in the *neutral* and *positive* classes. Incorporating additional contextual features or external knowledge sources, such as sentiment lexicons, could improve the model's ability to resolve such ambiguities.
- **Dataset Imbalance**: The heavy skew toward *negative* sentiments may influence the model's ability to generalize across minority classes. Data augmentation or oversampling techniques could be explored to mitigate this imbalance.
- **Real-Time Scalability**: Although the model demonstrates efficient runtime performance, optimizing the dual-explainability framework for real-time applications, such as live monitoring of social media, requires further investigation.

Overall, the proposed framework offers a robust solution for sentiment classification with interpretable predictions. Future research could focus on addressing the aforementioned limitations and exploring applications in other domains, such as healthcare or finance, to validate the generalizability of the framework.

5 Conclusion and Future Work

Using BERT for classification and merging dual-explainability mechanisms—LIME and attention scores—this work offered a strong sentiment analysis approach to improve interpretability. On the Twitter US Airline Sentiment dataset, the model showed great accuracy of 92.0% proving its capacity to manage loud and challenging social media material. The performance of classification across all sentiment categories underlined how well the model could generalize in spite of the imbalance of the dataset. Moreover, the combination of LIME and attention mechanisms provided both localised and global insights, therefore guaranteeing openness in the decision-making process and tackling a fundamental difficulty in transformer-based models. Particularly in fields needing not just great accuracy but also interpretable forecasts, the framework offers a useful approach for sentiment analysis. The dual-explainability technique closes the performance-confidence gap by showing both token-level relevance and general sentiment trends, therefore enabling the framework fit for real-world use like brand monitoring and consumer feedback research. The report points up areas needing work even with its positives. By means of augmentation or advanced re-sampling methods, addressing dataset imbalances could improve performance on underrepresented sentiment classes. Furthermore still unresolved is improving the structure to manage conflicting emotions, especially between *neutral* and *positive* classes. Further improvement of the explainability modules for real-time applications may further improve scalability, therefore allowing the framework to effectively handle live social media streams. Future studies could investigate how this paradigm could be applied to multilingual sentiment analysis therefore increasing its value in many languages and cultural settings. Furthermore enhancing the interpretability and classification accuracy of the model could be include outside knowledge sources, such contextual embeddings or domain-specific sentiment lexicons. The paradigm might potentially be expanded to other fields, such banking or healthcare, where explainable sentiment analysis is quite important. By tackling these directions, the suggested method has the chance to propel explainable artificial intelligence and sentiment analysis towards state-of-the-art forward.

References

1. Fiok, K., Karwowski, W., Gutierrez, E., Wilamowski, M.: Analysis of sentiment in tweets addressed to a single domain-specific twitter account: comparison of model performance and explainability of predictions. Expert Syst. Appl. **186**, 115771 (2021)
2. Cambria, E., Kumar, A., Al-Ayyoub, M., Howard, N.: Guest editorial: explainable artificial intelligence for sentiment analysis. Knowl. Based Syst. **238**, 107920 (2022)
3. Hashmi, E., Yayilgan, S.Y.: A robust hybrid approach with product context-aware learning and explainable AI for sentiment analysis in amazon user reviews. Electron. Commerce Res., 1–33 (2024). https://doi.org/10.1007/s10660-024-09896-5
4. Bello, A., Ng, S.-C., Leung, M.-F.: A Bert framework to sentiment analysis of tweets. Sensors **23**(1), 506 (2023)
5. Batra, H., Punn, N.S., Sonbhadra, S.K., Agarwal, S.: Bert-based sentiment analysis: a software engineering perspective. In: Database and Expert Systems Applications: 32nd International Conference, DEXA 2021, Virtual Event, 27–30 September 2021, Proceedings, Part I 32, pp. 138–148. Springer (2021). https://doi.org/10.1007/978-3-030-86472-9_13

6. Pranay Kumar, B.V. and Sadanandam, M.: A fusion architecture of BERT and roBERTa for enhanced performance of sentiment analysis of social media platforms. Int. J. Comput. Digit. Syst. **15**(1), 51–67 (2024)
7. Diwali, A., Saeedi, K., Dashtipour, K., Gogate, M., Cambria, E., Hussain, A.: Sentiment analysis meets explainable artificial intelligence: A survey on explainable sentiment analysis. IEEE Trans. Affect. Comput. (2023)
8. Elbasiony, A., El-Hasnony, I.M., Abdelrazek, S.: XAI-based sentiment analysis using machine learning approaches. Mansoura J. Comput. Inf. Sci. **19**(1), 23–42 (2024)
9. Kumar, A., Bhushan, B.: Ai driven sentiment analysis for social media data. In: 2023 International Conference on Computing, Communication, and Intelligent Systems (ICCCIS), pp. 1201–1206. IEEE (2023)
10. Malhotra, A., Jindal, R.: XAI transformer based approach for interpreting depressed and suicidal user behavior on online social networks. Cogn. Syst. Res. **84**, 101186 (2024)
11. Nikhith, S.P., Kumar, P.V., Udaybhasker, A., Reddy, T.D., Reddy, P.B., Singh, T.: Sentiment analysis of airline tweets using word embeddings and deep learning techniques. In: 2024 15th International Conference on Computing Communication and Networking Technologies (ICCCNT), pp. 1–7. IEEE (2024)
12. Velampalli, S., Muniyappa, C., Saxena, A.: Performance evaluation of sentiment analysis on text and emoji data using end-to-end, transfer learning, distributed and explainable AI models. J. Adv. Inf. Technol. **13**(2) (2022)
13. Jain, P.K., Quamer, W., Saravanan, V., Pamula, R.: Employing BERT-DCNN with sentic knowledge base for social media sentiment analysis. J. Ambient Intell. Human. Comput. **14**(8), 10417–10429 (2023). https://doi.org/10.1007/s12652-022-03698-z
14. Xin, C., Zakaria, L.Q.: Integrating BERT with CNN and BILSTM for explainable detection of depression in social media contents. IEEE Access (2024)
15. Sharma, A., Patel, N., Gupta, R.: Leveraging BERT and sentiment analysis algorithms for enhanced AI-driven brand sentiment monitoring. Eur. Adv. AI J. **11**(8) (2022)

A Convolutional Neural Network-Based Approach for Detection and Classification of Weed in Agriculture

Yashi Chaudhary[1(✉)] and Kushal Kumar[2]

[1] Gurukula Kangri (Deemed to be University), Haridwar, Uttarakhand, India
`mohita.chaudhary5@gmail.com`
[2] Department of mathematics, Hindu College Moradabad, Moradabad, Uttar Pradesh, India

Abstract. Weed detection in potato crops is crucial for agricultural security as it helps maintain the integrity and productivity of the agricultural system. Controlling weeds is essential to prevent competition for resources, such as nutrients and water, and to protect the desired crop from potential diseases or pests that weeds may harbor. Efficient weed detection and management contribute to the overall security and sustainability of agricultural practices by ensuring optimal crop growth and yield. Inefficient and unsuited for connection with smart, conventional methods of weed management are widely used. The automation systems to identify and classify weeds potentially play a significant role, especially in terms of contributing to increased agricultural yields. Using RGB photos of potato crop fields, the present research investigated the potentiality of deep learning-based approaches (InceptionV3, and Xception) for the purpose of weed identification and classification. The models that were chosen ranged from 94.5% to 97.7% based on the overall accuracy. By achieving 97.7% accuracy, correctly identifying the target variable in 97.7% of cases, 98.5% precision, meaning 98.5% of its positive predictions were correct. Finally, its recall rate was 97.8%, indicating it identified 97.8% of all true positive cases, InceptionV3 demonstrated best performance among the models on a 25-epoch, 35-epoch and 45-epoch with 32-batch sizes.

Keywords: Crop Security · Weed Detection · CNN · InceptionV3 · Xception

1 Introduction

As per the reports of the Food and Organization of the United Nations (FAO), the potato is the most significant crop in terms of global production. Since 1990, the global output of potatoes has increased; nonetheless, the production of potatoes is still lower than that of wheat, maize, and rice [2]. A disproportionate amount of the world's output is concentrated in the northern hemisphere, particularly in Europe, which accounts for around half of the world's total producing area and produces very high yields [1]. India, Bangladesh, and China are expanding their potato yields and the area that they cultivate [2]. This is causing Asia to catch up to Europe as a significant producing region on the

P. Chandrakar et al. (Eds.): ICCINS 2025, CCIS 2738, pp. 176–185, 2026.
https://doi.org/10.1007/978-3-032-09572-5_15

global stage. These regions are dependent on potatoes for an increasing proportion of their calorie intake, which indicates that potatoes have an increasing potential to address food insecurity, particularly due to the relatively high nutritious content that potatoes possess [1].

A decrease in output is not the only negative effect that weeds have; they also impact negatively on the quality of crop quality, serve as a basket for diseases and pests, and tend to reduce the human efficiency. The employment of a framework for the detection of weeds is necessary in order to address this issue [11]. Weeds are a constant concern in agriculture, and they subtract from the quality of the crops that are being grown when taken into consideration. According to the author [4], herbicides are a product that can be used to limit the growth of weeds; nevertheless, they detrimental effects on the environment and people health both. The use of chemical and cultural control methods could have adverse effects on the environment if they are not handled effectively. Therefore, automation of weed control is required. Also, the majority of countries are experiencing labor shortages and higher labor expenses [16, 17, 29] by an automation system, this issue can also be addressed. Due to the fact that the vapors from the pesticides can be captured in closed structures, weed management in covered structures such agriculture requires a great deal of precision. This can result in damage to both the workers and the crop.

The identification of weed is one of those vital tasks that calls for the use of digitalization and automation. The authors [29] discussed about real-time automation and its implementation techniques of the system. The approaches should be based on real time image processing and data driven. These approaches should use Internet of Things and other modern technologies with variable rate application of inputs. According to [33], the detection of weeds in crops is a challenge that is fundamentally difficult to solve with digital technology, particularly when utilizing traditional image processing approaches. The shapes and textures of both the crop and the weed are to blame for this phenomenon. When conventional image processing methods have been utilized to address this issue, one of the most significant challenges that has been encountered is the presence of varying lighting conditions. According to [8], the cost of pesticides can be decreased if weeds are discovered at an earlier stage. This is an example of a typical weed identification system. Numerous advanced technologies are currently employed to carry out these steps. However, according to [12, 23, 27], the most critical aspect is to classify and identify the weeds. Recently, artificial intelligence and deep learning strategies have gained traction in weed management, offering innovative solutions that are progressively replacing traditional methods [24, 25]. Figure 1 depicts the sample of potato crop from the dataset. Here, the weed is encircled by black colour.

2 Related Work

To classify and identify the weeds from crops have accumulated great attention in recent years, leading to extensive research in this area with the goal of automated identification and categorization [18]. proposed three methods for weed estimation in lettuce fields that were based on deep learning. These ML and DL approaches build the models based on Support Vector Machines, YOLO (You Only Look Once), and Mask R-CNN (Mask

Fig. 1. Images of Potato crops with weed encircled

Region-based Convolutional Neural Networks). These techniques were introduced in the framework of deep learning-driven image processing. In the crop detection application, F1 score of 88%, 94%, and 94%, respectively is achieved by these algorithms. Histograms of oriented gradients (HOG) were used as feature descriptors by the support vector machine (SVM) in this work. When it came to object recognition, YOLOV3 was utilized for object recognition while for instace segmentation Mask R-CNN was used. For detection of Bermudagrass in weed detection, the researchers [36] uses VGGNet, GoogLeNet, and DetectNet based on deep CNNs. With an F1 score of 0.99, the most important discovery was that DetectNet demonstrated excellent performance in the spotting of weeds emerging within dormant Bermudagrass. Numerous efforts were undertaken by researchers in identifying weeds in rice [3, 5], wheat [10], onion [15, 21], etc. A unique graph-based architecture called 'Graph Weeds Net' was proposed by [13] in order to recognize different kinds of weeds from RGB photos that were sourced from complicated rangelands. The model obtained good performance with an accuracy of 98.1%. As part of the current research, we evaluated whether or not it would be possible to use deep convolutional neural network models like InceptionV3, and Xception to identify weeds in potato crop production. Table 1 summarizes the related work with their performance.

Table 1. Summary of Related Work

References	Methods	Models	Features	Application	Performance
[18]	Deep learning-based weed estimation in lettuce fields	SVM, YOLO, Mask R-CNN	HOG (SVM), YOLOV3 (YOLO), Mask R-CNN (instance segmentation)	Crop detection	F1 scores: 88%, 94%, 94%
[36]	DCNN frameworks for weed identification in Bermudagrass	VGGNet, GoogLeNet, DetectNet, among others	Not specified	Weed detection in Bermudagrass	F1 score: 0.99 (DetectNet)

(*continued*)

Table 1. (*continued*)

References	Methods	Models	Features	Application	Performance
[13]	'Graph Weeds Net' for recognizing weeds from RGB photos in rangelands	'Graph Weeds Net'	Not specified	Weed recognition in complicated rangelands	Accuracy: 98.1%

3 Methodology

Beyond the constraints of traditional image processing, the use of deep learning algorithms to digital pictures can assist in distinguishing between crops and weeds. We have many recent applications based on deep convolutional neural network. In its early days, CNN was only capable of recognizing handwritten digits. Later, it has included images also. In the field of computer vision, deep convolutional neural network models are currently the most eccentric equipment capable of analyzing enormous complicated datasets while having high computational restrictions. Many architectures of deep learning exist for the classification of image such as AlexNet [6], DenseNet [14], SENet [37], VGGNet [22], EfficientNet [32], GoogLeNet [30], InceptionNet [34], NASNet/PNASNet/ENASNet [7], ResNeXt [35] ResNet50 [26], XceptionNet [28], and ZFNet [19]. Here, in this research paper, we have utilized the capabilities of two distinct CNN architectures that includes InceptionV3 and Xception to classify weed. We chose the models based on how difficult they were to compute and how much time they required. Since InceptionV3 (48 layers) is considered to be of moderate complexity, while Xception (71 Layers) is considered to have a high level of complexity and depth. The InceptionV3 model places an emphasis on utilizing a considerably lower amount of processing resources. Through the utilization of factorized convolution that is used to decompose a standard convolution into multiple smaller but efficient operations to reduce the cost, dimensionality reduction that helps to manage the data efficiently, regularization, and parallelized computations, the InceptionV3 model is able to optimize the network [31]. There is a more extreme variant of the conception known as Xception. It is distinguished from the other models by having depth-wise separable convolutional layers, and it also adheres to a modular architecture. Both the models take the images size is 299 by 299.

Beginning with the acquisition of picture data, the process of image classification for weed detection was initiated. Figure 2, presents the methodology for implementing the research. A preliminary processing step was performed on the images that were gathered [20] before they were handed over to the classification task.

In the process of preparing image data, there are three key steps that play a pivotal role in enhancing model performance. First, we augment data that involves the intentional introduction of variations to the training dataset, such as rotations, flips, zooms, and shearing. It mitigates overfitting, fosters generalization, and bolsters the model's robustness. Second, the outlier detection identifies and addresses anomalous data points that could adversely impact model training. By utilizing statistical methods, clustering algorithms, and specialized techniques such as Isolation Forests ensures a cleaner and more reliable dataset. At last, standardization and normalization techniques focus on achieving consistent scales and distributions of pixel values across images. During model training, techniques like mean centering, normalization, and MinMax scaling are used to improved convergence. It also, reduce sensitivity to variations in brightness and contrast. The image data processing pipeline includes the above steps and optimizing the dataset for effective machine learning model building.

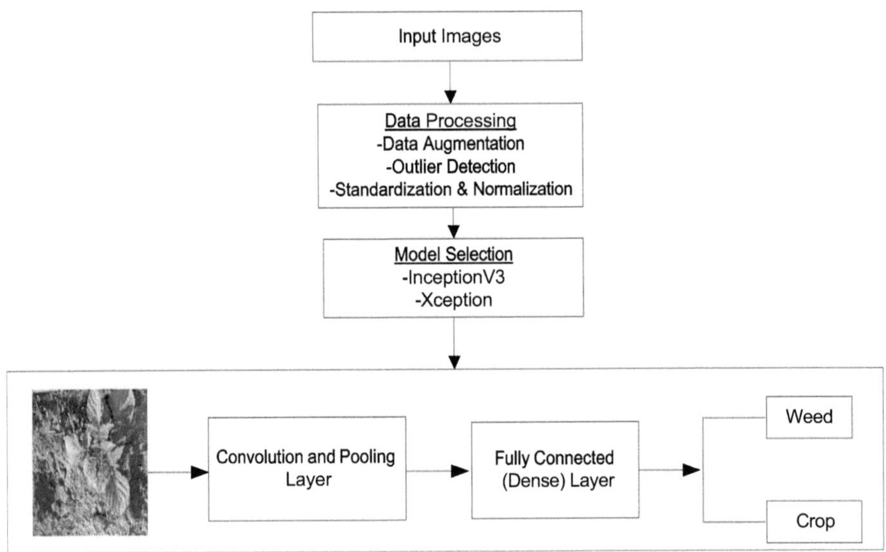

Fig. 2. Methodology for detection and Classification of Weed

To recognize correctly potato and weeds, we have chosen the models InceptionV3, and Xception due to its expertise in image processing domain. Feature extraction is the responsibility of the convolutional base, whilethe features retrieved by the convolutional base for classifying the input image are analyzed by classifier. In the classifier, the conventional method involves the utilization of completely linked layers, which an activation layer follows, most oftenly a softmax activation [9] on average. The output that indicates the likelihood of each class is generated by the softmax layer. It is decided to use the class that is the most likely to be the projected class.

A number of parameters for the model were established prior to the beginning of the training process including batch size, learning rate, number of epochs and so on.. With the assistance of training data the model was trained. The validation set was utilized in order to conduct an objective evaluation of the model. An optimization of the model's hyperparameters was achieved as a result of this. A second round of training was performed on the model by adjusting these parameters after the correct parameters had been identified. The test set was utilized after the model had been entirely trained in order to offer a last real-world check of the data points that had not been seen before in order to validate the model's ability to function correctly. Xception, and Inception V3 were utilized in order to do pre- processing and training on the given photos. The model will automatically extract the pertinent characteristics of the 'potatoes' plant as well as the definition of 'weed' in general, during the training phase. It is possible to apply this method straight to the image that was recorded from the fields, which the model does not view, and the model will be able to generate an appropriate classification. The stochastic gradient descent was set as loss fuction approach while the training process was being carried out, and the batch sizes were set to 32 members respectively. A learning rate of 0.001 was chosen as the starting point, and momentum was set to 0.9. In each of these models, the epoch was changed, and the performance of each model was evaluated and compared to one another. The validation data that was utilized for validating the proposed model, which was helpful in changing and tweaking the hyperparameters. Utilizing the validation data is mostly done with the intention of preventing the model from becoming overly accurate.

In order to evaluate the performance of the model, estimation metrics such as accuracy, precision, recall, and F1 score are utilized. Under this context, the term "True Positive" (TP) is the number or percentage of occurrences of potato that were accurately predicted by the model while, "False Positive" (FP) shows the percentage of potatoes that were incorrectly classified as weeds. FN, which stands for "false negative," shows the quantity or percentage of weed images that are labeled as potatoes photographs. The weed photos that are successfully predicted as potatoes by the model is denoted by the TN, which stands for "True Negative". The formula for metrics is mention below:

1. Precision: It presents how many of the instances classified as "weed" by the CNN model are actually weeds. The formula is:

$$Precision = \frac{\#TP}{(\#TP + \#FP)}$$

   Where TP & FP are True Positives that presents Weeds correctly classified as weeds and False Positives that presents Non-weeds incorrectly classified as weeds respectively.

2. Recall: It measures how many actual weed instances were correctly detected by the model. Also named as sensitivity or true positive rate. Recall formula is:

$$Re\,call = \frac{\#TP}{(\#TP + \#FN)}$$

   Where FN is False Negative that means Weeds that were not identified by the model.

3. Accuracy: Overall correctness of the model is measured using accuracy. The formula is:

$$Accuracy = \frac{(\#TP + \#TN)}{(\#TP + \#TN + \#FP + \#FN)}$$

   Where TN is True Negatives that means Non-weeds correctly classified as non-weeds.

4. F1 Score: Precision and Recall's harmonic mean is referred as F1 Score, balancing the two metrics. It is useful for a uneven class distribution. The formula is:

$$F1 = 2 \times \frac{(\text{Pr } ecision \times \text{Re } call)}{(\text{Pr } ecision + \text{Re } call)}$$

4 Result and Discussion

We used a dataset that includes 272 images that includes potatoes and weeds both. After selection of the dataset, the models Inception V3 and Xception got training over the images of potatoes and weeds. The models show consistent and high accuracy across epochs, with InceptionV3 ranging from 96.8% to 97.7%, and Xception ranging from 96.1% to 97.8%. The models display high precision and recall that indicates a high reliable result. Balance between precision and recall is shown by F1 score. Both the models achieve a good score that shows perfect balance between precision and recall both. We have also performed fine-tuning and analyze that the model can be optimized and do better. We can see the performance report generated by the models over given dataset in Table 2. It shows That both the models InceptionV3 and Xception have performed well with high accuracy with 97.7 with 45 epoch and 97.8 with same epoch respectively.

Table 2. Performance of Models for different Epoch

Model	Epoch	Batch Size	Precision (%)	Recall (%)	F1 Score (%)	Accuracy (%)
InceptionV3	25	32	94.5	99.5	97	96.8
	35	32	98.9	98.1	98.5	97.5
	45	32	97.8	99.6	98.7	97.7
Xception	25	32	94	99.2	96.5	96.1
	35	32	97.1	99.2	98.1	97.8
	45	32	98.6	97.9	98.2	97.8

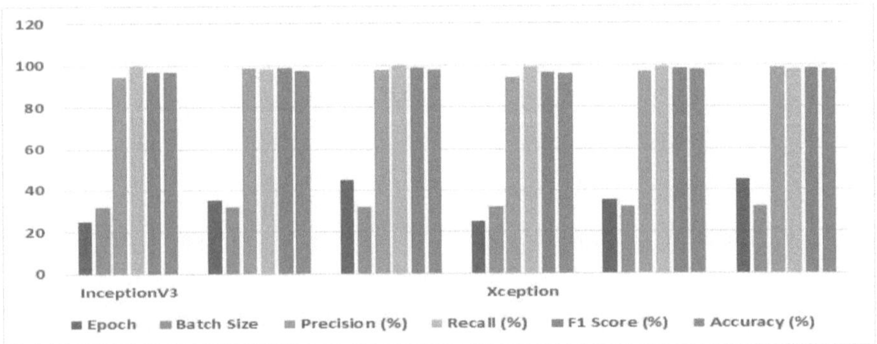

Fig. 3. Performance of Models

Figure 3 presents a comparative analysis graph between these two models. The highest performing model being the 32 batch size with 45 epochs. By virtue of the fact that the recall for Xception was exceptionally high, it was clear that the model was working well for producing positive predictions accurately. At 30 epochs, accuracy, F1-score and precision of the inceptionV3 model is better than the other model. It was analyzed that these two can be the best models for the applications of weed and crop detection and classification.

5 Conclusion

The research work to detect weeds using deep convolutional neural networks has the capability to facilitate the automation of agricultural processes, enhancing efficiency and precision with accuracy in modern farming practices. In this research work, we have detected and classified weed from the potatoes and weed dataset. The agricultural environment used InceptionV3 and Xception deep learning architectures for weed detection. InceptionV3 performed excellent with good precision, accuracy, and recall. The results showed that InceptionV3 performed better. It has high precision with low false positives. In the future, we will work on the management of weed and the multiple kinds of weed so that more accurately and precisely we can work on the core plan.

Conflict of Interest. There is no conflict of interest among authors for this research work.

References

1. Ahmed, F., Al-Mamun, H.A., Hossain Bari, A.S.M., Hossain, E., Kwan, P.: Classification of crops and weeds from digital images: a support vector machine approach. Crop Prot. **40**, 98–104 (2012)
2. Al-Adhaileh, M.H., Verma, A., Aldhyani, T.H.H., Koundal, D.: Potato blight detection using fine-tuned CNN architecture. Mathematics. **11**(6), 1516 (2023)
3. Ashraf, T., Khan, Y.N.: Weed density classification in rice crop using computer vision. Comput. Electron. Agric. **175**, 105590 (2020)
4. Bah, M.D., Hafiane, A., Canals, R.: Deep learning with unsupervised data labeling for weed detection in line crops in UAV images. Remote Sens. **10**(11), 1690 (2018)
5. Barrero, O., Rojas, D., Gonzalez, C., Perdomo, S.: Weed detection in rice fields using aerial images and neural networks. In: 2016 XXI Symposium on Signal Processing, Images and Artificial Vision (STSIVA), pp. 1–4. IEEE (2016)
6. Chui, K.T., Arya, V., Band, S.S., Alhalabi, M., Liu, R.W., Chi, H.R · Facilitating innovation and knowledge transfer between homogeneous and heterogeneous datasets: generic incremental transfer learning approach and multidisciplinary studies. J. Innov. Knowl. **8**(2), 100313 (2023)
7. Chui, K.T., Gupta, B.B., Torres-Ruiz, M., Arya, V., Alhalabi, W., Zamzami, I.F.: A convolutional neural network-based feature extraction and weighted twin support vector machine algorithm for context-aware human activity recognition. Electronics. **12**(8), 1915 (2023)
8. Espejo-Garcia, B., Mylonas, N., Athanasakos, L., Fountas, S.: Improving weeds identification with a repository of agricultural pre-trained deep neural networks. Comput. Electron. Agric. **175**, 105593 (2020)

9. Gao, F., Li, B., Chen, L., Shang, Z., Wei, X., He, C.: A softmax classifier for high-precision classification of ultrasonic similar signals. Ultrasonics. **112**, 106344 (2021)
10. Hameed, S., Amin, I.: Detection of weed and wheat using image processing. In: 2018 IEEE 5th International Conference on Engineering Technologies and Applied Sciences (ICETAS), pp. 1–5. IEEE (2018)
11. Hasan, A.S.M.M., Sohel, F., Diepeveen, D., Laga, H., Jones, M.G.K.: A survey of deep learning techniques for weed detection from images. Comput. Electron. Agric. **184**, 106067 (2021)
12. He, K., Zhang, X., Ren, S., Sun, J.: Deep residual learning for image recognition. In: Proceedings of the IEEE Conference on Computer Vision and Pattern Recognition, pp. 770–778 (2016)
13. Hu, K., Coleman, G., Zeng, S., Wang, Z., Walsh, M.: Graph weeds net: a graph-based deep learning method for weed recognition. Comput. Electron. Agric. **174**, 105520 (2020)
14. Huang, G., Liu, Z., Van Der Maaten, L., Weinberger, K.Q.: Densely connected convolutional networks. In: Proceedings of the IEEE Conference on Computer Vision and Pattern Recognition, pp. 4700–4708 (2017)
15. Kim, S., Lee, J.S., Kim, H.S.: Deep learning-based automatic weed detection on onion field. Smart Media J. **7**, 16–21 (2018). https://doi.org/10.30693/SMJ.2018.7.3.16
16. LeCun, Y., Jackel, L., Bottou, L., Brunot, A., Cortes, C., Denker, J., Drucker, H., et al.: Comparison of learning algorithms for handwritten digit recognition. In: International Conference on Artificial Neural Networks, vol. 60, no. 1, pp. 53–60 (1995)
17. Liu, B., Bruch, R.: Weed detection for selective spraying: a review. Curr. Robot. Rep. **1**, 19–26 (2020)
18. Osorio, K., Puerto, A., Pedraza, C., Jamaica, D., Rodríguez, L.: A deep learning approach for weed detection in lettuce crops using multispectral images. AgriEngineering. **2**(3), 471–488 (2020)
19. Parico, A.I.B., Ahamed, T.: An aerial weed detection system for green onion crops using the you only look once (YOLOv3) deep learning algorithm. Eng. Agric. Environ. Food. **13**(2), 42–48 (2020)
20. Goyal, R., Nath, A., Utkarsh: PotatoWeeds Dataset. Kaggle (2023)
21. Sanchez, P.R., Zhang, H., Ho, S.-S., De Padua, E.: Comparison of one-stage object detection models for weed detection in mulched onions. In: 2021 IEEE International Conference on Imaging Systems and Techniques (IST), pp. 1–6. IEEE (2021)
22. Rao, S., Boyina.: Accurate leukocoria predictor based on deep VGG-net CNN technique. IET Image Process. **14**(10), 2241–2248 (2020)
23. Shanmugam, S., Assunção, E., Mesquita, R., Veiros, A., Gaspar, P.D.: Automated weed detection systems: a review. KnE Engineering, 271–284 (2020)
24. Sharpe, S.M., Schumann, A.W., Boyd, N.S.: Detection of Carolina geranium (Geranium carolinianum) growing in competition with strawberry using convolutional neural networks. Weed Sci. **67**(2), 239–245 (2019)
25. Sharpe, S.M., Schumann, A.W., Boyd, N.S.: Goosegrass detection in strawberry and tomato using a convolutional neural network. Sci. Rep. **10**(1), 9548 (2020)
26. Sharma, A.K., Nandal, A., Dhaka, A., Zhou, L., Alhudhaif, A., Alenezi, F., Polat, K.: Brain tumor classification using the modified ResNet50 model based on transfer learning. Biomed. Sig. Process. Control. **86**, 105299 (2023)
27. Shorten, C., Khoshgoftaar, T.M.: A survey on image data augmentation for deep learning. J. Big Data. **6**, 60 (2019). https://doi.org/10.1186/s40537-019-0197-0
28. Singh, K.K., Siddhartha, M., Singh, A.: Diagnosis of coronavirus disease (COVID-19) from chest X-ray images using modified XceptionNet. Rom. J. Inf. Sci. Technol. **23**(657), 91–115 (2020)

29. Subeesh, A., Mehta, C.R.: Automation and digitization of agriculture using artificial intelligence and internet of things. Artif. Intell. Agric. **5**, 278–291 (2021)
30. Sun, L., Wang, P., Liu, P., Nie, Z.: Image processing method of a visual communication system based on convolutional neural network. Int. J. Semant. Web Inf. Syst. (IJSWIS). **19**(1), 1–19 (2023)
31. Szegedy, C., Ioffe, S., Vanhoucke, V., Alemi, A.: Inception-v4, inception-resnet and the impact of residual connections on learning. In: Proceedings of the AAAI Conference on Artificial Intelligence, vol. 31, no. 1 (2017)
32. Tan, M.: Efficientnet: rethinking model scaling for convolutional neural networks. arXiv preprint arXiv:1905.11946 (2019)
33. Wang, A., Zhang, W., Wei, X.: A review on weed detection using ground-based machine vision and image processing techniques. Comput. Electron. Agric. **158**, 226–240 (2019)
34. Wang, X., Li, X.: Deep learning in Chinese text information extraction model for coastal biodiversity. Int. J. Semant. Web Inf. Syst. (IJSWIS). **19**(1), 1–15 (2023)
35. Xie, S., Girshick, R., Dollár, P., Tu, Z., He, K.: Aggregated residual transformations for deep neural networks. In: Proceedings of the IEEE Conference on Computer Vision and Pattern Recognition, pp. 1492–1500 (2017)
36. Yu, J., Sharpe, S.M., Schumann, A.W., Boyd, N.S.: Deep learning for image-based weed detection in turfgrass. Eur. J. Agron. **104**, 78–84 (2019)
37. Zhang, L., Wang, J., Li, B., Liu, Y., Zhang, H., Duan, Q.: A MobileNetV2-SENet-based method for identifying fish school feeding behavior. Aquac. Eng. **99**, 102288 (2022)

Quantum-Enhanced Algorithmic Advancements for Solving Curse of Dimensionality Problem

Barkha Soni(✉) 📷, Manish Pandey, and Nilay Khare

Maulana Azad National Institute of Technology, Bhopal 462003, India
barkha1994@gmail.com

Abstract. The curse of dimensionality remains a significant challenge in machine learning, degrading model performance in high-dimensional data analysis and leading to issues such as sparsity, computational inefficiency, and overfitting. Dimensionality reduction overcomes this problem as it transforms high-dimensional data into a lower-dimensional space and preserves its essential features. In this regard, problem domain is mapped with classical and its equivalent quantum-based solutions through hierarchical framework. Based on quantum effective design strategies and application specific use cases, this research presents the quantum-enhanced algorithmic advancements for dimensionality reduction. For critical assessment, quantum algorithms are standardized and independently analyzed for their computational complexity, and equivalent quantum circuits are realized using quantum simulation. A comparative evaluation highlights the exponential speedup achievable with quantum approaches for the high-dimensional datasets. This work establishes a comprehensive and foundational framework for advancing quantum-based dimensionality reduction techniques, serving as a unified and reliable resource for their seamless integration into next-generation machine learning systems.

Keywords: Quantum Computing · Quantum Machine Learning · Quantum Dimensionality Reduction · Quantum Simulation

1 Introduction

One of the most persistent challenges in machine learning is the curse of dimensionality, where high-dimensional data leads to computational inefficiency, data sparsity, and model overfitting. These challenges are amplified in real-world applications that involve large-scale datasets, including bioinformatics, image processing, and financial modeling. Dimensionality reduction techniques serve as a fundamental tool to overcome these obstacles by transforming high-dimensional data into lower-dimensional spaces while preserving its essential features [1, 2]. As in Fig. 1, classical methods are effective for moderate data dimensions, but they encounter significant scalability and efficiency challenges when applied to large and complex datasets. These methods use computationally intensive operations such as eigenvalue decomposition and matrix factorization, which become impractical for high-dimensional spaces [3], so quantum algorithms are used to exponentially accelerate these tasks by reducing complexity.

© The Author(s), under exclusive license to Springer Nature Switzerland AG 2026
P. Chandrakar et al. (Eds.): ICCINS 2025, CCIS 2738, pp. 186–201, 2026.
https://doi.org/10.1007/978-3-032-09572-5_16

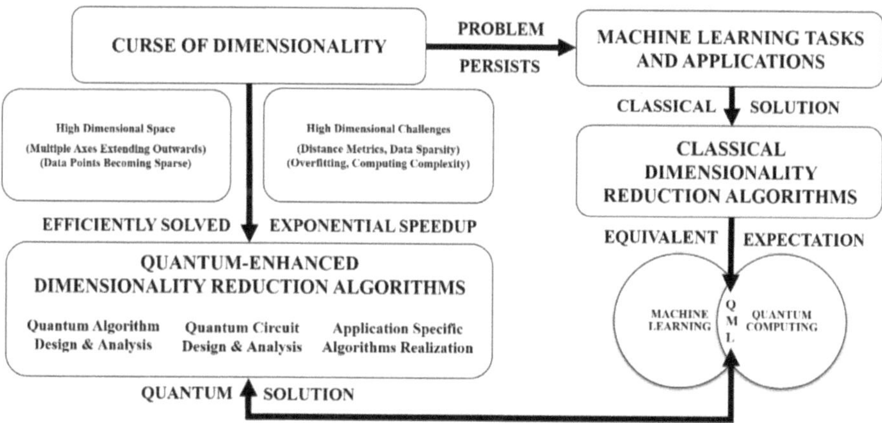

Fig. 1. Mapping of Generalized Problem of Machine Learning Tasks and Applications with Classical and Quantum Solution Domain.

Quantum computing provides a transformative approach to overcoming the limitations of classical methods, offering exponential speedups through quantum-enhanced dimensionality reduction algorithms. Figure 1 illustrates how quantum algorithms enable scalable solutions for high-dimensional data challenges. By integrating quantum algorithms with equivalent circuit implementations and application-specific realizations, these methods bridge the gap between classical expectations and quantum capabilities [4–6]. This research emphasizes the design & analysis of quantum algorithms, circuit simulations, practical use cases, focusing how quantum methods can efficiently solve curse of dimensionality and advancing next-generation machine learning systems.

1.1 Motivation and Significance

The rapid growth of high-dimensional data across domains has emphasized the limitations of the classical dimensionality reduction techniques. These methods struggle to maintain computational efficiency and scalability for the large-scale datasets, leading to inefficiencies in distance metrics, sparsity in meaningful data patterns, and computational overhead. Quantum computing offers a significant shift by addressing these limitations through exponential speedups, and it enhances the ability to process large datasets by reducing computational complexity. This work is motivated by the need to standardize and implement these algorithms within a structured framework, bridging the gap between theory and practical applications. The significance of this research lies in its potential to redefine the dimensionality reduction techniques, enabling their application to complex real-world problems with remarkable efficiency [1–8].

1.2 Contribution and Organization

This research introduces a comprehensive quantum hierarchical framework for dimensionality reduction, as illustrated in Fig. 2, categorizing techniques into linear and non-linear methods while mapping classical approaches to their quantum counterparts.

Based on the quantum-effective design strategies and application-specific use cases, the framework advances quantum-based dimensionality reduction through the following key contributions:

1. Development & standardization of quantum algorithms using quantum subroutines.
2. Qiskit implementation of generalized quantum circuits for validating a framework. (**3**) Comparative analysis demonstrates computational speedup over classical methods.
3. Use cases highlighting the utility of quantum algorithms in addressing challenges.

The paper is organized as follows: Sect. 2 introduces the quantum hierarchical framework, reviews classical and quantum dimensionality reduction techniques, and discusses about the methods and the design of linear and non-linear quantum techniques. Sect. 3 focuses on the development and standardization of quantum algorithms, generalized circuit representations for mapping quantum algorithms, comparative analysis of quantum and classical algorithms, highlighting computational complexity and application-specific use cases. Finally, Sect. 4 concludes with a summary of contributions and insights into future research directions of work.

2 Hierarchical Framework Based Review

Dimensionality reduction addresses the curse of dimensionality in machine learning by transforming high-dimensional datasets into lower-dimensional representations while preserving essential features [7, 9]. This research presents a hierarchical framework (Fig. 2) that categorizes methods into linear and non-linear approaches, bridging classical algorithms to quantum-enhanced counterparts. Based on the quantum principles like superposition and entanglement, the framework integrates quantum subroutines, emphasizing scalability, circuit realization, dimensionality preservation, and application-specific use cases. It offers a roadmap for advancing quantum algorithms, showcasing exponential and polynomial speedups for high-dimensional data analysis.

2.1 Classical Dimensionality Reduction Methods

Classical methods have been integral in handling high-dimensional datasets across various domains. Principal Component Analysis (PCA) and Singular Value Decomposition (SVD) are foundational techniques in linear dimensionality reduction. PCA is widely applied in genomics for analyzing gene expression data [4, 5, 7], while SVD is critical in collaborative filtering for the recommendation systems. Also, Linear Discriminant Analysis (LDA) is used in supervised classification tasks, such as sentiment analysis & document categorization, where it maximizes class separability. Independent Component Analysis (ICA) finds applications in separating mixed signals in domains like EEG analysis, and Canonical Correlation Analysis (CCA) is used to analyze correlations between multiple data modalities, such as in audio-visual datasets. Non-linear methods extend the capabilities of dimensionality reduction to datasets with non-linear relationships. Kernel-PCA (K-PCA) is used in image segmentation and pattern recognition, while Locally Linear Embedding (LLE) is effective in preserving local neighborhood structures, applicable to handwriting analysis tasks [8–17].

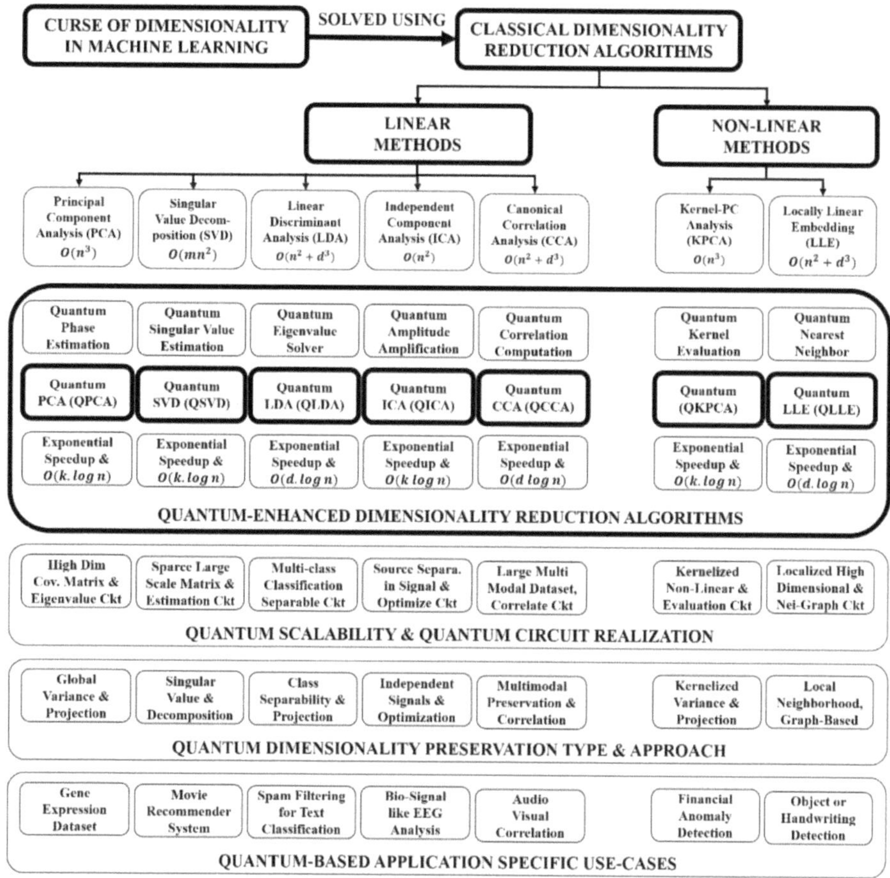

Fig. 2. Hierarchical Framework for Quantum-Enhanced Dimensionality Reduction: Mapping Classical and Quantum Methods with Algorithmic Complexities, Scalability, Preservation Approaches, and Domain-Specific Use Cases [1–23].

The classical dimensionality reduction methods are limited by their computational complexities, particularly in handling large-scale or high-dimensional datasets. Principal Component Analysis (PCA), a widely used method for linear dimensionality reduction, involves eigenvalue decomposition with a complexity of $O(n^3)$, where n is the number of features. Singular Value Decomposition (SVD), widely used for matrix factorization, incurs a computational cost of $O(mn^2)$ for dense matrices where m is number data samples, making it resource-intensive for high-dimensional data. Linear Discriminant Analysis (LDA), which maximizes class separability, has a complexity of $O(n^2 + d^3)$, where d is the no. of classes or dimensions to be reduced. This quadratic growth with d limits LDA applicability to datasets with a large number of classes or features. Independent Component Analysis (ICA) relies on iterative optimization algorithms with a computational complexity of $O(n^2)$. This quadratic dependency on the dimensionality makes ICA computationally expensive for high-dimensional data, such as in EEG signal analysis. Canonical Correlation Analysis (CCA) requires solving eigenvalue problems across multiple covariance matrices, leading to a complexity of $O(n^2 + d^3)$ where d is number of correlation dimensions, and this makes CCA challenging for multimodal datasets. Non-linear methods are even more computationally demanding. Kernel PCA, extends PCA to non-linear feature spaces, requires $O(n^2)$ memory to store the kernel matrix and $O(n^3)$ for its eigenvalue decomposition. Locally Linear Embedding (LLE), which preserves local neighborhood information, have $O(n^2 + d^3)$ complexity, where d is the number of neighbors, further limiting its use for large-scale applications. These complexities ensure limitations of classical methods, therefore quantum algorithms, to be discussed in next sub-section, promise to drastically reduce these complexities, enabling efficient handling of high-dimensional data while maintaining or improving accuracy [2, 4, 7, 9–18].

2.2 Quantum Dimensionality Reduction Methods

Quantum computing addresses the limitations of classical methods by introducing exponential speedups through the quantum-enhanced dimensionality reduction techniques. Quantum PCA (QPCA), based on the Quantum Phase Estimation (QPE), efficiently computes eigenvalues and eigenvectors, making it ideal for large-scale genomic studies. Similarly, Quantum SVD (QSVD) facilitates the matrix factorization in recommendation systems, enabling real-time predictions for the large user-item datasets. Quantum Linear Discriminant Analysis (QLDA) and Quantum Independent Component Analysis (QICA) extend linear methods to high-dimensional classification and signal separation tasks. Quantum Canonical Correlation Analysis (QCCA) is crucial for multimodal data analysis, such as correlating medical imaging data with genomic datasets. Non-linear methods, including Quantum Kernel PCA (QK-PCA), are instrumental in anomaly detection and fraud detection, where non-linear relationships dominate. Quantum Locally Linear Embedding (QLLE), retaining local neighborhood information, is effective in facial recognition and object detection tasks.

Quantum algorithms offer exponential reductions in computational complexity compared to classical methods, making them highly efficient for high-dimensional data analysis. In general, the reduced $O(log n)$ time be attained realizing the quantum operations; therefore, this factor must be included in complexities analysis of quantum algorithms. Quantum Principal Component Analysis (QPCA) utilizes the Quantum Phase Estimation (QPE) to compute eigenvalues and eigenvectors with a complexity of $O(k \log n)$ where k is the number of extracted principal components, significantly improving on the $O(n^3)$ complexity of classical PCA. Similarly, Quantum Singular Value Decomposition (QSVD), utilizing the Quantum Singular Value Estimation (QSVE) achieves $O(k \log n)$ compared to the $O(mn^2)$ cost in classical SVD where k is the number of singular vectors used for the singular value computations. Quantum Linear Discriminant Analysis (QLDA) reduces the complexity of solving generalized eigenvalue problems, from the classical $O(n^2 + d^3)$ to logarithmic time $O(d \log n)$ where d is the number of classes, by utilizing the quantum properties like amplitude amplification. Also, Quantum Independent Component Analysis (QICA) improves the complexity of iterative optimization algorithms, reducing the quadratic dependency from classical ICA $O(n^2)$ to $O(k \log n)$

where k is the number of independent components, facilitating scalable signal separation. Quantum Canonical Correlation Analysis (QCCA) benefits significantly using the Quantum Amplitude Amplification, reducing the computational complexity from the classical $O(n^2 + d^3)$ to $O(d \log n)$ where d is number of correlation dimensions. For the non-linear methods, Quantum Kernel PCA (QK-PCA) achieves efficient feature mapping by bypassing the $O(n^2)$ memory requirements and $O(n^3)$ computational costs of classical Kernel PCA, achieving $O(k \log n)$ complexity where k is the number of kernel dimensions. Finally, Quantum Locally Linear Embedding (QLLE) reduces the cost of preserving local neighborhood structures from classical $O(n^2 + d^3)$ to $O(d \log n)$ where d is number of nearest neighbors, making it scalable for large datasets. These quantum complexities are significant for quantum dimensionality reduction techniques in addressing bottlenecks faced by classical methods [2, 4, 7–23].

2.3 Quantum Subroutines for Dimensionality Reduction

The efficient design of quantum dimensionality reduction methods heavily relies on specialized quantum subroutines that exploit the unique capabilities of quantum computation. Quantum subroutines are essential as they utilize quantum mechanical properties like superposition, entanglement, and quantum interference to achieve remarkable efficiency and scalability in dimensionality reduction. These subroutines, defined below, enable the design of quantum algorithms that outperform classical counterparts by reducing computational complexities and ensuring precise data transformations.

Quantum Phase Estimation (QPE): This subroutine computes eigenvalues of unitary matrices, a critical step in dimensionality reduction methods such as Quantum PCA (QPCA) and Quantum Kernel PCA (QK-PCA). The process involves encoding the matrix as a quantum operator and using quantum Fourier transforms to extract eigenvalues with exponential precision. Now, QPE reduces the eigenvalue computation complexity from $O(n^3)$ to $O(k \log n)$ [4, 15].

Quantum Singular Value Estimation (QSVE): QSVE is a specialized technique for computing singular values of matrices, making it foundational for Quantum SVD (QSVD) and Quantum Canonical Correlation Analysis (QCCA). It combines QPE with matrix decompositions to achieve significant reductions in computational requirements. For example, QSVD handles large-scale matrix decomposition tasks efficiently by reducing classical complexity $O(mn^2)$ to $O(k \log n)$ [5, 11].

Quantum Eigenvalue Solver: This subroutine computes eigenvalues and eigenvectors of Hermitian matrices. It is integral to methods like Quantum Linear Discriminant Analysis (QLDA), where it ensures maximum class separability. The solver operates on Hamiltonian simulations and iterative amplitude estimation, achieving a complexity of $O(d \log n)$, suitable for multi-class classification tasks [6, 7].

Quantum Amplitude Amplification: This subroutine enhances the probability of success in quantum search and optimization algorithms. It is particularly valuable in Quantum ICA (QICA) for optimizing independent signal extraction and in QLDA for improving class separation. Amplitude amplification reduces the number of iterations required for optimization tasks, reducing a complexity from $O(n^2)$ to $O(\sqrt{n} \log n)$ providing a polynomial speedup [7, 9].

Quantum Correlation Computation: This subroutine calculates the correlations between two datasets, a crucial step in QCCA for multimodal data integration. It uses quantum state preparation and overlap measurements to compute correlation matrices with $O(d \log n)$ complexity, that is, reduces classical complexity of $O(n^2 + d^3)$ [12].

Quantum Kernel Evaluation: Quantum kernel methods transform data into higher-dimensional spaces to capture non-linear relationships. QK-PCA relies on this subroutine to compute kernel matrices efficiently, reducing the classical $O(n^3)$ complexity to $O(k \log n)$ for kernel feature mapping [8, 13].

Quantum Nearest Neighbor Search: This subroutine optimizes neighborhood graph construction, essential for Quantum Locally Linear Embedding (QLLE). By utilizing quantum search algorithms, it reduces complexity from $O(n^2 + d^3)$ to $O(d \log n)$, enabling preservation of local neighborhood structures in high-dimensional data [14].

These subroutines not only reduce computational burdens but also facilitate scalable quantum circuit designs. For example, QPE and QSVE are implemented using quantum Fourier transforms and controlled-unitary operations, while the Quantum Kernel Evaluation uses quantum matrix exponentiation techniques. This integration ensures that the proposed quantum dimensionality reduction methods are both theoretically sound and practically implementable using modern quantum simulation tools like Qiskit. By incorporating these subroutines into the hierarchical framework, the proposed methods achieve scalable, efficient, and precise dimensionality reduction, making them applicable to a wide range of machine learning tasks and applications to tackle the curse of dimensionality effectively [4–8, 11–13, 15–18, 21–23].

3 Quantum Dimensionality Reduction Algorithms

This main section includes the design and development of Quantum Dimensionality Reduction Algorithms, focusing on their transformative potential in addressing the challenges posed by high-dimensional datasets. As to maintain the uniformity in consecutive sub-sections, this paper initially standardizes each of the existing quantum algorithms of Fig. 2 using the general steps. Algorithm designs consider existence of the Quantum Memory (qRAM) for encoding classical data into quantum state efficiently. Next, generalized quantum circuits design is implemented using Qiskit due to paper length restriction, however, this generalized circuit gives details about the design of quantum circuits flow for quantum dimensionality reduction algorithms, and it provides clarity on the execution pipeline of the proposed framework for the quantum implementations of linear and non-linear dimensionality reduction algorithms. Then, subsequent section includes the comparative complexities analysis with the quantum resources estimation, as such analysis is required for the different possible designs of quantum algorithms and their equivalent circuits. Last sub-section discusses some specific use-cases of quantum algorithms as in accordance with the Fig. 2.

3.1 Standardization of Quantum Dimensionality Reduction Algorithms

---

Algorithm 1: Quantum Principal Component Analysis (QPCA)

Objective: Reduce data dimensionality by extracting principal components.

1.	Load covariance matrix C into quantum memory using Quantum RAM (qRAM)
2.	Use quantum phase estimation to find eigenvalues λ_i and eigenvectors $|v_i\rangle$ of C
3.	Construct diagonal matrix Λ with eigenvalues λ_i and apply it to eigenvectors $|v_i\rangle$
4.	Apply the controlled unitary rotations to amplify the significant eigenvalues λ_i
5.	Perform Inverse QFT (IQFT) to retrieve the reduced-dimensional quantum state
6.	Measure the reduced quantum state for obtaining the classical representation

Speedup: Exponential reduction compared to classical $O(n^3)$

End of QPCA Algorithm

The QPCA algorithm initializes qubits with the covariance matrix using qRAM, represented in the circuit as quantum state encoding $|\psi\rangle$. Hadamard gates create superposition, enabling controlled phase rotations for eigenvalue amplification. Inverse Quantum Fourier Transform (IQFT) extracts the principal components, while measurement maps quantum eigenstates to classical reduced-dimensional data. This process efficiently realizes QPCA through quantum circuit implementation.

Algorithm 2: Quantum Singular Value Decomposition (QSVD)

Objective: Decompose a matrix into singular vectors and singular values.

1.	Represent matrix A as a quantum state $|A\rangle$ using Quantum RAM (qRAM)
2.	Apply QPE to compute singular values σ_i and corresponding vectors $|u_i\rangle$, $|v_i\rangle$
3.	Construct the diagonal matrix operator Σ in superposition of states using σ_i
4.	Construct the singular value decomposition $A = U\Sigma V^\dagger$ using $|u_i\rangle$, $|v_i\rangle$, and σ_i
5.	Project the data onto the reduced singular values and corresponding vector space
6.	Measure the result as reduced quantum states for the classical analysis

Speedup: Exponential reduction compared to classical $O(mn^2)$

End of QSVD Algorithm

The QSVD quantum circuit represents matrix decomposition into singular values and vectors. qRAM encodes the matrix, Hadamard gates create superpositions, controlled rotations amplify eigenvalues, and IQFT extracts singular values. Measurements provide classical outputs of the quantum singular value decomposition.

Algorithm 3: Quantum Linear Discriminant Analysis (QLDA)

Objective: Maximize class separability in reduced-dimensional space.

1.	Load S_B (between-class) and S_W (within-class) scatter matrices into qRAM
2.	Compute $S_W^- S_B$ using Quantum Value Solver (QES) using QPE method
3.	Identify significant eigenvalues λ_i & eigenvectors $|v_i\rangle$ for maximal separability
4.	Project data onto the eigenvector space $|v_i\rangle$ to achieve reduced dimensions
5.	Measure the quantum state for obtaining reduced classical classification tasks

Speedup: Exponential reduction compared to classical $O(n^2 + d^3)$

End of QLDA Algorithm

The QLDA employs qRAM for encoding covariance matrices and QPE for extracting discriminant directions. Controlled rotations and IQFT ensure efficient class separability with exponential speedup, enabling scalable multi-class classification tasks.

Algorithm 4: Quantum Independent Component Analysis (QICA)

Objective: Separate independent sources from mixed signals.

1.	Load mixed signals X into quantum memory via Quantum Memory (qRAM)
2.	Use QPE to compute eigenvalues and eigenvectors for independent components
3.	Apply Quantum Amplitude Amplification for Independent Sources Separation
4.	Transform the data into independent components for dimensionality reduction
5.	Measure the quantum state for the classical output of independent signals

Speedup: Exponential reduction compared to classical $O(n^2)$

End of QICA Algorithm

The QICA algorithm uses qRAM and amplitude amplification to separate mixed signals. QPE optimizes signal independence, while quantum circuits reconstruct components efficiently, achieving mainly complexity reduction in signal separation tasks.

Algorithm 5: Quantum Canonical Correlation Analysis (QCCA)

Objective: Identify relationships between two datasets.

1.	Represent datasets X and Y as quantum states $|X\rangle$ and $|Y\rangle$ using the qRAM
2.	Use quantum phase estimation to compute covariance matrices C_{XX}, C_{YY}, C_{XY}
3.	Use QPE to calculate canonical correlations ρ_i and their corresponding vectors
4.	Apply Quantum Amplitude Amplification for finding significant correlations
5.	Combine ρ_i with canonical vectors and measure them for classical analysis

Speedup: Exponential reduction compared to $O(n^2 + d^3)$

End of QCCA Algorithm

The QCCA algorithm encodes covariance matrices in quantum states and computes correlations via QPE and controlled rotations. Amplitude amplification & IQFT ensure efficient correlation analysis with exponential speedup for multimodal datasets.

Algorithm 6: Quantum Kernel Principal Component Analysis (QKPCA)

Objective: Extract nonlinear features using kernel-based mapping.

1.	Represent the kernel matrix K in quantum memory using Quantum Memory
2.	Use QPE to compute eigenvalues λ_i and eigenvectors $|v_i\rangle$ of kernel matrix K
3.	Transform data into kernel space in quantum superposition for kernel mapping
4.	Apply controlled rotations for amplification of significant principal components
5.	Retrieve the reduced-dimensional quantum state for the feature set extraction
6.	Measure the quantum state for classical representation of nonlinear features

Speedup: Exponential reduction compared to classical $O(n^3)$

End of QKPCA Algorithm

The QKPCA algorithm utilizes quantum kernel evaluation and QPE to extract nonlinear principal components. Controlled rotations amplify key features, enabling scalable dimensionality reduction with exponential efficiency.

Algorithm 7: Quantum Locally Linear Embedding (QLLE)

Objective: Compute local neighborhood relationships for embedding.

1.	Construct the nearest neighborhood graph G as a quantum adjacency matrix
2.	Compute weight matrix W by quantum phase estimation for local relationships
3.	Apply quantum SVD to compute eigenvalues and eigenvectors of $I - W$ matrix
4.	Choose the smallest k eigenvalues & corresponding eigenvectors for embedding
5.	Use quantum amplitude amplification to optimize the weights and embeddings
6.	Measure the reduced-dimensional embedding for the classical interpretation

Speedup: Exponential reduction compared to $O(n^2 + d^3)$

End of QLLE Algorithm

The QLLE algorithm encodes adjacency graphs for local preservation using QSVD and amplitude amplification. IQFT retrieves reduced embeddings, achieving efficient manifold learning with exponential complexity reduction.

3.2 Generalized Quantum Circuit for Mapping Quantum Algorithms

A generalized quantum circuit of Fig. 3 represents a unified framework for quantum dimensionality reduction. It begins with data encoding via qRAM, applies controlled unitary operations for feature extraction (e.g., QPE), and uses the Inverse Quantum

Fourier Transform (IQFT) to prepare the state for measurement. This adaptable design supports both linear and non-linear algorithms, ensuring efficient processing of high-dimensional datasets for diverse applications. All the quantum dimensionality reduction algorithms can be implemented, based on the generalized circuit, and simulated using the **Qiskit** framework. However, to execute a circuit of Fig. 3, Qiskit Version 1.3.1 installed on host computer with Intel(R) Core(TM) i5-9300HF running at CPU @ 2.40GHz and 8 GB RAM. The experiments were conducted on the **AerSimulator** backend, which simulates ideal quantum hardware without noise.

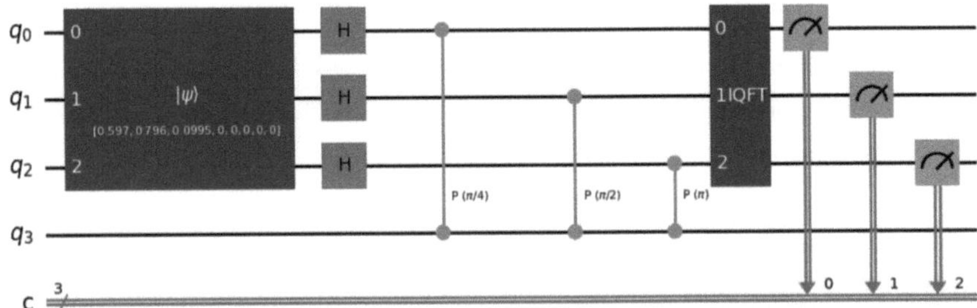

Fig. 3. Qiskit Simulated Generalized Quantum Circuit for Dimensionality Reduction: A framework presents a detailed quantum circuit flow for quantum dimensionality reduction algorithms, highlighting the sequence of matrix encoding, controlled unitary operations, Inverse Quantum Fourier Transform and measurement steps. This detailed circuit view provides clarity on execution pipeline of the proposed framework for the quantum implementations of linear & non-linear dimensionality reduction algorithms.

The framework starts with the qRAM-based matrix encoding, efficiently transforming classical data into quantum states. It adapts to diverse representations like covariance matrices (QPCA), singular value matrices (QSVD), and kernel matrices (QK-PCA), ensuring compatibility across linear and non-linear dimensionality reduction methods.

Controlled unitary operations are used for each algorithm. QPE handles eigenvalue extraction for linear methods (QPCA, QSVD), while Quantum Amplitude Amplification enhances feature separation (QICA, QLDA). This modularity ensures specific algorithmic needs are effectively met. IQFT converts quantum states from the Fourier to computational basis for the direct eigenvalue or singular value measurement. This standardized step, followed by measurement, ensures accurate classical data output for all algorithms, linear or non-linear.

The framework's modular design ensures scalability for large datasets while maintaining resource efficiency. Its unified structure, integrating encoding, controlled operations, inverse QFT, and measurement, provides a consistent and scalable approach to quantum dimensionality reduction algorithms.

3.3 Comparative Analysis of Algorithms & Circuits with Resource Estimation

The complexities of classical and quantum algorithms are discussed already in the previous Sect. 2. However, referring to the Table 1, this section includes comparative analysis among the presented algorithms and in reference to a generalized quantum circuit of Fig. 3, Table 2 includes resource estimation of quantum algorithms. In quantum computing, resource estimation refers to the following key factors: (1) Number of Qubits for total number of qubits required to implement the algorithm; (2) Circuit Depth for total number of gates (including controlled gates) used in the quantum circuit; (3) Gate Complexity for number and type of quantum gates used (e.g., Hadamard, CNOT, QPE); (4) Measurement Overhead for number of shots or repetitions required for getting reliable output. Now, further ahead to this paragraph, subsequent paragraphs are written for the summarized facts noted about the algorithms.

QPCA addresses the problem of covariance matrix decomposition, which classically scales as $O(n^3)$. A QPE reduces this to $O(k \log n)$. The quantum circuit includes Hadamard gates for superposition, controlled rotations for eigenvalue extraction, and an Inverse QFT for measurement. The resource estimation indicates that QPCA requires $n + 1$ qubits, where n is the number of input features, achieving a circuit depth of $O(n^2)$ due to the matrix multiplication involved in QPE and controlled rotations. The measurement overhead is $O(logn)$ which reflects logarithmic reduction in classical cubic complexity. This results in exponential speedup over PCA using QPE.

QSVD solves eigenvalue problems for singular vectors, which classically scale as $O(mn^2)$, and Quantum Singular Value Estimation (QSVE) reduces it to $O(k \log n)$. The circuit requires $m + n$ qubits where m for row encoding and n for column encoding, and it uses QSV estimation and controlled rotations. The circuit depth is $O(n^2)$ due to controlled operations & inverse QFT, with measurement overhead of $O(logn)$ It ensures exponential complexity reduction in handling singular value computation.

QLDA reduces the classical complexity of generalized eigenvalue problems from $O(n^2 + d^3)$ to $O(d \log n)$ with Quantum Eigenvalue Solver (QES). The circuit needs $n + d$ qubits where n for input dimensions and d for class projections. Circuit depth is $O(d^3)$ due to solving covariance matrix problems. QES extracts eigenvalues efficiently, and controlled operations refine the state. Measurement overhead is $O(logd)$, enabling improved classification in high-dimensional datasets.

QICA reduces classical complexity from $O(n^2)$ to $O(k \log n)$ using the Quantum Amplitude Amplification (QAA). The circuit requires $n + k$ qubits where n for input features and k for component estimation. Circuit depth is $O(n^2)$ due to state preparation and amplitude amplification. QAA enhances feature separation, achieving logarithmic complexity scaling. Measurement overhead is $O(logn)$, ensuring exponential speedup in feature extraction.

QCCA reduces classical complexity from $O(n^2 + d^3)$ to $O(d \log n)$ using Quantum Correlation Computation (QCC). The circuit requires $n + d$ qubits where n for input matrices and d for correlation state projection. Circuit depth is $O(d^3)$ due to the correlation extraction and state refinement. QCC extracts canonical correlations efficiently, reducing complexity exponentially. Measurement overhead is $O(logd)$, and this is ensuring improved scalability in correlation analysis.

QKPCA reduces classical complexity from $O(n^3)$ to $O(k \log n)$ using QPE and Kernel Evaluation (KE). The circuit requires $n + k$ qubits where n for input dimensions and k for kernel state projection. Circuit depth is $O(n^2)$ due to kernel matrix encoding and QPE-based refinement. KE enhances feature separation in kernel spaces, reducing

Table 1. Comparative Classical-Quantum Complexities Analysis for Algorithms

Quantum Algorithms	Classical Complexity	Quantum Complexity	Contextual Justification and Achieved Speedup
QPCA	$O(n^3)$	$O(k\ log\ n)$ $k=$ No. of the principal components	PCA involves covariance matrix decomposition, which is classically cubic in matrix size. QPE reduces this to logarithmic scaling based on no. of principal components. *(Exponential Speedup using QPE)*
QSVD	$O(mn^2)$	$O(k\ log\ n)$ $k=$ No. of the singular vectors	SVD solves eigenvalue problems for singular vectors, classically quadratic in matrix size. QSVE reduces complexity to a logarithmic scale using no. of singular values. *(Exponential Speedup using QSVE)*
QLDA	$O(n^2 + d^3)$	$O(d\ log\ n)$ $d=$ No. of classes	LDA requires solving a generalized eigenvalue problem, classically involving cubic complexity for within-class and between-class covariance matrices. Quantum Eigenvalue Solver reduces this to the logarithmic scale with respect to the number of classes. *(Exponential Speedup using QES)*
QICA	$O(n^2)$	$O(k\ log\ n)$ $k=$ No. of independent components	ICA estimates the independent components based on second-order statistics. Quantum Amplitude Amplification reduces classical complexity to logarithmic scaling in terms of independent components. *(Exponential Speedup using QAA)*
QCCA	$O(n^2 + d^3)$	$O(d\ log\ n)$ $d=$ Number of correlation dimensions	CCA solves a pair of eigenvalue problems for correlation matrices. Quantum Correlation Computation reduces this to logarithmic factor in correlation dimensions terms. *(Exponential Speedup using QCC)*
QKPCA	$O(n^3)$	$O(k\ log\ n)$ $k=$ Number of kernel dimensions	KPCA computes the kernel matrix and its eigenvalues, which scales cubically with a sample size. The QPE + Kernel Evaluation reduces this to logarithmic scaling based on the kernel dimensions. *(Exponential Speedup using QPE & KE)*
QLLE	$O(n^2 + d^3)$	$O(d\ log\ n)$ $d=$ Number of nearest neighbors	LLE computes the neighborhood graph and its Laplacian matrix, classically involving cubic complexity. Quantum Nearest Neighbor Search reduces this to logarithmic scaling based on nearest neighbors. *(Exponential Speedup using QNN)*

Table 2. Resource Estimation Analysis for Quantum Algorithms

Quantum Algorithms	No. of Qubits	Circuit Depth	Used Quantum Gates/Circuits	Measurement Overhead
QPCA	$n + 1$	$O(n^2)$	Hadamard, QPE, Controlled Rotation	$O(logn)$
QSVD	$m + n$	$O(n^2)$	QSV Estimation, Controlled Rotation	$O(logn)$
QLDA	$n + d$	$O(d^3)$	Quantum Eigenvalue Solver	$O(logd)$
QICA	$n + k$	$O(n^2)$	Quantum Amplitude Amplification	$O(logn)$
QCCA	$n + d$	$O(d^3)$	Quantum Correlation Computation	$O(logd)$
QKPCA	$n + m$	$O(n^2)$	QPE, Kernel Evaluation	$O(logn)$
QLLE	$n + d$	$O(d^3)$	Quantum Nearest Neighbor Search	$O(logd)$

cubic complexity. Measurement overhead is $O(logn)$, ensuring scalable kernel-based processing.

QLLE reduces classical complexity from $O(n^2 + d^3)$ to $O(d \log n)$ using Quantum Nearest Neighbor Search (QNN). The circuit requires $n + d$ qubits where n for input dimensions and d for nearest neighbor projection. Circuit depth is $O(d^3)$ due to nearest neighbor state preparation and projection. QNN reduces pairwise computation complexity, improving graph-based scalability. Measurement overhead is $O(logn)$, ensuring efficient handling of complex neighbor structures.

This section concludes with the facts that the classical equivalent quantum presented algorithms achieve at least an exponential speedup, and these algorithms can be declared as effective based on the estimated quantum computing resources.

3.4 Specific Use-Cases of Dimensionality Reduction Algorithms

Based on vertical of quantum dimensionality reduction, Fig. 2, suitability and significance of algorithms for specific use-cases are evident here to address complex high-dimensional challenges effectively. Linear algorithms such as QPCA and QSVD are well-suited for global variance analysis in gene expression datasets and singular value decomposition for movie recommender systems. Algorithms like QLDA and QICA validate their value in multi-class classification tasks, including spam filtering and independent signal optimization for bio-signal analysis like EEG. Non-linear methods such as QKPCA and QLLE excel in capturing intricate relationships, making them ideal for the financial anomaly detection, localized object, or handwriting recognition tasks. These quantum algorithms streamline dimensionality reduction, ensuring scalability, precision, and computational efficiency in application-specific domains.

4 Concluding Remarks

This research addresses the persistent challenge of the curse of dimensionality in machine learning by proposing quantum-enhanced solutions for dimensionality reduction. Through a hierarchical framework, the classical algorithms were mapped to their quantum equivalents, by utilizing quantum subroutines such as quantum the phase estimation, quantum singular value estimation, and quantum amplitude amplification. These standardized quantum algorithms demonstrated exponential speedups, reducing complexities from classical to quantum-efficient. Equivalent quantum circuits were realized using qiskit simulation based generalized frameworks, ensuring scalable and efficient implementations. The study about a comprehensive framework bridges theoretical advancements and practical realization, aligning the quantum algorithms with application-specific use cases across bioinformatics, finance, and image processing. In addition, the comparative analysis with resource estimation justifies that quantum algorithms are achieving exponential speedups. This work establishes a foundational resource for integrating quantum-based dimensionality reduction into next-generation machine learning systems. Future research can explore optimized quantum circuit designs and real-world deployment on quantum hardware to further advance the field.

References

1. Nielsen, M.A., Chuang, I.L.: Quantum Computation and Quantum Information, 10th anniversary ed. Cambridge University Press (2010)
2. Wittek, P.: Quantum Machine Learning—What Quantum Computing Means to Data Mining, 1st edn. Elsevier Publication (2014)
3. Lloyd, S., Mohseni, M.: Quantum principal component analysis. Nat. Phys. Lett. **10**, 631–633 (2014)
4. Zhang, Y., Ni, Q.: Recent advances in quantum machine learning. Quantum Eng. **2**, e34 (2020)
5. Lin, Z., Liu, H., et al.: Hardware-efficient quantum principal component analysis for medical image recognition. Front. Phys. **19**(5), 51202 (2024)
6. Ozpolat, Z., Karabatak, M.: Performance evaluation of quantum-based machine learning algorithms for cardiac arrhythmia classification. Diagnostics. **13**, 1099 (2023)
7. Mancilla, J., Pere, C.: A preprocessing perspective for quantum machine learning. Entropy. **24**, 1656 (2022)
8. Cong, I., Duan, L.: Quantum discriminant analysis for dimensionality reduction & classification. New J. Phys. **18**, 073011 (2016)
9. Xiao-Fan, X., et al.: An efficient quantum algorithm for independent component analysis. New J. Phys. **26**, 1–18 (2024)
10. Lin, J., Buo, W.-S., Zhang, S., Li, T., Wang, X.: An improved quantum principal component analysis algorithm. Phys. Lett. A. **383**, 2862–2868 (2019)
11. He, C., et al.: A low-complexity quantum principal component analysis algorithm. IEEE Trans. Quantum Eng. **3**, 3100713 (2022)
12. Song, C.-D., et al.: Quantum canonical correlation analysis algorithm. Laser Phys. Lett. **20**, 105203 (2023)
13. Vasques, X., Paik, H., Cif, L.: Application of quantum machine learning using quantum kernel algorithms on multiclass neuron M-type classification. Sci. Rep. **13**, 11541 (2023)
14. He, X., Sun, L., Lyu, C., Wang, X.: Quantum locally linear embedding for nonlinear dimensionality reduction. Quantum Inf. Process. **19**, 309 (2020)

15. Soni, B., Khare, N.: Advancement of quantum methods in principal component analysis. In: 12th IEEE International Conference on Computing, Communication, and Networking Technologies, pp. 1–7. IIT Kharagpur, India (2021)
16. Daskin, A.: Obtaining a linear combination of the principal components of a matrix on quantum computers. Quantum Inf. Process. **15**, 4013–4027 (2016)
17. Soni, B., Khare, N.: Quantum principal component analysis based on the dynamic selection of eigenstates. Europhys. Lett. **139**(4), 48002 (2022)
18. Suzuki, Y., et al.: Amplitude estimation without phase estimation. Quantum Inf. Process. **19**(2), 1–20 (2020)
19. Maronese, M.: Quantum amplitude estimation algorithms on near-term devices. Quantum Rep. **6**, 1–13 (2024)
20. Grinko, D., et al.: Iterative quantum amplitude estimation. npj Quantum Inf. **7**, 52 (2021)
21. Maheshwari, D., et al.: Quantum machine learning applications in the biomedical domain: a systematic review. IEEE Access. **10**, 80463 (2022)
22. Li, H.-S., et al.: Quantum implementation circuits of quantum signal representation and type conversion. IEEE Trans. Circuits Syst.–I. **1**, 341–354 (2019)
23. Wei, L., et al.: Quantum machine learning in medical image analysis: a survey. Neurocomputing. **525**, 42–53 (2023)

Intelligent Networking and Communication Systems

Optimizing Energy Efficiency Through Power Factor Monitoring: A Smart Approach with IoT Sensors

Krishnakant Singh[1]([✉])(iD), Manju Pandey[1], and Sunil Pandey[2]

[1] Department of Computer Applications, National Institute of Technology Raipur, Raipur, India
{ksingh.phd2024.ca,mpandey.mca}@nitrr.ac.in
[2] Central Computer Center, NIT Raipur, Raipur, India
sys_admin@nitrr.ac.in

Abstract. Power Factor (PF) plays a critical role in determining the efficiency of electrical systems, directly impacting energy losses, system stability, and operational costs. Poor PF conditions lead to increased reactive power, reduced efficiency, and potential penalties in industrial and commercial setups. Traditional PF correction techniques, such as fixed capacitor banks, suffer from inefficiencies due to their inability to dynamically adjust to load variations. This paper presents a smart, IoT-driven PF monitoring and correction system that integrates real-time sensor data acquisition with adaptive capacitor switching to optimize energy efficiency. The system utilizes ESP32 microcontrollers, SCT013 current sensors, and ZMPT101B voltage sensors to provide continuous monitoring, while relay-controlled capacitor banks ensure dynamic correction. Through an experimental setup, the proposed method demonstrates a notable improvement in PF to 0.98, significantly reducing energy losses and improving system reliability. Compared to traditional correction methods, the system achieves faster response times and greater adaptability to fluctuating load conditions, making it suitable for residential, commercial, and industrial applications. This research further explores the mathematical modeling, hard-ware architecture, and experimental validation of the proposed system while benchmarking its efficiency against existing PF correction techniques. The findings underscore the importance of IoT and automated control strategies in modern smart grid and energy management solutions.

Keywords: Power Factor · Internet of Things · Energy Monitoring · Energy Efficiency

1 Introduction

Energy efficiency is essential in modern power systems, where PF significantly impacts power quality and optimization. Defined as the ratio of real power (P)

P. Chandrakar et al. (Eds.): ICCINS 2025, CCIS 2738, pp. 205–217, 2026.
https://doi.org/10.1007/978-3-032-09572-5_17

to apparent power (S), PF reflects how efficiently electrical power is used. A PF close to 1.0 is ideal, while lower values indicate excessive Reactive Power (Q), leading to higher losses, increased costs, and inefficiencies. Industrial facilities with inductive loads—motors, compressors, transformers—often experience PF deterioration, requiring correction for improved stability and reduced energy waste [1].

Traditional PF correction methods, like fixed capacitor banks and manual switching, lack adaptability, often causing overcorrection or undercorrection. IoT-driven solutions offer real-time monitoring, automated control, and adaptive capacitor switching, ensuring optimal PF maintenance. Recent advancements in IoT-based power management systems demonstrate that integrating microcontroller-based correction mechanisms enhances PF stability and minimizes energy loss [1]. Studies have also explored adaptive capacitor switching using relay-controlled compensation units to address fluctuating loads in industrial and commercial settings [2].

This study introduces a novel IoT-enabled PF monitoring and correction system, overcoming conventional Power Factor Correction (PFC) limitations with real-time sensor feedback and adaptive control strategies. It integrates SCT013 current sensors, ZMPT101B voltage sensors, and an ESP32 microcontroller, which continuously analyzes PF deviations and activates capacitor banks via relays. Using threshold-based correction, the system ensures stable, efficient power compensation, eliminating static capacitor bank drawbacks. A 24-hour monitoring study as shown in Fig. 1, tracked real-time PF fluctuations under varying loads. Results show significant PF degradation during high inductive load usage, necessitating real-time compensation. The proposed system dynamically engages the required capacitance, optimizing power usage and minimizing energy losses. Compared to conventional static compensation, the proposed

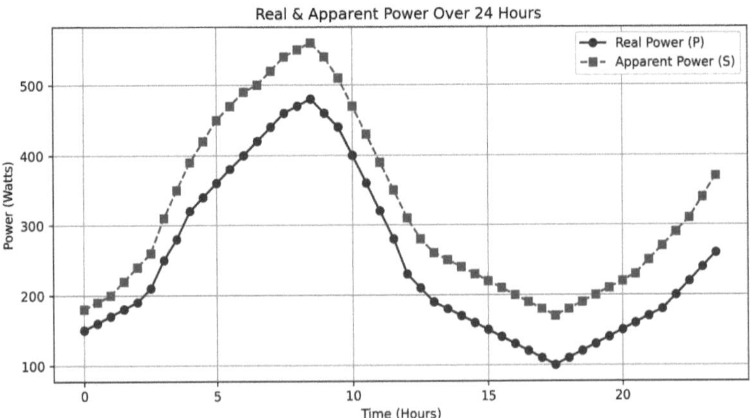

Fig. 1. Graph of Real & Apparent Power Over 24 h (Illustrating PF fluctuations and the necessity of real-time correction).

system achieves real-time correction, preventing excessive power wastage and improving overall PF efficiency.

The paper is structured as follows: Sect. 2 reviews existing PFC techniques, their strengths, and limitations. Section 3 details the proposed system, including mathematical models, hardware integration, and dynamic capacitor switching. Section 4 presents experimental results and performance comparisons. Section 5 summarizes key findings, applications, and future research directions.

2 Literature Review

PFC has been widely researched due to its critical role in enhancing energy efficiency and reducing reactive power (Q) losses in electrical systems. Traditionally, power factor (PF) improvement has been achieved through fixed capacitor banks, switched capacitor controllers, and manual compensation strategies. However, these techniques although effective under steady-state load conditions fail to accommodate dynamic load variations, often resulting in either over-compensation or undercompensation of Q losses [1]. This limitation necessitates real-time, adaptive PF correction systems capable of responding to fluctuating power demands while maintaining optimal compensation efficiency.

The integration of IoT in PF monitoring has significantly improved the accuracy and responsiveness of correction mechanisms. In [3] the authors demonstrated that real-time voltage and current sensing using IoT-enabled controllers can dynamically regulate capacitor switching, ensuring improved PF stabilization. Similarly, [4] highlighted that, unlike traditional correction systems that operate on predefined compensation thresholds, IoT-based solutions utilize continuous monitoring and automated decision-making algorithms to enhance correction efficiency. Furthermore, studies in [5] and [6] emphasized that relay-controlled capacitor switching presents a practical alternative to conventional correction techniques, reducing response time while ensuring optimal capacitor engagement.

Advancements in embedded system based PF correction have also introduced microcontroller-controlled compensation strategies that employ sensors to continuously analyze phase deviations and trigger corrective actions in real time. Comparative analyses in [7] and [8] demonstrated that using ESP32 controllers significantly enhances both the speed and precision of Q compensation. Additionally, the integration of SCT013 and ZMPT101B sensors enables real-time tracking of current and voltage phase shifts, facilitating instantaneous capacitor switching to maintain PF levels within the optimal range [9]

The economic implications of PF correction have been extensively explored, with studies indicating that inefficient compensation leads to higher electricity costs and increased system losses. In [6] it was noted that traditional capacitor bank solutions often suffer from high operational costs due to excessive capacitor wear and tear, necessitating frequent maintenance. This argument is further supported by [3] and [10], which contend that adaptive, automated correction systems using IoT-based monitoring frameworks can significantly reduce power wastage while minimizing long-term maintenance costs.

While machine learning (ML) based PF correction has been investigated as a potential alternative for predictive compensation, its real-time implementation in industrial environments remains challenging. Studies in [11] and [12] argued that ML-driven correction mechanisms demand high computational power, extensive training datasets, and frequent model updates factors that render them impractical for low-cost, real-time PF correction applications. Instead, [13] suggested that relay-controlled capacitor switching with threshold-based compensation logic offers a more practical, scalable alternative for industrial and commercial power management.

This research builds upon prior IoT-enabled PF correction studies by developing a dynamic, real-time capacitor switching mechanism that ensures an instantaneous response to fluctuating load conditions. Unlike conventional static capacitor banks which suffer from delayed response times the proposed relay-controlled correction system continuously adjusts capacitor engagement based on live PF readings. By eliminating ML dependencies, the system maintains low computational overhead, making it cost-effective, scalable, and adaptable to diverse electrical environments.

3   System Design and Methodology

The proposed real-time PFC system dynamically adjusts reactive power (Q) compensation to optimize energy efficiency under varying load conditions. Maintaining a high power factor (PF) is essential for achieving optimal energy performance, particularly in environments with dynamic load variations. Studies have shown that inefficient PF correction leads to increased Q losses, which adversely affect power quality and system stability [14]. Similarly, research on static compensation mechanisms reveals that their delayed response times contribute to unnecessary energy losses, rendering them unsuitable for applications with rapidly fluctuating load conditions [2].

To address these challenges, the proposed system adopts a real-time correction strategy that continuously monitors PF deviations and automatically adjusts compensation levels. Unlike traditional capacitor banks that engage fixed capacitance values based on predefined assumptions, this system dynamically adapts to changing conditions, optimizing power usage and minimizing energy wastage. It leverages an IoT-enabled approach integrating real-time monitoring, automated capacitor switching, and adaptive decision-making. The ESP32 microcontroller continuously analyzes voltage and current data from SCT013 and ZMPT101B sensors, as shown in Fig. 2. A capacitor bank (10 µF, 20 µF, 40 µF) provides compensation, with relay-controlled switching ensuring precise engagement. An LCD module displays system status and PF values. By responding to instantaneous PF readings, the system applies compensation as needed, avoiding overcorrection or undercorrection while maintaining low computational overhead.

Capacitor selection plays a crucial role in ensuring efficient PFC. Table 1 outlines the capacitance values required for different load conditions, optimizing Q

Fig. 2. Hardware Setup for PF Monitoring.

Table 1. Capacitor Selection & Required Capacitance for PF Correction

Load Type	PF Before Correction	Target PF	Required Compensation (VAR)	Capacitor Value (μF)	Capacitors Engaged
Light Load	0.93	0.98	250	10 μF	1 × 10 μF
Medium Load	0.87	0.95	670	20 μF	1 × 20 μF
Heavy Load	0.82	0.90	1280	40 μF	1 × 40 μF
Mixed Load	0.85	0.95	850	40 μF + 10 μF	1 × 40 μF + 1 × 10 μF

compensation. For light loads (initial PF = 0.93), a 10 μF capacitor provides 250 VAR, achieving a PF of 0.98. In medium-load conditions (initial PF = 0.87), a 20 μF capacitor delivers 670 VAR, improving PF to 0.95. Heavy loads (initial PF = 0.82) require 1280 VAR, corrected using a 40 μF capacitor to raise PF to 0.90. For mixed loads (initial PF = 0.85), a combination of 40 μF and 10 μF capacitors supplies 850 VAR, achieving a final PF of 0.95. This structured approach ensures that only the necessary capacitance is engaged, preventing both undercompensation and overcompensation, while maximizing energy efficiency. The workflow of the correction mechanism, illustrated in Fig. 3, follows a structured decision-making process that allows for dynamic capacitor switching. The system continuously monitors PF in real time, analyzing voltage and current waveforms. If the PF drops below 0.95, additional capacitors are engaged to improve efficiency. When the PF exceeds 0.98, capacitors are disengaged to prevent overcorrection. If the PF remains within the optimal range of 0.95 to 0.98, no changes are made to the capacitor configuration, thereby maintaining system stability.

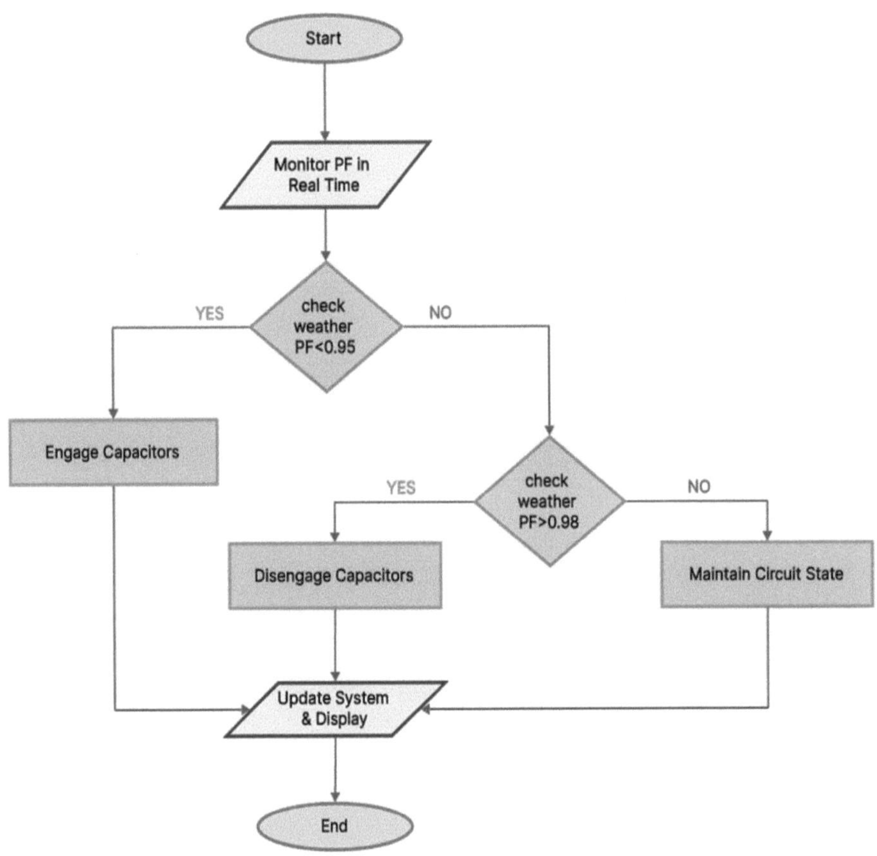

Fig. 3. System Workflow Diagram for PFC.

PF is mathematically expressed as the ratio of real power (P) to apparent power (S), given by:

$$PF = \cos(\theta) = \frac{P}{S} = \frac{VI \, \cos(\theta)}{VI} \tag{1}$$

where P represents real power and S is apparent power. To determine the compensation required, the system computes the reactive power Q_c as:

$$Q_c = P\left(\tan\left(\theta_{\text{before}}\right) - \tan\left(\theta_{\text{after}}\right)\right) \tag{2}$$

where θ_{before} and θ_{after} are the phase angles before and after correction. The required capacitive reactance X_c is given by:

$$X_c = \frac{1}{2\pi f C} \tag{3}$$

where f is the supply frequency (50 Hz) and C is the required capacitance. The optimal capacitance for correction is computed as:

$$C_{\text{required}} = \frac{Q_c}{2\pi f V^2} \tag{4}$$

By utilizing these equations, the system precisely determines the amount of capacitance needed to compensate for Q fluctuations, ensuring efficient PF correction and this is implemented in the system as Algorithm 1 which is the Dynamic PF Correction.

Algorithm 1. Dynamic PF Correction

Require: PF_threshold $\leftarrow$ 0.98, PF_overcomp $\leftarrow$ 1.0, $t \leftarrow$ 1s
Require: C_available = $\{10\mu F, 20\mu F, 40\mu F\}$
Ensure: Automatic capacitor switching for correction
1: **while** system is active **do**
2:     Read $PF_{current}$
3:     **if** $PF_{current} < PF_{threshold}$ **then**
4:         Compute $Q_c \leftarrow f(P, \theta_{before}, \theta_{after})$
5:         Compute $C_{required} \leftarrow f(Q_c, V, f)$
6:         Select $C_{engaged} \in$ C_available where $C_{engaged} \geq C_{required}$
7:         Activate relay for $C_{engaged}$
8:     **else**
9:         Maintain $C_{engaged}$
10:     **end if**
11:     Wait t seconds
12:     Read updated $PF_{current}$
13:     **if** $PF_{current} > PF_{overcomp}$ **then**
14:         Compute $C_{adjust} = C_{engaged} - C_{last}$
15:         Deactivate relay for C_{last}
16:     **end if**
17: **end while**

The Dynamic PF Correction process starts with Algorithm 1, which takes input parameters: PF_current, PF_threshold = 0.98, PF_overcomp = 1.0, P, θ_{before}, θ_{after}, C_available = $\{C1, C2, C3\}$, t = 1s. The algorithm continuously monitors PF in real time and automatically switches capacitors for correction. It first reads PF_current to check if correction is needed. If PF_current < PF_threshold, it calculates the required reactive power compensation Q_c = Function($P, \theta_{before}, \theta_{after}$) to determine the correction needed based on real power and phase shift. The required capacitance is then computed as $C_{required}$ = Function(Q_c, voltage, frequency) to counteract Q imbalance. Next, the system selects the optimal capacitor(s) using $C_{engaged}$ = select from $\{10\mu F, 20\mu F, 40\mu F\}$ where $C_{engaged} \geq C_{required}$, ensuring neither under- nor overcompensation. The selected capacitor(s) are activated via relay integration. If no correction is needed, $C_{engaged}$ remains unchanged. After t seconds for stabilization, the system rechecks PF_current. If PF_current $>$ PF_overcomp, indicating overcompensation, it adjusts capacitance using $C_{\text{adjust}} = C_{\text{engaged}} - C_{\text{last}}$, deactivating C_{last} via relay to restore optimal PF. This process runs dynamically in real time, repeating steps every t seconds for continuous load-based adjustments. The algorithm stops when monitoring is no longer required, ensuring efficient, automated PF correction with minimal relay operations.

4   Results and Discussions

The proposed real-time adaptive PFC system was evaluated across various load conditions to assess its effectiveness in improving PF, response time, and energy efficiency. Unlike conventional capacitor banks that operate on fixed compensation principles, this system dynamically engages capacitors based on instantaneous PF variations, ensuring optimal correction while minimizing unnecessary relay activations. The study focused on quantifying the system's capability to stabilize PF, reduce energy losses, and enhance response time, demonstrating its applicability in industrial and commercial environments. To evaluate the system's response under different load categories, various household and industrial appliances were considered. Light load conditions included small electrical devices such as LED bulbs and ceiling fans, exhibiting relatively stable PF characteristics. Medium loads, represented by appliances such as refrigerators, and air coolers, introduced moderate Q variations. Heavy load conditions, including motors and air conditioning units, significantly affected PF due to high inductive components, necessitating rapid correction. The ability of the system to adaptively engage and disengage capacitors in response to these fluctuations highlights its superior performance over traditional fixed capacitor banks.

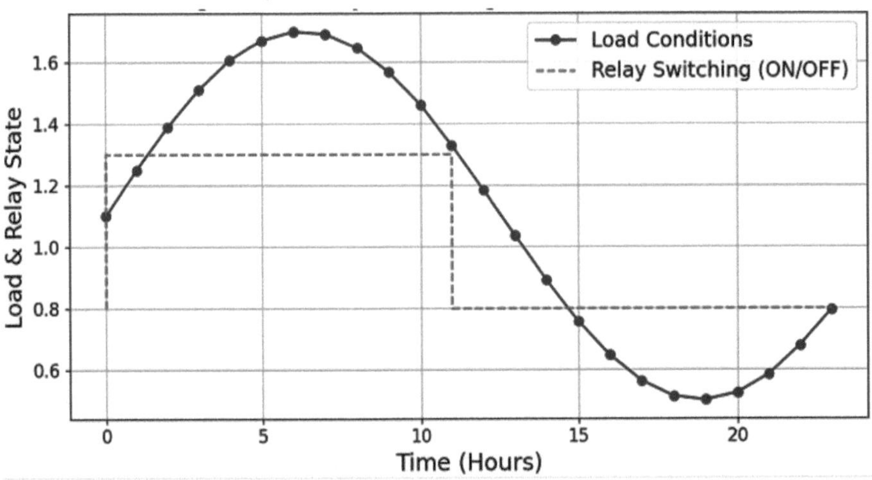

Fig. 4. Relay Switching vs. Load Conditions Graph illustrating capacitor engagement over time.

The dynamic nature of capacitor engagement is illustrated in Fig. 4, which showcases relay switching activity in response to varying load demands. Unlike conventional capacitor banks that operate on a predetermined threshold, the proposed system continuously monitors PF and engages capacitors only, when necessary, thereby optimizing Q compensation while preventing excessive relay activations. The graph clearly depicts how the system dynamically adjusts capacitance levels, ensuring that PF remains within the desired range without causing overcorrection or unnecessary energy consumption.

Table 2. Load Conditions & PF Variation Before & After Correction

Load Condition	Voltage (V)	Current (A)	Real Power (W)	Apparent Power (VA)	PF Before Correction	PF After Correction
No Load	230	0.1	23	23	1.00	1.00
Light Load	230	1.0	215	230	0.93	0.98
Medium Load	230	3.5	700	805	0.87	0.95
Heavy Load	230	6.0	1140	1380	0.82	0.90

The system's ability to improve PF across different load scenarios is further validated by Table 2, which presents the recorded values of voltage, current, real power, apparent power, and PF before and after correction. The results indicate a consistent improvement in PF, with heavy loads experiencing the most significant enhancement from 0.82 to 0.90. Medium and light loads also exhibited notable improvements, confirming the effectiveness of the real-time correction mechanism. These findings align with those reported by [15], who demonstrated that dynamic capacitor engagement strategies significantly improve PF stability in fluctuating industrial environments.

The system's response time and energy efficiency were further analyzed to determine its effectiveness in minimizing Q losses and improving overall system performance. Table 3 presents a combined analysis of response time and energy loss reduction across different load conditions, comparing the traditional capacitor bank approach with the proposed system.

The results confirm that while conventional capacitor banks required 4.8 to 8.5 s to adjust, the proposed system achieved correction within 1.2 to 2.1 s, demonstrating a notable enhancement in response speed.

Additionally, the system's energy-saving capability was evident in the significant reductions observed in power losses across all load categories, with up to 36% energy savings in mixed-load scenarios. These improvements contribute directly to cost reductions and enhanced power quality, making the system highly effective for industrial applications.

The system's impact on PF stability over time is demonstrated in Fig. 5, which illustrates how the proposed system maintains optimal PF levels even when load variations occur. The ability to rapidly adjust capacitance levels

Table 3. Response Time and Energy Loss Reduction Before & After PF Correction

Load Condition	Response Time of traditional Capacitor bank (s)	Proposed System Response time (s)	Energy losses before correction (kWh)	Energy losses after correction (kWh)	Reduction (%)
No Load	0.0	0.0	–	–	–
Light Load	4.8	1.2	1.2	0.9	25%
Medium Load	6.2	1.5	3.8	2.5	34%
Heavy Load	8.5	2.1	7.5	4.8	36%

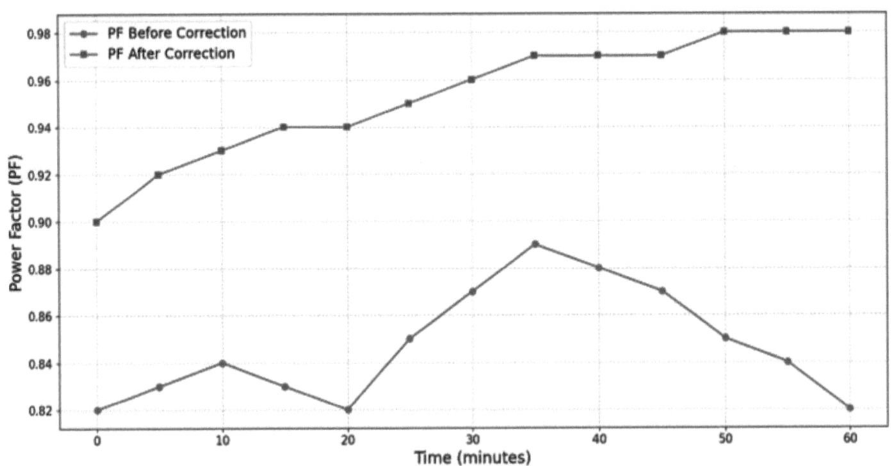

Fig. 5. PF Correction Impact Over Time demonstrating system stabilization after load changes.

ensures that PF fluctuations are immediately corrected, reducing energy inefficiencies and improving voltage stability. This contrasts with conventional systems, which often exhibit delayed responses, leading to momentary voltage instability.

The comparative analysis in Table 4 further highlights that the proposed real-time PFC system provides faster response time, real-time adaptability, and higher energy efficiency under dynamically changing load conditions. While the Bridgeless Zeta Converter [16] improves PF through an efficient topology, it lacks real-time adjustments and IoT automation. The Dynamic Capacitor Bank Control [15] works well for industrial-scale optimization but is slower to respond to rapid load variations. The proposed IoT-based real-time PFC system ensures superior energy efficiency and power quality by dynamically adjusting capacitance levels in real-time.

Table 4. Comparative Analysis of the Proposed System with other PFC Models

Sr. No.	Aspect	Proposed Real-Time PFC System	Dynamic PowerFactor Correction [15]	Bridgeless Zeta-Derived PFC Converter [16]
1.	Approach	Real-time monitoring & correction of PF using IoT-enabled capacitor banks	Automated capacitor bank control for staged Q compensation	Bridgeless topology to reduce losses and improve efficiency
2.	Correction Mechanism	Automated capacitor switching based on instantaneous PF readings	Pre-determined capacitor stages for correction rather than dynamic response	Diode-free design to lower conduction losses but lacks adaptive control
3.	Response Time	Immediate (within 1.2-2.1s response)	Slower than proposed, as MATLAB-based decisions require com-putational cycles	Fixed compensa-tion , not responsive to instantaneous variations
4.	PF Enhancement	Actively improves PF from 0.82 → 0.90 (heavy loads)	Adjusts PF gradually but not optimized for fast-changing loads	Provides better PF but lacks real-time adjustments
5.	Industrial Application	Suitable for real-world industrial & household loads with dynamic correction	Best suited for large-scale industrial settings with controlled environments	More applicable for single-phase applications with structured loads

The study confirms that the real-time adaptive PFC system provides substantial advantages in efficiency, response time, and energy savings, making it a highly practical and scalable solution for modern electrical power networks. The ability to rapidly adapt to load variations, reduce energy wastage, and maintain voltage stability demonstrates the system's reliability for large-scale deployment in both industrial and commercial applications. Beyond performance advantages, the proposed system introduces several practical benefits that make it a viable solution for large-scale deployment. Its real-time adaptation to load variations ensures continuous system optimization, reducing the likelihood of overcompensation or undercorrection, which can negatively impact power quality. Additionally, the system minimizes relay activations, which enhances component longevity and reduces maintenance costs associated with frequent capacitor switching failures. This aspect is particularly beneficial in high-load industrial setups, where switching reliability and minimal downtime are crucial for uninterrupted operation. The findings validate that the real-time adaptive PFC system offers a significant improvement over existing correction methods, making it a scalable, efficient, and practical solution for modern power management applications. Its ability to deliver instantaneous correction, higher energy efficiency, and long-term reliability reinforces its potential to redefine power quality optimization in industrial and commercial settings.

5   Conclusion

The proposed real-time adaptive PFC system offers significant improvements over conventional methods by dynamically engaging capacitors based on real-time load variations. Results show enhanced PF stabilization, faster response times, and greater energy efficiency, making it highly effective for industrial and commercial applications. Comparative analysis reveals a PF improvement upto 0.98, with the response time of 1.2–2.1 s, and up to 36% reduction in energy loss. A key advantage of this system is its instantaneous response to load fluctuations, ensuring optimal power compensation without unnecessary capacitor engagement. Reduced relay switching cycles enhance system reliability, extend component lifespan, and lower maintenance costs. Additionally, minimized energy losses translate to lower operational expenses, making the system a cost-effective power management solution. Maintaining a high-PF under varying load conditions also prevents voltage instability and avoids utility penalty charges. Despite these benefits, some limitations exist. The system's accuracy depends on sensor precision, meaning deviations in real-time voltage or current readings can affect compensation effectiveness. Additionally, initial setup costs may be higher than traditional fixed capacitor banks due to the need for microcontroller-based monitoring and relay switching. However, long-term advantages, including reduced energy wastage and improved power distribution efficiency, outweigh these initial costs. Future improvements could focus on predictive control algorithms that anticipate load variations and adjust capacitor engagement proactively. Machine learning integration could enhance adaptability and further reduce. Incorporating renewable energy sources into the correction mechanism would support sustainable power management, ensuring compatibility with modern smart grids. The study confirms that real-time adaptive PFC is a robust, scalable, and efficient approach to power quality optimization. By eliminating fixed compensation drawbacks, improving reactive power management, and reducing energy losses, it presents a technologically advanced solution for modern industrial and commercial power networks.

Acknowledgments. The authors did not receive support from any organization for the submitted work.

References

1. Pawanr, S., Garg, G.K., Routroy, S.: Prediction of energy efficiency, power factor, and associated carbon emissions of machine tools using soft computing techniques. Int. J. Interact. Des. Manuf. **17**, 1165–1183 (2023)
2. Patoliya, K., Sant, A.V., Patel, A., Sinha, A.: Smart meter for the estimation of power quality indices based on second-order generalized integrator. Environ. Sci. Pollut. Res. (2024)
3. Saeed Qazi, H., Ullah, Z., Alferidi, A., Alsolami, M., Lami, B., Muhammad Abrar Akber, S.: Stability analysis and voltage improvement in DG-integrated distribution networks using VCPI-based critical buses and lines detection considering uncertain power factor. Ain Shams Eng. J. (2024)

4. Beheshti Asl, M., Fofana, I., Meghnefi, F.: Review of various sensor technologies in monitoring the condition of power transformers. Energies (Basel) **17** (2024)
5. Setayesh, H., Kasaeian, A., Najafi, M., Madani Pour, M., Akbari, M.: The effects of geometric factors on power generation performance in solar chimney power plants. Energy **310** (2024)
6. Kharade, P.A., Jeyavel, J., Ingale, N.R., Jadhav, S.D.: Design and control of high-power density converters with Power Factor Correction using multilevel rectifiers. e-Prime – Adv. Electr. Eng. Electr. Energy **11** (2025)
7. Chaiwas, S., Kasayapanand, N., et al.: Enhanced thermoelectric power factor of magnetron Co-sputtered Pd-added Bi^2Te^3 thin films. Vacuum **230** (2024)
8. Uematsu, Y., et al.: Anomalous enhancement of thermoelectric power factor in multiple two-dimensional electron gas systems. Nat. Commun. **15** (2024)
9. Masci, A., Dimaggio, E., Neophytou, N., Narducci, D., Pennelli, G.: Large increase of the thermoelectric power factor in multi-barrier nanodevices. Nano Energy **132** (2024)
10. Dadhaniya, P., Maurya, M., Vishwanath, G.M.: A bridgeless modified boost converter to improve power factor in EV battery charging applications. IEEE J. Emerging Selected Topics Indust. Electr. **5**, 553–564 (2024)
11. Wang, X., Leng, M., He, L., Lu, S.: An improved LCC-S compensated inductive power transfer system with wide output voltage range and unity power factor. IEEE Trans. Trans. Electrifi. **10**, 2342–2354 (2024)
12. Wei, J., Liu, D., Wang, P., Lei, X., Liu, Y.: Thermoelectric power factor of cementitious composites significantly reinforced with boron/nitrogen co-doped reduced graphene oxide. Ceramics Inter. (2025)
13. Sun, P., Song, Q., Wang, W.: Minimum loss modulation method for various power factor angles. Energy **310** (2024)
14. Wu, S., et al.: Large transverse thermoelectric power factor in topological semimetal $NbAs^2$. Adv. Energy Mater. **14** (2024)
15. Ayaz, M., Rizvi, S.M.H., Akbar, M.: Dynamic power factor correction in industrial systems: an automated capacitor bank control approach. In: 2023 2nd International Conference on Emerging Trends in Electrical, Control, and Telecommunication Engineering (ETECTE 2023) – Proceedings. IEEE (2023)
16. M. V., Venkadesan, A., Sakthivel, S.S.: Design and analysis of bridgeless zeta-derived Power Factor Correction converter with reduced magnetic components. Comput. Electr. Eng. **123** (2025)

Describing the Properties of Privacy Tools to Non-expert Users

Shumaila Khan[✉]

MCB Bank, Lahore, Pakistan
`shumaila.khan@mcb.com.pk`

Abstract. In this research, we examine the gap between non-expert customers and their inability to comprehend or engage with privacy tools that include, policies, securities, and responsibilities of customers in digital banking services. The study, which involves an analysis of privacy and security statements of the UK (UNITED KINGDOM)'s major banking institutions, including HSBC, Barclays, Lloyds, and NatWest, with non-expert users, show that users have limited understanding of some of the most crucial policies such as data retention policies, PIN, and other legal phrases such as 'grossly negligent' and 'reasonable care'. Based on the findings of this paper, the following suggestions have been provided in a bid to facilitate easier changes to the current policies, and improve user understanding of the policies through creative educative campaigns as well as proper user interface designs. Therefore, this research which seeks to investigate the aforementioned challenges will go ahead add to the pool of knowledge on how to come up with better methods of enabling users to protect their information, thereby leading to better protection of data by all users, especially those who are most vulnerable.

Keywords: Privacy tools · digital banking · cybersecurity · data protection · user perception

1 Introduction

Since digital services are gradually becoming a part of life and work, cybersecurity tools are necessary to protect personal data. Regardless whether a user or a company bank online or use social networks, or participate in any other type of project that involves interaction with the personal data, all the privacy tools are an essential protective barrier against such attempts as hacking, data breaches, identity thefts and others. However, it is not very easy for non-expert users to work with these tools—one can mention encryption protocols, password management systems, two-factor authentication, and others, since they are frequently quite complicated and are described in legal terms. Such a gap could result in misuse of privacy tools or insufficient protection measures for personal information and increase users' risks.

Another common problem is that users cannot read and understand privacy policies. Such policies are used by organizations such as banks to determine the role and obligations of the supplier and customer. Even though these policies are intended to safeguard

P. Chandrakar et al. (Eds.): ICCINS 2025, CCIS 2738, pp. 218–233, 2026.
https://doi.org/10.1007/978-3-032-09572-5_18

accounts from unauthorized access, most of the users either do not read or find it very hard to comprehend the terms and conditions that they sign for. Research shows that some of the terms used when allocating blame in the event of a data breach such as 'gross negligence' or 'reasonable care' go unnoticed or misinterpreted.

This ambiguity can lead to serious problems especially when users demand their money back or some compensation after a breach.

Scholars have paid considerable attention to understanding how ordinary people approach privacy tools and policies. For instance, Pattnaik et al. [1] conducted a study of the discussion forum in order to understand how the common people think about cybersecurity and privacy. Their research shows that the use of unfamiliar technical terms hampers privacy practices. In the same vein, Distler et al. [4] looked into how to demystify encryption technologies which include the use of graphics and plain language. In the study, the authors found that the level of user engagement increases dramatically when the users are provided with simple and clear privacy tools. The issue of sound security measures in digital banking is more crucial than in other sectors of banking. Components such as encryption, multi-factor authentication, and biometric identification are crucial for the safety of financial and personal information. Nevertheless, a review of the privacy policies of four major UK (UNITED KINGDOM) banks: HSBC, Barclays, Lloyds, and NatWest showed that many of these documents are filled with legal language that would be difficult for most users to understand. This difference between expected and real users' awareness can lead to missed security threats, lack of adherence to even the most elementary security measures such as changing default PINs, or not reporting suspicious activity. As a result, even the most sophisticated technical measures may not be sufficient, if the users themselves are able to protect their privacy [5].

1.1 Problem Statement

Experts report that non-expert users struggle to protect their privacy due to difficult and intricate terms in privacy tools and policies, as well as a lack of guidance on how to shield their information. The use of technical terminology and information overload makes it hard for users to make informed decisions about their privacy and security. Phrases like "reasonable care" and "grossly negligent" contribute to privacy breaches and are often misunderstood or ignored by non-professionals [6]. Many users are unaware of these risks, making them vulnerable to identity fraud, scams, and phishing.

1.2 Objective

Therefore, the purpose of this paper is to bring awareness to non-expert users about privacy tools and categorize these tools based on their language and functionality in the context of digital banking. This paper aims to discover factors that hinder ordinary users from protecting their data through analysing the privacy policies employed by the largest banks in the UK (UNITED KINGDOM) and conducting interviews with non-specialist individuals.

2 Literature Review

Privacy tools have become important in the current digital environment especially since normal users are more in touch with sensitive information in various applications. But, it is important to note that several studies have stresses that revealed the fact that even though there are numbers of privacy tools available, the average users lack the ability or knowledge to understand let alone use the privacy tools in the best way possible. The following review includes major works concerning these problems to outline how various scholars have attempted to overcome the challenge for instance, Pattnaik et al. (2023) revealed that over 40% of the participants must go through a lot of struggle to determine simple privacy-related decisions and more so their privacy settings [1]. This shows how hard it is for non-expert users to understand certain simple terms on privacy to enable them make right decisions concerning their data. On the same note, Distler et al. (2022) opined that by presenting encryption technologies in simple graphics and words, users are likely to make better decisions about their privacy, perhaps taking more autonomy over the decision-making process. According to their studies, the intended audience was more active and made smarter choices especially when security technologies where presented in a basic graphic form [2].

To this effect, Barth et al. (2022) recently reiterated this aspect by stating that, at the very least, privacy tools are poorly understood by experts and non-experts due to the fact that even when experts attempt to explain the availability of privacy tools, the language they use includes technical terms, acronyms, and complex legal phrases that are incomprehensible to the common user. This results in users ignoring essential privacy features, allowing potentially their data to be exposed [3]. Furthermore, the implementation of privacy tools such as encryption, multi-factor authentication, password management and others is difficult, consequently, people do not fully engage in these measures, despite the fact that their focus is on the security [4]. Ebbers et al. (2020) reported that users stay away from the use of such tools when they are complex and time consuming. Abu-Salma (2020) also pointed that privacy tools should be congruent and aligned to the average user aiming to decrease the cognitive load and making them easy to use and comprehend. This makes the usability of the privacy tools a principle which is very important in the extension of the privacy tools to other more people [5].

Of these privacy policies are especially difficult for non-expert users in the context of digital banking. A study by Hosseini et al. (2021) also showed that many users were actually unaware of, for example, the legal distinction between 'gross negligence' and 'reasonable care', of which the former puts the burden of responsibility for a security breach on the party that was negligent. Such misunderstanding can offend the users and further diminish their confidence in digital banking services, besides making them reluctant to engage with the privacy polices [8]. Concerning these difficulties, several suggestions have been made, meaning easier language in privacy policies and better definitions of the privacy terms. Moreover, giving users understandable and relevant information about the function of the AI increases their trust in the system, a concept previously researched by Larasati et al., 2023 that may also apply to privacy tools [7].

Several suggestions have been made to enhance the usage and understanding of tools for fashioning user's privacy, as follows: These are referring to actions such as presenting the information in the privacy policies in ordinary language, using pictures or graphics to illustrate facets that may be hard to understand, and giving users fast or 'how-to' guides to his/her privacy settings. Additional, it is necessary to update users and inform about any changes in privacy practices constantly [9, 13, 22]. Thus, following the recommendations of Rodak et al. (2020), it is necessary to minimize cognitive load in privacy policies: using traffic light platforms to increase visibility of privacy risks and privacy features for users. By distilling privacy tools and policies, and by making sure that users of those systems are informed about their privacy rights and duties, this problem can be solved—non-expert users can more effectively code their digital privacy.

3   Methodology

This research employs a qualitative research strategy to analyze non-expert users' perception of privacy policies and the effectiveness of privacy tools in the context of digital banking. The research hence aims at identifying comprehension issues, usage patterns and difficulties that participants, who are users in major UK (UNITED KINGDOM) banks, encounter while engaging with privacy policies and security terms (Fig. 1).

3.1   Research Design

The study employs a qualitative, cross-sectional research approach to explore the perceptions and attitude of non-professional users towards the use of privacy tools. The design is based on the knowledge of other meanings that people attribute to privacy policies, their behavior when dealing with security practices such as passwords, and how they understand legal and technical terms. This design was chosen to ensure more elaborate and comprehensive answers than could be offered by quantitative research only.

3.2   Participant Selection

Participants were selected through purposive sampling, focusing on non-expert users of digital banking services from four major UK (UNITED KINGDOM) banks: HSBC, Barclays, Lloyds, and NatWest.

Be currently using an active digital banking account with one of the four targeted local banks.

Lack sophistication by not being experts in cyber security, banking or data privacy.

Must have one year of experience in Internet banking services.

Participants were attracted through emails and via communications through the WhatsApp application, specifically to university customers of banks. The study involved 20 participants, of which 7 were from HSBC, 7 from Barclays, 4 from Lloyds, and 2 from NatWest.

3.3   Pilot Phase

To reduce the potential variability, a pilot phase of the main study with 3 participants was used to fine-tune the interview protocol. The pilot identified that the participants had difficulties in understanding some of the privacy-related terms (for instance 'grossly negligent' and 'reasonable care'. Therefore, to avoid any misunderstanding, the interview questions were modified to provide a brief definition of these terms.

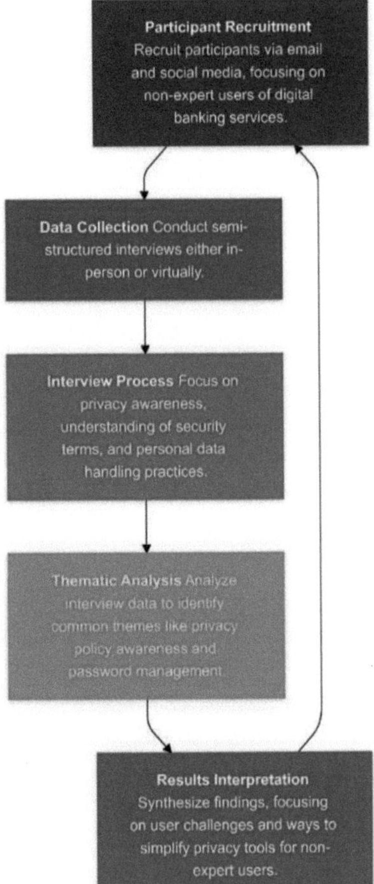

Fig. 1. Research Methodology Flowchart

3.4 Data Collection

Information was gathered during the survey using both closed questions as well as face-to-face interviews lasting 30 min on average to provide the participants with an opportunity for elaboration. The semi-structured approach provided a level of standardization across the interviews while allowing participants to provide additional details on their perceptions.

- Awareness of privacy policies
- Understanding of security practices
- Perceptions of privacy terminologies
- Behavioural responses to security risks

Before the interview, participants were given privacy policy statements from their banks to study in advance. This helped in ensuring that the discussions revolved around their real-life encounter with the policies and how they understood them. These privacy policies were obtained from the banks' official website to ensure the correctness and applicability to the participant's context.

3.5 Ethical Considerations

The study got clearance from the College Research Ethics Committee thus conforming to the highest ethical standards. The study participants were offered a consent form in which they were informed on the goals and objectives of the study, their rights regarding their involvement in the research, and measures that would be taken to maintain their anonymity.

3.6 Data Analysis

Thematic Analysis was employed as the method of choice for analysing the interview data since it is a common method of analysing qualitative data and it enables the determination of patterns (themes) in the data. The following steps were taken during the analysis process:

Familiarization with data
The researcher carefully read all the interview transcripts several times to ensure he or she was familiar with the review contents.

Coding
Consistent with the coding of data, important words and phrases from participant's responses were highlighted. Coding was specific in aspects such human's knowledge of privacy policies, security measures, and legal stuff.

Theme Development
In order to enhance intercoder reliability, codes were aggregated into more general categories based on similarities of participants' accounts. The central concerns assessed were the general awareness of the privacy policy, perceptions of security measures taken, management of the PIN number and understanding of security terms used.

Reviewing Themes
Each of the identified themes was discussed and examined in order to establish that they are informative of the data and the research questions. Defining themes
The final themes are then described and labelled and interview data samples are provided for each of these themes.

3.7 Tools and Frameworks

Computer aided qualitative data analysis was used throughout the study with specific reference to the NVivo software, which was used to manage the data and code system. This software served as a way of categorizing the codes and the themes in a way that all facets of the data were analysed systematically. The study also relied on theories from user-centered design (UCD) to explain the results of the study by looking at PTPs through the lens of user interfaces.

4 Results

The research done makes it clear that customers especially those with non-expert knowledge lack adequate information regarding actual privacy and security of digital banking services.

4.1 Awareness of Privacy Policies

The work conducted showed that non-expert users generally do not comprehend privacy policies, nine out of ten individuals revealed that they have never read their bank's privacy policy or read it superficially at best. Some of them had no idea that such policies exist or the importance they hold when it comes to protection of their information.

In order to make a coherent analysis, participants who had come into contact with these policies referred to them as "thick" and "just comprehensibly opaque, cluttered with terms they themselves could not fully understand". Similarly, a meager 10% of the users said that they had at some point made an attempt to understand the terms and conditions but they were too many to read through properly. This is an indication that for all the legal requirement of putting up privacy policies, Fig. 2 shows the Percentage of participants who are aware of the privacy related concept and practice such as data sharing, data retention, password management, legal terms and conditions including "reasonable care "and "gross negligence".

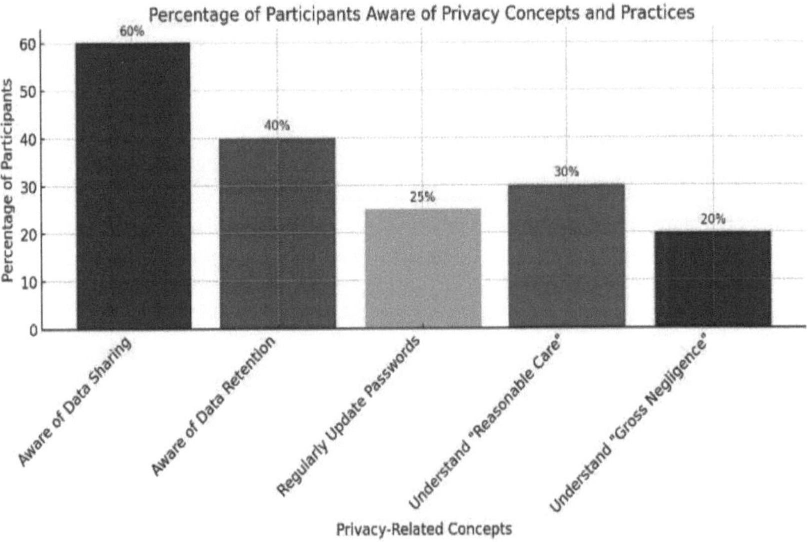

Fig. 2. Percentage of participants who are aware of the privacy related concept and practice such as data sharing, data retention, password management, legal terms and conditions including "reasonable care "and "gross negligence".

4.2   Understanding of Data Protection Regulations

About regulation concerning data protection in general, and the GDPR (general data protection regulation) in particular, a significant number of the Participants indicated that they were not familiar with the law and its impact on the banking data. Social media poll showed that only 30% of the respondents had some knowledge of GDPR (general data protection regulation) or how their information is protected by this law. Of the rest 30%, although they had heard something about data protection, their level of knowledge was very low, and they thought that it means only encryption. Evidently, most could not explain how GDPR (general data protection regulation) applied to their usage of banking services or how their rights were impacted on data collection, storage, and utilization.

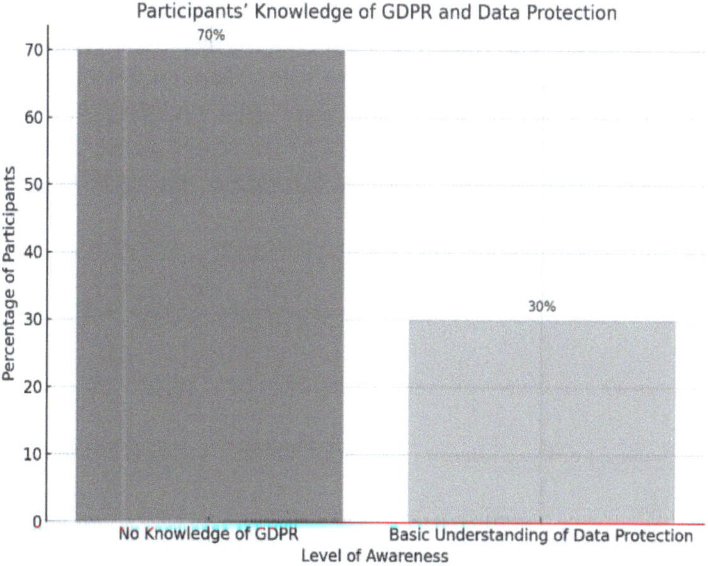

Fig. 3. GDPR (general data protection regulation) Awareness among participants which indicates that a considerable number of the respondents did not know about the GGDPR (general data protection regulation) and the rights they have over their personal data.

This Bar chart in Fig. 3 shows that exactly half of the participants had no idea about the GDPR (general data protection regulation) and its protection of banking data, while the other half had only slight knowledge of data protection which included encryption, even though only a third of them knew that GDPR (general data protection regulation) protected banking data.

4.3   Handling of Passwords and PINs

While 80% of the participants stated that they created a default PIN or password when opening the banking accounts, only a few of the participants stated that they changed them or updated them regularly. A small percentage of the participants claimed they protected their accounts by changing passwords every few weeks, and other participants admitted that they used easily discernible passwords or kept them in a place where others could have access to them.

4.4 Reasonable Care and Gross Negligence

All participants found terms like the 'reasonable care', 'gross negligence' that often appear in privacy policies confusing. None of the respondents were able to provide a clear definition of these concepts, although these concepts are important in an assessment of the degree of negligence of an organization when it comes to authorizing people to conduct transactions on behalf of an organization.

4.5 Security Breach Awareness

There were aware to different capacities regarding security breaches among the participants. Whereas everyone might easily identify reckless signs like fraudulent transactions, they hardly identified the less apparent risks, like phishing attempts or having a malware.

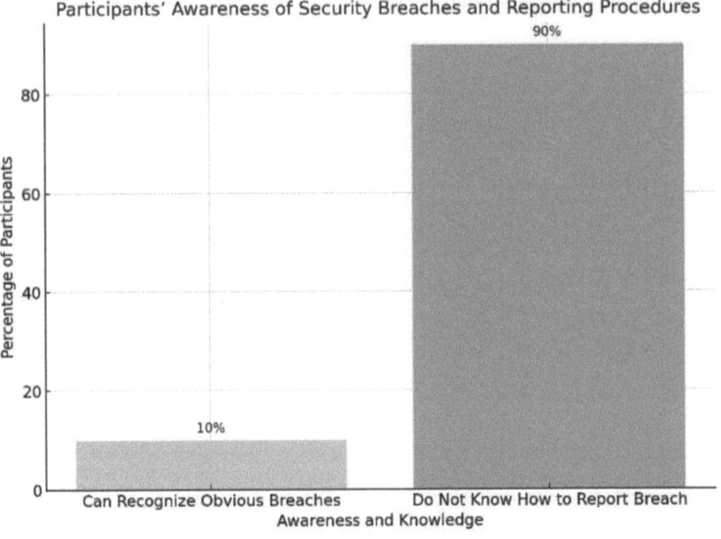

Fig. 4. Security breaches awareness by the participant and how the participant reported

The Fig. 4 the bar chart representing Participants Awareness of Security Breaches and the Reporting Procedures are best represented in this bar chart. This proves that 90% of participants claimed they had no idea how one could report a breach or suspected fraudulent transaction whereas only 10% were able to identify apparent breaches.

4.6 Lack of Transparency on Data Transfers

Of all the participants, none were informed their data could be transferred to another country to fulfill their bank's operational requirements. Should they be informed of this possibility, participants were worried, especially about the differences in data protection laws by different jurisdictions (Fig. 5 and Table 1).

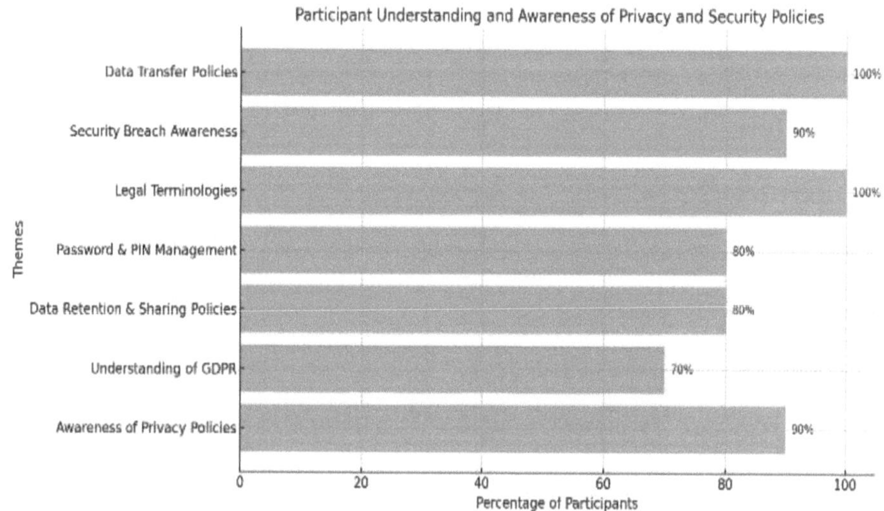

Fig. 5. Privacy and Security knowledge among Participants on Privacy and Security aspects that include: Awareness of privacy policy, Understanding of GDPR (GENERAL DATA PROTECTION REGULATION), Beliefs on data retention, and comfort with legal terms and conditions concerning Privacy and Security.

Table 1. Data Summary Table

Theme	Key Insights
Awareness of Privacy Policies	90% of the participants stated that they had never read their bank's policy on the privacy of the customers.
Understanding of GDPR (GENERAL DATA PROTECTION REGULATION)	They found out that 70% of the participants were not aware of GDPR (GENERAL DATA PROTECTION REGULATION) protection measures.
Data Retention and Sharing Policies	As to data retention, only 20% of the respondents stated that they knew how long their data was stored and with whom.
Password and PIN Management	80% modified it by changing the password by default, but not many did so frequently.
Legal Terminologies	All the respondents had no prior knowledge about the meaning of the terms "reasonable care" and "gross negligence: respectively.

(*continued*)

Table 1. (continued)

Theme	Key Insights
Security Breach Awareness	Ninety percent were unaware of how to report a security breach or how to identify clandestine threats.
Data Transfer Policies	Consequent to this, none of the participants had an understanding of international data transfer.

5 Discussion

This section discusses the findings with reference to the research questions and the existing literature on privacy tools and non-expert users' interactions with digital banking services. The findings highlight a serious discrepancy between the existence of the privacy-enhancing technologies and the understanding and application of these technologies by the members of the general public. This work has therefore described a worrying chasm between the existence of privacy-preserving technologies and the knowledge and adoption by consumers of digital banking services not trained in computer science or engineering. Participants' knowledge of privacy policies was generally inadequate due to poor or no reading or understanding of the policy. In the same regard, there was poor knowledge about data protection regulations such as GDPR (general data protection regulation), which needs improved educating by the banks and the policymakers. The other mini-findings that emerged from the study are that non-expert users tend to consider Ease of use as a priority over security when dealing with passwords PIN numbers and other security features thus creating insecurity in the handling of data. Moreover, user awareness regarding important legal vocabulary, which can cause ambiguity in case of violations of security, was low. Little of such was known regarding how to report such incidents more concerning numerous participants were compromised. Finally, existing consumer ignorance of cross-border transfers might lower the public's confidence in financial organizations and indicate the need to address how they process and transfer data. To overcome these challenges, banks should step up on promoting a more personal user experience superintending, the simplification of security, aspects together with a clearer procedure concerning the privacy of personal data.

5.1 Limitations

The first limitation of this study is the number of participants chosen, which although appropriate for qualitative research may not be diverse enough to include all nonexperts. However, it is essential to note that the study was conducted only on four UK (UNITED KINGDOM) banks, and thus, the results would not be universally generalizable to other financial firms or countries. Also, the use of the self-completion technique may have led to some biases where the findings reflect either underestimation of difficulties experienced or overestimation of comprehension of privacy policies and tools.

6   Conclusion

This study helps to untangle the severe problems in privacy tools and digital banking policies that nonexpert users encounter. The findings showed that despite the fact that banks do offer clear and detailed policy statements regarding the privacy of consumers' information and measures for information security, such statements and measures are quite often concealed by complex and technical language. Most of the users either never read the privacy policies or became overwhelmed and frustrated at the rudimentary level when trying to comprehend what it means, for example, retention of data, sharing of data, and legal obligations regarding terms such as reasonable care and gross negligence. As far as the results of this study are concerned, one of the most striking observations made is the gross ignorance prevailing in the subject domain concerning fundamental legal and security related terms. A significant number and proportion of non-expert users who avail the services of digital banking are, perhaps, not cognizant with the ways in which their data is processed, stored and/or disseminated. Due to this lack of understanding, users' privacy is at risk and their decision-making concerning data is constrained. The study also highlighted other bad practices some of which are poor password and PIN management and these put the users to more risks.

The findings can be useful to the existing debate on the usefulness of PETs and stress the necessity to provide the wider community with tools and policies that are easy to understand and apply. It appears that today's digital banking privacy features are more effective in meeting legal requirements than they are in helping the user understand them. As such the same tools are compliant with the legal requirements of GDRP for instance but the actual effectiveness is derailed by the fact that most users do not understand how to use them let alone implement the best practice as expected. In its summary despite finding that the privacy tools being employed in digital banking are fully functional from the technical standpoint, it is necessary to note that their primary aim is not achieved because these tools are virtually unreachable to people who have no sufficient background to work with computers.

In this regard, there is a need for banks to move from the more standard compliance to the general knowledge and the need of the general public. These findings are useful in understanding the need for the simplification of privacy tools and policies and the need for financial institutions particularly to encourage more active pursuit of creating awareness among their users on how to safely conduct their affairs online.

7   Future Work

This research has provided directions for subsequent research on factors that influence the usability of privacy tools for ordinary users. Future research needs to improve their sample and range, demographic and financial institutions, and cultural backgrounds to learn about the global consumer behavior in an online bank. While comparing nonexpert users to the expert we could be able to spot all the areas that need to be improved concerning the privacy tools and the training of users on the same. Moreover, the investigation of other approaches in presenting privacy policies, for example in the form of an entertaining tutorial or in graphic visualization, could enhance understanding among users and their interest in it.

Knowledge exchange activities related to timely user training and security notifications are limited, such as pop-up tips during suspicious activities might increase users' awareness. Additional research on privacy by design principles, including default privacy setting and simplified consent forms may help to enhance the effectiveness of the privacy tools. Future research should shift its attention toward the ways that financial institutions inform its members or customers regarding data handling and regarding their perceptions about international data transfer as it becomes the new norm for organizations. Thus, further developments in the field of user-oriented design, digital competencies as well as cybersecurity research will turn out to be vital to improve data security and users' confidence in digital banking.

Acknowledgments. This Special thanks go to all the participants and other stakeholders who offered their valuable time and input during the research. We also wish to express our gratitude to the reviewers, whose helpful comments helped to improve the quality of this paper. Technical and academic help has also been acknowledged which the researcher received throughout the course of the project.

References

1. Pattnaik, N., Li, S., Nurse, J.R.C.: Perspectives of non-expert users on cyber security and privacy: an analysis of online discussions on Twitter. Comput. Secur. **125**, 103008 (2023). https://doi.org/10.1016/j.cose.2022.103008
2. Distler, V., Gutfleisch, T., Lallemand, C., Lenzini, G., Koenig, V.: Complex, but in a good way? How to represent encryption to non-experts through text and visuals—evidence from expert co-creation and a vignette experiment. Comput. Hum. Behav. Rep. **5**, 100161 (2022). https://doi.org/10.1016/j.chbr.2021.100161
3. Barth, S., de Jong, M.D.T., Junger, M.: Lost in privacy? Online privacy from a cybersecurity expert perspective. Telemat. Inform. **68**, 101782 (2022). https://doi.org/10.1016/j.tele.2021.101782
4. Ebbers, F., Zibuschka, J., Zimmermann, C., Hinz, O.: User preferences for privacy features in digital assistants. Electron. Mark. (2020). https://doi.org/10.1007/s12525-020-00418-2
5. Abu-Salma, R.: Designing User-Centered Privacy-Enhancing Technologies. UCL (University College London) (2020)
6. Bove, C., Aigrain, J., Lesot, M.-J., Tijus, C., Detyniecki, M.: Contextualization and exploration of local feature importance explanations to improve understanding and satisfaction of non-expert users. In: 27th International Conference on Intelligent User Interfaces, pp. 233–245 (2022). https://doi.org/10.1145/3490099.3511122
7. Larasati, R., De Liddo, A., Motta, E.: Meaningful explanation effect on user's trust in an AI medical system: designing explanations for non-expert users. ACM Trans. Interact. Intell. Syst. **13**(4), 1–39 (2023). https://doi.org/10.1145/3571711
8. Hosseini, M.B., Breaux, T.D., Slavin, R., Niu, J., Wang, X.: Analyzing privacy policies through syntax-driven semantic analysis of information types. Inf. Softw. Technol. **138**, 106608 (2021). https://doi.org/10.1016/j.infsof.2021.106608
9. Oesch, S., et al.: Understanding user perceptions of security and privacy for group chat: a survey of users in the US and UK (UNITED KINGDOM). In: Annual Computer Security Applications Conference, pp. 234–248 (2020). https://doi.org/10.1145/3427228.3427243

10. Rodak, A., Jamson, S., Kruszewski, M., Pedzierska, M.: User requirements for autonomous vehicles—a comparative analysis of expert and non-expert-based approach. In: 2020 AEIT International Conference of Electrical and Electronic Technologies for Automotive (AEIT AUTOMOTIVE), pp. 1–6 (2020). https://doi.org/10.23919/AEITAutomotive50086.2020.930 7345

11. Tadic, B., Rohde, M., Randall, D., Wulf, V.: Design evolution of a tool for privacy and security protection for activists online: cyberactivist. Int. J. Hum. Comput. Interact. **39**(1), 249–271 (2023). https://doi.org/10.1080/10447318.2022.2127940

12. Schaffhauser, T., et al.: A framework for the broad dissemination of hydrological models for non-expert users. Environ. Model. Softw. **164**, 105695 (2023). https://doi.org/10.1016/j.env soft.2023.105695

13. Agrawal, N., Binns, R., Van Kleek, M., Laine, K., Shadbolt, N.: Exploring design and governance challenges in the development of privacy-preserving computation. In: Proceedings of the 2021 CHI Conference on Human Factors in Computing Systems (2021). https://doi.org/ 10.1145/3411764.3445185

14. Luan, K., Halvorsrud, R., Boletsis, C.: Evaluation of a tool to increase cybersecurity awareness among non-experts (SME employees). In: Proceedings of the 9th International Conference on Information Systems Security and Privacy, vol. 509, pp. 509–518 (2023). https://doi.org/ 10.5220/0012078700003410

15. Alshamsan, A.R., Chaudhry, S.A.: A GDPR (General Data Protection Regulation) compliant approach to assign risk levels to privacy policies. Comput. Mater. Contin. **74**(3), 4631–4647 (2023). https://doi.org/10.32604/cmc.2023.027989

16. Miyata, S., Chang, C.-M., Igarashi, T.: Trafne: a training framework for non-expert annotators with auto validation and expert feedback. In: Lecture Notes in Computer Science, pp. 475–494. Springer, Cham (2022). https://doi.org/10.1007/978-3-030-97992-4_27

17. Fainchtein, R.A., Aviv, A.J., Sherr, M.: User perceptions of the privacy and usability of smart DNS. In: Proceedings of the 38th Annual Computer Security Applications Conference, pp. 591–604 (2022). https://doi.org/10.1145/3564625.3564629

18. Velykoivanenko, L., Niksirat, K.S., Zufferey, N., Humbert, M., Huguenin, K., Cherubini, M.: Are those steps worth your privacy?: fitness-tracker users' perceptions of privacy and utility. Proc. ACM Interact. Mob. Wearable Ubiquitous Technol. **5**(4), 1–41 (2021). https://doi.org/ 10.1145/3494961

19. Oesch, S., et al.: User perceptions of security and privacy for group chat. Digit. Threat. **3**(2), 1–29 (2022). https://doi.org/10.1145/3487064

20. Shen, H., Jin, H., Cabrera, Á.A., Perer, A., Zhu, H., Hong, J.I.: Designing alternative representations of confusion matrices to support non-expert public understanding of algorithm performance. Proc. ACM Hum. Comput. Interact. **4**(CSCW2), 1–22 (2020). https://doi.org/ 10.1145/3415208

21. Chalhoub, G., Kraemer, M.J., Flechais, I.: Useful shortcuts: using design heuristics for consent and permission in smart home devices. Int. J. Hum. Comput. Stud. **182**, 103177 (2024). https:// doi.org/10.1016/j.ijhcs.2023.103177

22. Lavalle, A., Maté, A., Trujillo, J., Teruel, M.A., Rizzi, S.: A methodology to automatically translate user requirements into visualizations: experimental validation. Inf. Softw. Technol. **136**, 106592 (2021). https://doi.org/10.1016/j.infsof.2021.106592

23. Virkar, S., Alexopoulos, C., Tsekeridou, S., Novak, A.-S.: A user-centred analysis of decision support requirements in legal informatics. Gov. Inf. Q. **39**(3), 101713 (2022). https://doi.org/ 10.1016/j.giq.2022.101713

24. Jia, H., et al.: Truth in a sea of data: adoption and use of data search tools among researchers and journalists. Inf. Commun. Soc. **26**(16), 3237–3256 (2023). https://doi.org/10.1080/136 9118X.2023.2176695

25. Schufrin, M., Reynolds, S.L., Kuijper, A., Kohlhammer, J.: A visualization interface to improve the transparency of collected personal data on the internet. IEEE Trans. Vis. Comput. Graph. **27**(2), 1840–1849 (2021). https://doi.org/10.1109/TVCG.2020.3030384
26. Breve, B., Desolda, G., Deufemia, V., Greco, F., Matera, M.: An end-user development approach to secure smart environments. In: Lecture Notes in Computer Science, pp. 36–52. Springer, Cham (2021). https://doi.org/10.1007/978-3-030-72632-0_3
27. Yaji, S., Bayyapu, N.: Reinforcement technique for classifying quasi and non-quasi attributes for privacy preservation and data protection. In: Communications in Computer and Information Science, pp. 3–17. Springer, Singapore (2023). https://doi.org/10.1007/978-981-199 9899_1
28. Porcelli, L., Mastroianni, M., Ficco, M., Palmieri, F.: A user-centered privacy policy management system for automatic consent on cookie banners. Computers. **13**(2), 43 (2024). https://doi.org/10.3390/computers13020043
29. Krsek, I., Wenzel, K., Das, S., Hong, J.I., Dabbish, L.: To self-persuade or be persuaded: examining interventions for users' privacy setting selection. In: CHI Conference on Human Factors in Computing Systems (2022). https://doi.org/10.1145/3491102.3517580
30. Stojkovski, B., Lenzini, G., Koenig, V.: I personally relate it to the traffic light: a user study on security & privacy indicators in a secure email system committed to privacy by default. In: Proceedings of the 36th Annual ACM Symposium on Applied Computing, vol. 149, pp. 1235–1246 (2021). https://doi.org/10.1145/3412841.3442036
31. Bertrand, A., Eagan, J.R., Maxwell, W.: Questioning the ability of feature-based explanations to empower non-experts in robo-advised financial decision-making. In: 2023 ACM Conference on Fairness, Accountability, and Transparency (2023). doi:https://doi.org/10.1145/359 3013.3594050
32. Müller, J.: Evaluation methods for citizen design science studies: how do planners and citizens obtain relevant information from map-based E-participation tools? ISPRS Int. J. Geoinf. **10**(2), 48 (2021). https://doi.org/10.3390/ijgi10020048
33. Barth, C.-M., Schmid, J., Al-Hazwani, I., Sachdeva, M., Cibulski, L., Bernard, J.: How applicable are attribute-based approaches for human-centered ranking creation? Comput. Graph. **114**, 45–58 (2023). https://doi.org/10.1016/j.cag.2023.06.003
34. Frik, A., Kim, J., Sanchez, J.R., Ma, J.: Users' expectations about and use of smartphone privacy and security settings. In: CHI Conference on Human Factors in Computing Systems (2022). https://doi.org/10.1145/3491102.3517504
35. Al-Momani, A., Bösch, C., Wuyts, K., Sion, L., Joosen, W., Kargl, F.: Mitigation lost in translation: leveraging threat information to improve privacy solution selection. In: Proceedings of the 37th ACM/SIGAPP Symposium on Applied Computing, pp. 1236–1247 (2022). https://doi.org/10.1145/3477314.3507077
36. Schulze-Weddige, S., Zylowski, T.: User study on the effects explainable AI visualizations on non-experts. In: Lecture Notes of the Institute for Computer Sciences, Social Informatics and Telecommunications Engineering, pp. 457–467. Springer, Cham (2022). https://doi.org/10.1007/978-3-030-93982-9_32
37. Nuñez von Voigt, S., Mehner, L., Tschorsch, F.: From theory to comprehension: a comparative study of differential privacy and k-anonymity. In: Proceedings of the Fourteenth ACM Conference on Data and Application Security and Privacy, pp. 221–232 (2024). https://doi.org/10.1145/3577923.3592957
38. Becker, I., et al.: International comparison of bank fraud reimbursement: customer perceptions and contractual terms. J. Cybersecur. **3**(2), 109–125 (2017). https://doi.org/10.1093/cybsec/tyx008
39. Beautement, A., Sasse, M.A., Wonham, M.: The compliance budget: managing security behaviour in organisations. Nspw.org. Available: https://www.nspw.org/papers/2008/nspw2008-beautement.pdf. Accessed 15 Sept 2024

40. Adams, A., Sasse, M.A.: Users are not the enemy. Commun. ACM. **42**(12), 40–46 (1999). https://doi.org/10.1145/322796.322806

41. Dev, J., Rashidi, B., Garg, V.: Models of applied privacy (MAP): a persona based approach to threat modeling. In: Proceedings of the 2023 CHI Conference on Human Factors in Computing Systems, vol. 12, pp. 1–15 (2023). https://doi.org/10.1145/3544549.3585751

42. Shankar, A., Waldis, A., Bless, C., Rodriguez, M.A., Mazzola, L.: PrivacyGLUE: a benchmark dataset for general language understanding in privacy policies. Appl. Sci. (Basel). **13**(6), 3701 (2023). https://doi.org/10.3390/app13063701

43. Easley, W.B., Asgarali-Hoffman, S.N., Hurst, A., Mentis, H.M., Hamidi, F.: Using a participatory toolkit to elicit youth's workplace privacy perspectives. In: Proceedings of the 2021 European Symposium on Usable Security, vol. 10, pp. 211–222 (2021). https://doi.org/10.1145/3481390.3481392

44. Bugeja, J., Jacobsson, A., Davidsson, P.: PRASH: a framework for privacy risk analysis of smart homes. Sensors (Basel). **21**(19), 6399 (2021). https://doi.org/10.3390/s21196399

45. Rossi, A., Palmirani, M.: Can visual design provide legal transparency? The challenges for successful implementation of icons for data protection. Des. Issues. **36**(3), 82–96 (2020). https://doi.org/10.1162/desi_a_00625

46. Buchanan, K., Healy, C.: Analyzing cybersecurity definitions for non-experts. Nist.gov. Available: https://tsapps.nist.gov/publication/get_pdf.cfm?pub_id=936618. Accessed 15 Sept 2024

47. Souza, J., Leung, C.K.: Explainable artificial intelligence for predictive analytics on customer turnover: a user-friendly interface for non-expert users. In: Explainable AI Within the Digital Transformation and Cyber Physical Systems, pp. 47–67. Springer, Cham (2021). https://doi.org/10.1007/978-3-030-79418-3_4

48. Rodrigues, A., Villela, M.L., Feitosa, E.: PTMOL: a suitable approach for modeling privacy threats in online social networks. In: Proceedings of the 21st Brazilian Symposium on Human Factors in Computing Systems, vol. 01, pp. 1–12 (2022). https://doi.org/10.1145/3562427.3564962

49. Klymenko, A., Meisenbacher, S., Favaro, L., Matthes, F.: On the integration of privacyenhancing technologies in the process of software engineering. In: Proceedings of the 26th International Conference on Enterprise Information Systems, pp. 41–52 (2024). https://doi.org/10.5220/0012078700003410

50. Dorafshanian, M., Aitsam, M., Mejri, M., Di Nuovo, A.: Beyond data collection: safeguarding user privacy in social robotics. In: 2024 IEEE International Conference on Industrial Technology (ICIT), pp. 1–6 (2024). https://doi.org/10.1109/ICIT57183.2024.1234567

A Method to Optimally Discover the Controller Locations in Software-Defined Networks

Mili Dhar[1]([✉]) [iD], Pooja[2], Vartika Mishra[1] [iD], and Suyash Shukla[2] [iD]

[1] Galgotias University, Greater Noida, India
{mili.dhar,vartika.mishra}@galgotiasuniversity.edu.in
[2] Bennett University, Greater Noida, India
pooja@bennett.edu.in

Abstract. In this paper, we have shown a method for placing multiple controllers in a network. It is based on the minimization of distances between the controllers and the switches. The location of the controller has been found in a way that the sum of distances between these said switches will have a minimum distance by balancing the switch loads. Our method mainly focused on distributing an almost equal number of switches among controllers to balance the load, improve fault tolerance, and optimize the average switch-to-controller (NC) latency. We have compared our results with K-Means clustering algorithm and found that our method always generates a consistent result of the average NC latency, accompanied by reduced load disbalance and a lower coefficient of variation.

Keywords: Controller Placement Problem (CPP) · Software Defined Network · Centre of Mass · latency · load disbalance

1 Introduction

Software Defined Network (SDN) technology is one kind of network management that facilitates dynamic and efficient network configuration, improving both performance and monitoring [1]. Current networks require more flexibility and easy troubleshooting instead of having decentralized and complex traditional network architecture. SDN is a favorable architecture to get rid of the challenges faced by traditional networks [2]. It permits network administrators to control, modify, and manage network behavior effectively [3]. Traditional network architecture is hardware-based, where the control and data planes are bundled together within a network device. Each network device makes its own decisions about traffic flow. To maintain a global network view, network devices need to communicate with each other [4]. Any modifications required in the network by operators need manual intervention, which is time-consuming, especially for large-scale networks.

P. Chandrakar et al. (Eds.): ICCINS 2025, CCIS 2738, pp. 234–243, 2026.
https://doi.org/10.1007/978-3-032-09572-5_19

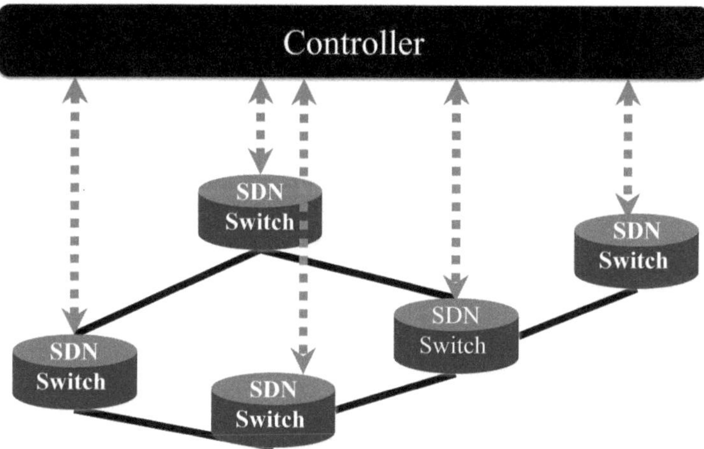

Fig. 1. SDN architecture with single controller

The fundamental idea of SDN is disassociating [5] the data-plane from the control-plane. Instead of communicating with multiple controllers to maintain a global network view while forwarding packets, a single controller is now sufficient to handle all network devices while maintaining the global network view[6]. A SDN architecture with a single controller is shown in Fig. 1. A SDN with a single controller is favorable as it comes up with centralized management and a global network view. When a packet arrives at the SDN-controlled forwarding device, it checks its flow table to find the corresponding flow entries. The forwarding device first processes the packet header and it either knows what to do with that packet or it will ask its controller for the updated flow-table. Execution of flow entries in the data-plane originated by the control-plane [7]. So, a controller determines the routing policies and makes decisions dependent on the broad view of the entire network. Forwarding devices are commonly known as switches in SDN. Figure 2 shows the communication between a controller and switches whenever a new packet arrives. If there is a miss in the flow table, the switch will query its controller by sending a packet-in message to determine the appropriate action—whether to forward or drop the packet. The controller will respond by sending a packet-out message and simultaneously update the flow tables of each affected switch along the path to the destination [8].

However, this technique remarkably increases the latency, if switches are far away from the controller. The largest part of latency is called "Propagation Latency (PL)", in the process of data forwarding between the controller and its assigned switches. Hence, clearly, there is a bottleneck in using a single controller within a SDN network related to scalability and performance [9].

Depending on the PL, the Controller Placement Problem (CPP) becomes the most prevalent research area in today's network. CPP has been first introduced in SDN by Heller et al. [10]. The CPP is a key responsibility in SDN architecture

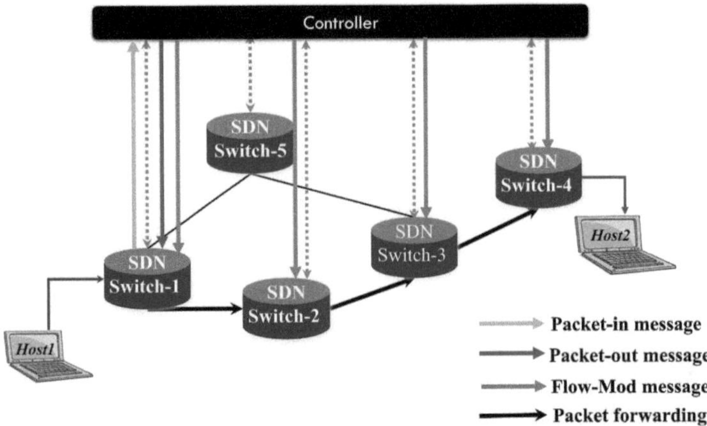

Fig. 2. setup to forward a new packet

placing multiple entities within a network which will be beneficial to improve various requirements such as latency, failure tolerance, and load balancing. CPP in SDN mainly depends on the controller numbers required for a network, the location of those controllers, and the roles of the controller to control assigned switches. Controllers and switches employ a protocol like 'OpenFlow' [11] which provides an open standard method to transmit and exchange information and has drawn significant interest from both industry and academics.

Inspired by this, we propose a mathematical algorithm for deploying controllers and assigning switches to those controllers so that it can optimize the propagation latency by minimizing the path lengths.

Main contributions:

- We have proposed a centroid-based mathematical model. Using this model, we divide the network into clusters and place one centroid in each cluster. These centroids represent the locations for placing controllers.
- This centroid-based model has been developed by minimizing the path distance between network components (NC). Reducing the path distance helps to get a minimum average NC latency.
- We have also proposed a multi-controller placement strategy that improves network load balancing. Our method finds appropriate controller location while distributing switches evenly among controllers, which ensures better network resilient in case of controller failure.
- Simulations have been done in MATLAB on two different networks. Performance analysis shows that our proposed method consistently achieves optimal results irrespective of the number of iterations. Our method generates a lower average NC latency while maintaining traffic loads and improves network resilience.

2   Existing Works

Benzekki et al. [3] presented some issues related to SDN networks and gives an insight view of the challenges faced by the future network model. It also presented distinct mitigation techniques and solutions based on SDN dependability, scalability, reliability, elasticity, security, resiliency, performance, and high availability concerns. This paper also shows a brief comparison with other existing techniques depending on these performance matrices.

In the paper [12], authors considered a backup controller for each switch to reduce the NC latency once any main controller fails. Authors proposed a method where they predefined how many controller may fail and how those switches will be operated during failure time. It reduces worst case delay one could have during failure.

There is a research proposed by authors of [13] where they have divided a network into clusters while reducing ene-to-end latency between NC. This multi-controller placement method considered single controller in each cluster. Their proposed method is useful to reduce worst case end=to-end latency into two different topologies [14] called OS3E and ChainaNet.

Paper [7] presented a comprehensive survey on the classification of SDN controller placement problems. This paper analyses the existing SDN controller placement problem and shows some limitations and future scope related to open research issues of controller placement problem.

The approach by paper [10] addressed a heuristic-based solution by using a simulated annealing genetic algorithm for the wireless SDN CPP. They presented a CPP solution that shields the failure probability of links, transparency, and latency on the wireless southbound interface.

An adaptive approach for CPP in SDN networks is proposed in the paper [15]. They presented network architecture-related issues and gave a SDN-based solution. This solution solved CPP i.e., the positioning of several resources in a network to meet network requirements (latency, failure tolerance, and load balancing).

However, paper [16] first proposed a CPP to address the fact of the resilience properties of a network. The author used the term, called min-cut algorithm by calculating the probability of link failure between network nodes which characterizes the resilience of a network. The whole network is partitioned into two domains and the centroid of each domain is selected as a controller in the network for better performance in terms of the resilience of the network.

The authors in [17] addressed a CPP strategy that considers reliability and controller capacity as well as makes plans for controller failure to avoid latency, repeated network intervention, and disconnection.

3   Problem Formulation

Our proposed algorithm formulates the optimization problem of controller placement by minimizing the average NC latency. We followed [18], the method developed by A. Debnath et al. while talking about the vehicular Network. We expanded the algorithms for determining the minimum distances in any complex network A graph G (N_s, E_s, C_s) is represented as a physical network, where N_s is the set of switches and E_s is the set of connections between the switches and the controllers. C_s is the set of controllers which means $|Cs| = C$ and $|Ns| = N$ represent the total number of controllers and nodes in a network respectively. Network switches have been distributed into multiple clusters and each cluster is controlled by a different controller.

Here, a mathematical algorithm has been introduced that minimizes the distance between the controller and switches and successfully deploys C number of controllers. For each controller, k number of switches is selected, where $k = 1, 2, 3...N(k \in N)$. All these switches are selected out of the total N number of switches, such that $N = \sum_{i=0}^{C} \sum_{j=1}^{k} S_j^i$.

At first, we assume that a single controller controls at least k number of switches, and the co-ordinate locations of the controllers are represented as (X_m^C, Y_m^C) where $[m = 1$ to $C]$. These co-ordinates (X_m^C, Y_m^C) are the locations found using the following Eqs. (1) and (2). Here, $k = floor(\frac{N}{C})$.

$$X_C = \frac{\sum_{i=1}^{N} \frac{a_i x_i}{R_i}}{\sum_{i=1}^{N} \frac{1}{R_i}} \tag{1}$$

$$Y_C = \frac{\sum_{i=1}^{N} \frac{a_i y_i}{R_i}}{\sum_{i=1}^{N} \frac{1}{R_i}} \tag{2}$$

In the above equation, the value of a_i indicates the activity level of each switch. In our model, we have assumed that all switches are equally active; thus, $a_i = 1$. R_i is the distance of the ith switch from the centroid coordinate. Centroid coordinates are the locations found by using equation (1) which are also denoted by (X_m^C, Y_m^C) here $[m = 1, 2, 3...., C]$.

At the beginning of the algorithm, we don't know the centroid location. Thus, we will assume an arbitrary centroid location until it converges. For example, at first, we assumed that our centroid location is $(X_m^C, Y_m^C) = (0, 0)$. Then, we calculate the actual value of the centroid point using (1). After that, we select only k number of switch points among N which are at a minimum distance from the first centroid point (X_m^C, Y_m^C). After getting our first k number of switches, we shall again find a centroid location. We consider the nearest node of this centroid location as our first controller location. In the next step, we take the remaining $(N - k)$ number of switches and do the same until we have all the C number of controller locations, namely (X_m^C, Y_m^C) where $[m = 1$ to $C]$. After getting centroid locations for each k number of switches, if there are any switches left to assign to any controller then it will be assigned to its nearest controllers. The procedure described above is shown in Fig. 3.

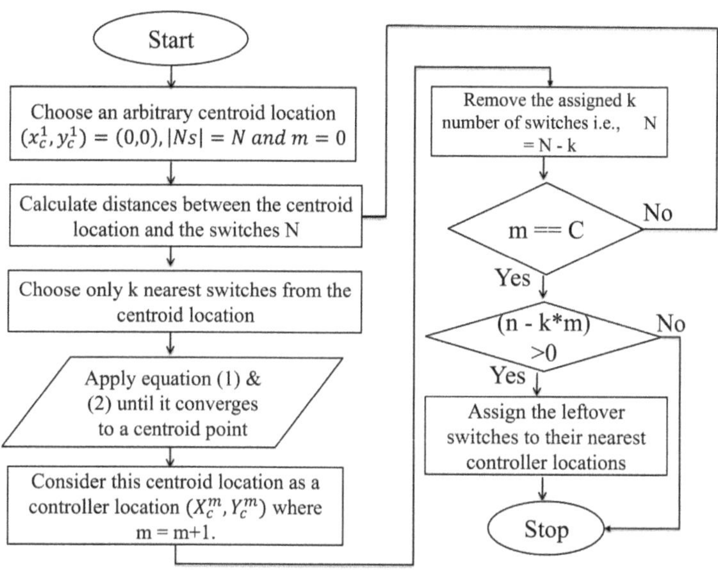

Fig. 3. Flowchart of our proposed method

4   Result Analysis

We considered two networks AT&T and Random with 25 and 100 switches to analyze the results. To analyze our results, we have also considered K-Means method. Simulations of both the methods (proposed method and K-means) have been done in MATLAB.

4.1   Analysis Based on Load Disbalance

Balancing loads among controllers is responsible for making a network fault-tolerant. To improve the survivability of a network against failures, one should have a lower load disbalance. Load disbalance refers to the difference between the minimum and the maximum number of switches controlled by a controller in a network [8,19]. Equation (3) is used calculate load disbalance among clusters in a network. Here, s indicates load of a controller.

$$Load_d = \max_{i \in C}\left(s_i\right) - \min_{j \in C}\left(s_j\right) \tag{3}$$

In Fig. 4, we show the load disbalance of different networks (AT&T and Random). Here, Blue color represents our proposed method and Red color represents the K-means clustering method. From Fig. 4, one can easily observe that our method generates a lower value of load disbalance compared to K-means. To reduce the NC latency K-Means sometime assigns a large number of switches to a single controller. As a result, packets may be discarded due to the limited capacity of a controller. It means using our method, a network controller can utilize their maximum capacity over K-means.

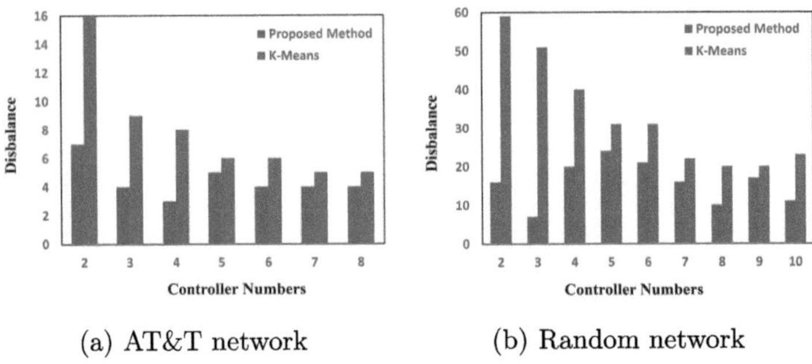

(a) AT&T network (b) Random network

Fig. 4. Load Disbalance

4.2 Analysis Based on Average NC Latency

Minimization of average NC latency is an important and mostly used performance matric because controllers are placed far from its data plane. Equation (4) is used to calculate this parameter.

$$L = \frac{1}{v|N|} \sum_{i \in N} \min_{j \in C} (NC_{ij}) \tag{4}$$

In the equation (4), v is the velocity of a wired medium and NC_{ij} is the shortest path between NC.

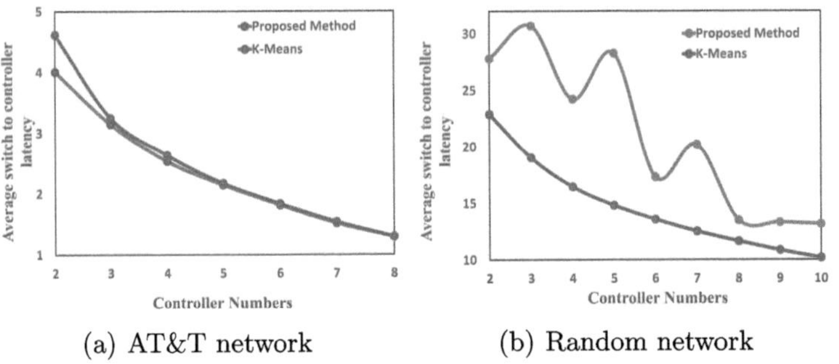

(a) AT&T network (b) Random network

Fig. 5. Average NC latency

In Fig. 5, we show the results of average NC latency in different networks (AT&T and Random). Here, the blue represents our proposed method and the red represents the K-means clustering method. We have selected the lowest result of K-means after executing the program 100 times because K-means randomly selected the starting center location. This random selection has a higher standard deviation. Thus, we run the K-means program 100 times and select the lowest value whereas our method always generates the same result which means our method has zero standard deviation. In Fig. 5, we can see that our proposed method can generate the lowest average NC latency in the case of AT&T but K-means generates the lowest result compared to ours in the case of Random network. This is because we balanced the switch loads among controllers by distributing an almost equal number of switches among controllers. For balancing the loads, we need to compromise latency to be minimized. Though we have not achieved the minimum average NC latency for all the networks, our method is still more consistent with zero standard deviation and better load distribution.

4.3   Analysis Based on Coefficient of Variation (CV)

Here, we have compared the Coefficient of variation ()CV once a controller fails. CV is a ratio of standard deviation (σ) and average NC latency (L). Lower CV specifies lower dispersion which means fairness of switch distribution [20]. Equation (6) shows the mathematical formulation to calculate this parameter. In equation (5), s_c is the average cased NC latency at each individual clusters. c_j is the total number of switches allocated to a cluster using our centroid based model and d_{ij} is the shortest distance between i^{th} switch to j^{th} controller. In equation (6), L is the average NC latency of a network calculated using equation (4).

$$s_c = \frac{1}{vc_j} \sum_{i=1}^{c_j} d_{ij} \tag{5}$$

$$CV = \frac{\sigma(s_c)}{L} \tag{6}$$

Figure 4 shows an analysis based on CV, considering when a controller will fail switches of that failed controller will be controlled by the nearest controller. From Fig. 6, one can easily observe proposed method has a lower value of CV compared to K-means. Our method finds appropriate controller location while distributing switches evenly among controllers which further shows that the network is more resilient in case of controller failure.

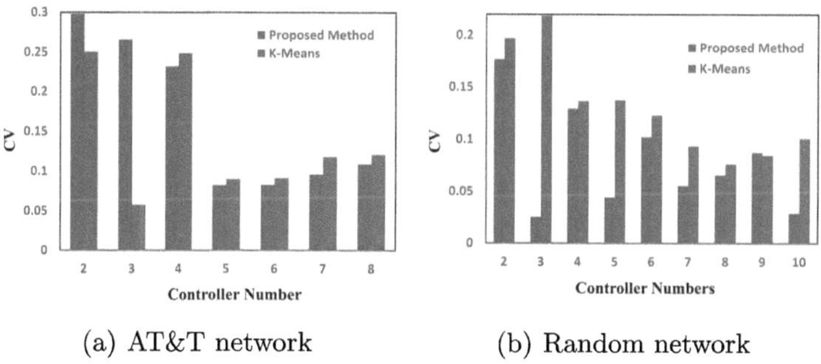

(a) AT&T network   (b) Random network

Fig. 6. Coefficient of variation

5   Conclusion

SDNs are supposed to eliminate all the traditional networks because of their logical, centralized, and programmable paradigm. Locating controllers in their appropriate positions is one of the more challenging tasks in SDN as network traffic and data transfer are increasing daily. In this paper, we have shown a technique to find the controller locations while balancing the loads. It has been found that to get the minimum average NC latency, one needs to compromise in load balancing. Our method always generates a consistent value irrespective of the number of iterations. We always need to run k-means multiple times to get the minimum result resulting higher standard deviation. It is also found that the proposed method also gives better fault tolerance in case of failure. Thus, we can say that our proposed method is much more suitable compared to K-means to get better performances in load balancing, fault tolerance, and average NC latency.

References

1. Wang, H.-C., Doong, H.-S.: Validation in internet survey research: Reviews and future suggestions. In: 2007 40th Annual Hawaii International Conference on System Sciences (HICSS'07). IEEE, pp. 243c–243c (2007)
2. Hu, Y.-N., Wang, W.-D., Gong, X.-Y., Que, X.-R., Cheng, S.-D.: On the placement of controllers in software-defined networks. J. China Univ. Posts Telecommun. **19**, 92–171 (2012)
3. Benzekki, K., El Fergougui, A., Elbelrhiti Elalaoui, A.: Software-defined networking (SDN): a survey. Secur. Commun. Netw. **9**(18), 5803–5833 (2016)
4. Li, D., Wang, S., Zhu, K., Xia, S.: A survey of network update in SDN. Front. Comp. Sci. **11**, 4–12 (2017)
5. Scott-Hayward, S., O'Callaghan, G., Sezer, S.: SDN security: a survey. In: 2013 IEEE SDN For Future Networks and Services (SDN4FNS), pp. 1–7. IEEE (2013)
6. Rana, D.S., Dhondiyal, S.A., Chamoli, S.K.: Software defined networking (SDN) challenges, issues and solution. Int. J. Comput. Sci. Eng. **7**(1), 884–889 (2019)
7. Singh, A.K., Srivastava, S.: A survey and classification of controller placement problem in SDN. Int. J. Netw. Manage **28**(3), e2018 (2018)
8. Dhar, M., Debnath, A., Bhattacharyya, B.K., Debbarma, M.K., Debbarma, S.: A comprehensive study of different objectives and solutions of controller placement problem in software-defined networks. Trans. Emerg. Telecommun. Technol. **33**(5), e4440 (2022)
9. Dvir, A., Haddad, Y., Zilberman, A.: The controller placement problem for wireless SDN. Wirel. Netw. **25**, 4963–4978 (2019)
10. Heller, B., Sherwood, R., McKeown, N.: The controller placement problem. ACM SIGCOMM Comput. Commun. Rev. **42**(4), 473–478 (2012)
11. McKeown, N., et al.: OpenFlow: enabling innovation in campus networks. ACM SIGCOMM Comput. Commun. Rev. **38**(2), 69–74 (2008)
12. Killi, B.P.R., Rao, S.V.: Towards improving resilience of controller placement with minimum backup capacity in software defined networks. Comput. Netw. **149**, 102–114 (2019)

13. Wang, G., Zhao, Y., Huang, J., Wu, Y.: An effective approach to controller placement in software defined wide area networks. IEEE Trans. Netw. Serv. Manage. **15**(1), 344–355 (2017)
14. Knight, S., Nguyen, H.X., Falkner, N., Bowden, R., Roughan, M.: The internet topology zoo. IEEE J. Sel. Areas Commun. **29**(9), 1765–1775 (2011)
15. Talhar, P., Bhagat, A.P.: An adaptive approach for controller placement problem in software defined networks. In: 2018 International Conference on Research in Intelligent and Computing in Engineering (RICE), pp. 1–11. IEEE (2018)
16. Zhang, Y., Beheshti, N., Tatipamula, M.: On resilience of split-architecture networks. In: 2011 IEEE Global Telecommunications Conference-GLOBECOM 2011, pp. 1–6. IEEE (2011)
17. Killi, B.P.R., Rao, S.V.: Capacitated next controller placement in software defined networks. IEEE Trans. Netw. Serv. Manage. **14**(3), 514–527 (2017)
18. Debnath, A., Basumatary, H., Tarafdar, A., DebBarma, M.K., Bhattacharyya, B.K.: Center of mass and junction based data routing method to increase the QoS in VANET. AEU-Int. J. Electron. Commun. **108**, 36–44 (2019)
19. Dhar, M., Bhattacharyya, B.K., Debbarma, M.K., Debbarma, S.: A cost-effective and load-balanced controller placement method in software-defined networks. Int. J. Netw. Manage **32**(5), e2199 (2022)
20. Dhar, M., Bhattacharyya, B.K., Kanti Debbarma, M., Debbarma, S.: A new optimization technique to solve the latency aware controller placement problem in software defined networks. Trans. Emerg. Telecommun. Technol. **32**(10), e4316 (2021)

Graeco-Latin Square-Based Modified Moss Growth Optimization (GLS-mMGO): A Triumvirate of Optimization Paradigms for Resource Allocation in 6G Network

Subrat Kumar Sethi[✉] and Arunanshu Mahapatro

Veer Surendra Sai University of Technology Burla, Odisha, India
subrat.vssutburla@gmail.com
http://www.vssut.ac.in

Abstract. The forthcoming 6G wireless networks will be characterized by unprecedented network densification, driven by the explosive growth of human-type and machine-type devices. This paradigm shift introduces new challenges, rendering conventional optimization techniques ineffective. To address these challenges, this paper proposes the Graeco-Latin Square-Based Modified Moss Growth Optimization (GLS-mMGO) technique, a novel paradigm for solving complex optimization problems in 6G network resource allocation. Inspired by the intricate growth patterns of moss, GLS-mMGO integrates a Graeco-Latin Square Design to orchestrate the placement of moss individuals, ensuring a balanced exploration of the solution space. The algorithm leverages the orthogonal properties of Graeco-Latin Squares to create a structured framework for moss placement, enabling efficient exploration and avoiding premature convergence. Furthermore, GLS-mMGO incorporates a Modified Spore Dispersal Search mechanism to adaptively adjust the dispersal range and optimize the trade-off between exploration and exploitation. To evaluate the effectiveness of GLS-mMGO, we compare its performance with the existing Moss Growth Optimization (MGO) technique using 23 standard benchmark functions. The results demonstrate the GLS-mMGO technique's superior performance, robustness, and efficacy in solving complex optimization problems in 6G network resource allocation.

Keywords: Exploration · Exploitation · Graeco-Latin Square · Moss Growth Optimization · 6G Wireless Network

1 Introduction

The optimization of resource allocation in 6G wireless networks presents significant challenges due to the complex, dynamic nature of these systems. Some of the network activities like allocating power to enhance spectral and energy efficiency [1], assigning of subcarriers in OFDMA and multi-carrier NOMA systems

[2], enabling cognitive routing in VANETs [3], and selecting remote radio heads in C-RANs [4] are generally modeled as mixed-integer non-linear programming (MINLP) problems.

These problems are NP-hard, making them difficult to solve using traditional methods. While global optimization techniques such as branch-and-bound and dynamic programming can provide optimal solutions, their exponential computational complexity limits their practical application in real-time, especially with the increasing number of devices and dynamic conditions in future 6G networks. Heuristic methods and machine learning (ML) approaches [5,6], though more scalable, often fail to guarantee optimal performance or struggle with the challenges of high-dimensional, real-time problems.

To address these limitations, metaheuristic algorithms (MAs) have emerged as powerful tools for solving complex optimization problems, owing to their ability to explore large solution spaces and balance computational efficiency with solution quality. Among these, the Moss Growth Optimization (MGO) algorithm [7] stands out as a bio-inspired approach, drawing idea from the way moss develops and expands, characterized by its adaptability and intricate patterns of growth in nature. MGO categorizes the solution population into key individuals based on problem dimensions. It determines the evolutionary trajectory of the population by analyzing the disparity between the top-performing individual and the major group, providing a unique approach to guiding optimization. Its primary mechanisms include spore dispersal search, dual propagation search, and cryptobiosis, allow it to effectively perform both global exploration and local exploitation, crucial for optimizing resource allocation in dynamic environments like 6G networks.

However, while MGO has shown promise in optimizing complex problems, there are areas for improvement in its ability to explore solution spaces efficiently. To enhance its performance, we propose an improvisation of the MGO algorithm by integrating a Graeco-Latin Square Design (GLSD) to orchestrate the placement of moss individuals. This modification ensures a more balanced and systematic exploration of the feasible region, enhancing the algorithm's capability to search effectively across high-dimensional solution spaces typical of resource allocation problems in 6G networks.

The traditional MGO mechanism is based on the "monitoring of wind trajectory," which simulates the dispersal of moss spores [7]. This monitoring of wind trajectory determine the direction of wind, which plays a crucial role in shaping the population's overall progression, steering the search process in a manner that helps avoid local optima. The spore dispersal search mechanism mimics the way moss spores spread under stable and turbulent wind conditions, enabling the algorithm to conduct a broad search in stable conditions and a more localized search during turbulent conditions. A binary propagation strategy, rooted in the moss's unique reproductive methods, combines sexual and vegetative reproduction to refine solutions through local exploitation, ensuring that the algorithm quickly converges to high-quality solutions. Finally, the cryptobiosis mechanism avoids stagnation at locally optimal points by modifying the traditional greedy selection mechanism, fostering a more robust search process.

By incorporating the GLSD, our enhanced MGO algorithm (mMGO) ensures that the population is distributed in a more structured and balanced manner across the solution space, facilitating better coverage and reducing the likelihood of premature convergence. Our work investigates the fusion of the GLSD with MGO, evaluating its performance in solving complex resource allocation problems in 6G networks. We demonstrate how mMGO can efficiently optimize the allocation of resources such as bandwidth, power, and time slots in highly dynamic and massive IoT environments, and compare its performance with existing optimization techniques.

The rest of this article is divided into the following sections: Sect. 2 explores the biological inspiration behind the growth mechanism of moss, providing insights into its adaptive and efficient growth strategies. Section 3 delves into the implications of these mechanisms for 6G resource allocation, highlighting the potential for bio-inspired approaches in optimizing network resources. In Sect. 4, the Graeco-Latin Square approach is introduced as a n5ovel method for improving the efficiency of resource distribution. Experimental results and analysis are presented in Sect. 5, offering a thorough evaluation of the proposed strategies. The paper ends with Sect. 6, which recapitulates the key takeaways and identifies areas for future exploration.

2   Biological Inspiration: The Growth Mechanism of Moss

Grey-cushion edge grimmia [8], also known as pulvinated dry rockmoss (Grimmia pulvinata), is a bryophyte moss species found in temperate regions globally [9]. The lifecycle of mosses, like that of plants, consists of two distinct phases: the gametophyte and the sporophyte. This type of lifecycle, where two different generations alternate, is referred to as alternation of heteromorphic generations. In vascular plants, the diploid generation (sporophyte) is the dominant phase of the lifecycle. However, in mosses, the dominant phase is the haploid generation, which is the gametophyte. This means that the green, leafy structures of moss, which are commonly seen, are haploid, containing only one set of chromosomes. The sporophyte, in contrast, is the diploid generation of the moss and remains attached to the gametophyte throughout its lifecycle. It is considered parasitic because it depends entirely on the gametophyte for nutrients and water, as it cannot independently perform photosynthesis. The sporophyte is essentially a reproductive structure that grows on and is nourished by the gametophyte, highlighting the interdependence between these two stages (Figs. 1, 2 and 3).

The dispersal of moss spores occurs predominantly in the early morning, allowing them to travel farther due to the calmer winds. The more turbulent winds present later in the day limit the dispersal distance of spores released at that time. Moreover, mosses undergo sexual reproduction through the migration of motile sperm between male and female gametophytes, where fertilization occurs and zygotes are formed. These zygotes develop into sporophytes, which remain attached to the gametophyte and rely on it for nourishment. Many moss species also reproduce vegetatively, often shedding fragments that can grow into new individuals when dispersed by the wind. Additionally, mosses have the remarkable ability to enter a cryptobiotic state, where metabolic activity halts, enabling them to survive under harsh conditions and rehydrate when environmental conditions improve.

Fig. 1. Grimmia pulvinata, flourishing on a concrete divider [8]

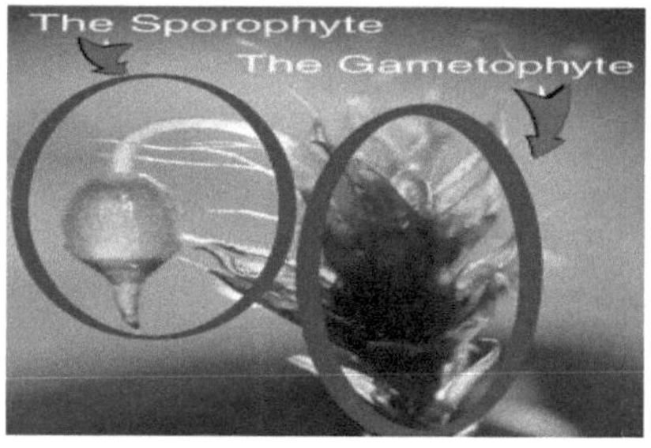

Fig. 2. Grimmia pulvinata: illustrating the gametophyte and sporophyte stages [8]

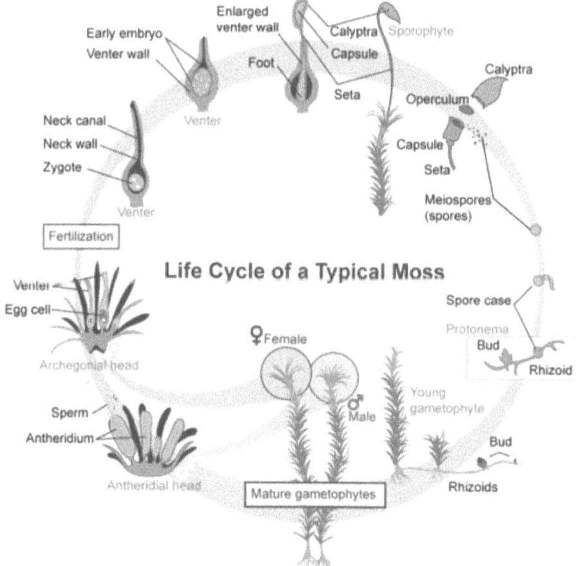

Fig. 3. Life cycle of a Typical moss Grimmia pulvinata [9].

2.1   Optimization Framework: Mathematical Model and Algorithm

The wind conditions significantly influence moss growth, particularly in spore dissemination. The MGO algorithm employs a novel 'wind direction determination' method. This approach informs the evolutionary trajectory of population members by analyzing spatial interactions between dominant and optimal individuals, helping the algorithm avoid becoming stuck in local optima. In the MGO algorithm, every moss entity is treated as a distinct search agent M, collectively forming the moss population X. To model wind direction, the algorithm makes the following assumptions:

- Throughout every cycle of the algorithm, the wind direction remains unchanged.
- The moss individuals are positioned in the solution space, with the best individual, denoted as M_{best}, representing the optimal solution found so far.
- The wind flows from regions with higher concentrations of moss toward the optimal individual, facilitating the movement of other agents toward the best solution.

The most optimal individual in the population X is labeled as M_{best}. The algorithm uses the j-dimensional value of M_{best} as a threshold and compares the j-dimensional values of all individuals to it. Based on this comparison, the individuals are divided into two sets.

$$DX_{j1} = \{M_i = (M_{i1}, M_{i2}, \ldots, M_{i\text{dim}}) \mid M_{ij} \geq M_{\text{best},j}, M_i \in X\} \tag{1}$$

$$DX_{j2} = \{M_i = (M_{i1}, M_{i2}, \ldots, M_{i\text{dim}}) \mid M_{ij} < M_{\text{best},j}, M_i \in X\} \tag{2}$$

The group with the higher number of members is selected as follows:

$$\text{divX}_j = \begin{cases} DX_{j1}, & \text{if count}(DX_{j1}) \geq \text{count}(DX_{j2}) \\ DX_{j2}, & \text{if count}(DX_{j2}) < \text{count}(DX_{j2}) \end{cases} \tag{3}$$

For sets obtained through multiple divisions, the algorithm defines a subset of moss individuals based on specific criteria:

$$\text{divX} = \{M_i = (M_{i1}, M_{i2}, \ldots, M_{i\text{dim}}) \mid M_i \in \bigcap_{j=1}^{\text{dn}} \text{divX}_{p_j}, M_i \in X\} \tag{4}$$

Here, dn represents the number of divisions, calculated as $\lfloor \frac{\text{dim}}{4} \rfloor$ and is at least 1. The floor function is denoted by $\lfloor \cdot \rfloor$, and p_j is the j-th random number within the range $(1, 2, \ldots, \text{dim})$, ensuring:

$$\bigcap_{j=1}^{\text{dn}} p_j = \emptyset \tag{5}$$

A simplified wind simulation is conducted, where the wind consistently blows from the region divX towards the optimal individual M_{best}. The calculation of dirX is given by:

$$\text{dirX} = \{M_{\text{best}} - M_i \mid M_i \in \text{divX}\} \tag{6}$$

In this context, dirX represents the collection of distances between individuals within divX and M_{best}.

The precise computation of wind direction is expressed as:

$$D_{\text{wind}} = \frac{1}{\text{num}} \sum_{i=1}^{\text{num}} dM_i, \quad dM_i \in \text{dirX} \tag{7}$$

Here, D_{wind} denotes the calculated wind direction, having the same dimension as the individuals. The variable num indicates the total number of individuals in dirX. Calculating the mean distance between major individuals and M_{best} helps smooth the path of individuals approaching M_{best}, thereby enhancing the optimization capability of MGO.

2.2 Exploration Phase of the MGO

During the exploration phase of the MGO, the dispersal of spores is crucial. Wind plays a significant role in this process, as its strength and stability determine how far the spores are dispersed. In conditions with stable winds, spores travel over greater distances, whereas in turbulent conditions, the dispersal range is shorter. The majority of spores are dispersed under stable wind conditions, with a smaller proportion spreading during turbulent periods. As the wind weakens, the spores tend to settle closer to the moss.

In this study, the spore positions are treated as potential solutions. The dispersal characteristics of the spores are modeled according to wind conditions. The position of a spore is updated based on the following equation:

$$M_i^{\text{new}} = \begin{cases} M_i + \text{step1} \cdot D_{\text{wind}}, & \text{if } r_1 > d_1 \\ M_i + \text{step2} \cdot D_{\text{wind}}, & \text{if } r_1 \leq d_1 \end{cases} \tag{8}$$

Here, M_i^{new} represents a new moss position derived from the dispersal of spores from the i-th moss individual M_i. r_1 is a random number between 0 and 1, and d_1 is a constant set to 0.2. If $r_1 > d_1$, spores are dispersed under stable wind conditions; otherwise, they are dispersed in turbulent conditions. The distance for dispersal in stable wind conditions is represented by step1, defined as:

$$\text{step1} = w \cdot (r_2 - 0.5) \cdot E \tag{9}$$

where w is a constant parameter set to 2, r_2 is a random vector within the range (0,1) with the same dimension as D_{wind}, and E represents the strength of the wind, which decreases with each iteration, as shown in Equation (9).

The distance for dispersal under turbulent wind conditions is represented by step2, calculated as:

$$\text{step2} = 0.1 \cdot w \cdot (r_3 - 0.5) \cdot E \cdot \left[1 + \frac{E}{2} \cdot \left(1 + \tanh\left(\beta\sqrt{1 - \beta^2}\right)\right)\right] \qquad (10)$$

where r_3 is a random vector in the range (0,1) with the same dimension as D_{wind}, and E is the wind strength as defined earlier.

The value of E is calculated as:

$$E = 1 - \frac{\text{FEs}}{\text{MaxFEs}} \qquad (11)$$

where FEs is the current evaluation count, and MaxFEs is the maximum number of iterations.

Finally, the value of β is given by:

$$\beta = \frac{\text{count}(\text{divX})}{\text{count}(X)} \qquad (12)$$

where β represents the proportion of the population in divX relative to the total population in X.

2.3 Exploitation Phase of the MGO

In the exploitation phase of the MGO, the *dual propagation search* plays a crucial role. This process mimics both sexual and vegetative reproduction, generating new individuals that are positioned close to the existing ones. For this mechanism to work correctly, the condition $c < 0.8$ must hold, where c is a random value between 0 and 1. During *sexual reproduction*, the genes of individuals are treated as potential solutions, allowing the creation of new individuals that inherit genes from both the current individual and the optimal one. In contrast, *vegetative reproduction* involves the generation of new individuals from fragments of existing moss, each of which represents a new solution. The movement of these fragments, much like spore dispersal, is influenced by the wind. However, unlike spore dispersal, the *dual propagation search* restricts the spread of new individuals to a smaller area, which accelerates the process of identifying optimal environments for moss growth.

In context of dual propagation search, the position of a new moss individual is defined by Equation (11). This technique improves local exploration by increasing the likelihood of modifying only one dimension of an individual, as opposed to traditional MAs.

$$M_i^{\text{new}} = \begin{cases} (1 - \text{act}) \cdot M_i + \text{act} \cdot M_{\text{best}}, & \text{if } r_4 > d_2 \\ M_{(i,j)}^{\text{new}} = M_{(best,j)} + \text{step3} \cdot D_{\text{wind},j}, & \text{if } r_4 \leq d_2 \end{cases} \qquad (13)$$

Here, M_i^{new} is the new individual derived from the i-th moss individual, and $M_{(i,j)}^{\text{new}}$ is the j-th particle of M_i^{new}, with j being a random number not exceeding the maximum dimension. M_{best} denotes the current optimal individual, while $M_{(best,j)}$ represents the j-th particle of M_{best}. $D_{\text{wind},j}$ is the j-th particle in D_{wind}, and r_4 is a random number between 0 and 1. If $r_4 > d_2$, the dual propagation search simulates sexual reproduction; otherwise, vegetative reproduction is applied. The variable act determines whether the particles in M_{best} are utilized, as shown in Equation (14). Lastly, the calculation for step3 is given in Equation (15).

$$\text{act} = \begin{cases} 1, & \text{if } \frac{1}{1.5 - 10 \cdot r_5} \geq 0.5 \\ 0, & \text{if } \frac{1}{1.5 - 10 \cdot r_5} < 0.5 \end{cases} \tag{14}$$

where r_5 is a random vector within the range (0,1) having the same dimension as M_{best}.

$$\text{step3} = 0.1 \cdot (r_6 - 0.5) \cdot E \tag{15}$$

where r_6 is a random number in the range (0,1), and E represents the wind strength.

2.4   Cryptobiosis Mechanism

To enhance the greedy selection process, this work uses cryptobiosis mechanism. The historical data of moss individuals are recorded using this method. Instead of directly modifying individuals, the mechanism maintains a record of each moss individual. When the maximum record count (set to 10) is reached or after a population iteration, the mechanism revives the optimal individual, replacing the current one. This approach ensures global exploration while maintaining population quality.

The MGO algorithm integrates biologically inspired mechanisms for optimization. The spore dispersal search, based on moss spore movement by wind, is utilized for global exploration. The dual propagation search, inspired by moss's sexual and vegetative reproduction, aids in local exploitation. Finally, the cryptobiosis mechanism enables multiple explorations from the same individual, avoiding local optima traps and improving population quality.

3   Optimization Model for Resource Allocation in a 6G Network

This optimization model for resource allocation in a **6G network** leverages the **MGO** algorithm, which is inspired by the biological behavior of moss, such as **spore dispersal, reproduction**, and **growth** mechanisms. By adopting these natural principles, the model efficiently addresses dynamic network management and resource distribution. Specifically, it incorporates the **Graeco-Latin Square Matrix**, which ensures diversity and balance in exploration and exploitation phases, alongside evolutionary direction, spore dispersal search, dual propagation search, and cryptobiosis mechanisms.

3.1 Graeco-Latin Square Matrix

A **Graeco-Latin Square** is a matrix of order n, where each cell contains a pair of distinct symbols. Two **Latin squares** L_1 and L_2 are considered **orthogonal** if, when superimposed, every pair of symbols from the two squares appears exactly once [10,11]. This orthogonality ensures that the individuals' search paths (solutions) are distinct and balanced, facilitating diverse exploration and exploitation strategies in the **MGO** algorithm.

For example, two orthogonal Latin squares of order 3 can be represented as:

$$
L_1 = \begin{pmatrix} A_1 & A_2 & A_3 \\ A_2 & A_3 & A_1 \\ A_3 & A_1 & A_2 \end{pmatrix} \quad \text{and} \quad L_2 = \begin{pmatrix} B_1 & B_2 & B_3 \\ B_2 & B_3 & B_1 \\ B_3 & B_1 & B_2 \end{pmatrix} \tag{16}
$$

Where:

- A_1, A_2, A_3 represent the first set of individuals (solutions).
- B_1, B_2, B_3 represent the second set of individuals.

These two matrices are **orthogonal**, meaning that every pair (A_i, B_j) is unique across both matrices. This orthogonality is central to the search process in the **MGO algorithm**, where the values of L_1 and L_2 are used to guide the placement of individuals in the search space.

In the context of the **MGO algorithm**, the **Graeco-Latin square matrix** ensures that each individual's path through the search space is distinct, enabling better exploration (via the spore dispersal search) and balanced exploitation (via the dual propagation search) strategies.

The individuals' placements in the search space are represented by the values $L_{i,j}$, where:

$$
L_{i,j} = (A_i, B_j) \tag{17}
$$

Thus, during the optimization process, for each individual M_i in the population, the placement is determined by the combination of symbols from L_1 and L_2 at index (i, j), ensuring diversity in search directions while keeping the individuals' paths balanced.

3.2 Evolutionary Direction in Resource Allocation

The **evolutionary direction** mechanism, inspired by the moss growth pattern, governs how resources are dynamically allocated across the **6G network**. Just as moss growth is influenced by environmental conditions (e.g., wind direction), the evolutionary direction adjusts resource allocation based on network conditions, such as signal strength, bandwidth demand, and interference.

The evolutionary direction is mathematically represented as:

$$M_i^{\text{new}} = L_{i,j} \cdot D_{\text{wind},j} \tag{18}$$

Where:

- $L_{i,j}$ represents the orthogonal pair from the Graeco-Latin square at position (i, j), guiding the resource allocation.
- $D_{\text{wind},j}$ is a vector representing the **wind direction** or the evolutionary force at position j, which adapts the individual's movement through the solution space.

This mechanism helps in avoiding local optima by ensuring that resource allocation adapts dynamically, guiding the system towards more efficient configurations while maintaining a diverse set of search directions.

3.3   Spore Dispersal Search (Exploration Phase)

The **spore dispersal search** phase models the **exploration** of resource configurations in a **6G network**. It simulates the dispersal of moss spores by wind, where spores spread over large distances under stable conditions and are more confined during turbulent conditions. This concept is applied to the search for optimal configurations of resources across the network.

The mathematical formulation for spore dispersal search can be written as:

$$M_i^{\text{new}} = \begin{cases} L_{i,j} \cdot D_{\text{wind},j} + \text{step1}, & \text{if } r_1 > d_1 \\ L_{i,j} \cdot D_{\text{wind},j} + \text{step2}, & \text{if } r_1 \leq d_1 \end{cases} \tag{19}$$

Where:

- r_1 is a random number in the range [0,1].
- d_1 is a threshold value (e.g., 0.5).
- step1 and step2 represent the dispersal steps for long-range and short-range exploration, respectively.

This mechanism ensures that the network explores a wide range of potential resource configurations during favorable conditions, while it adjusts more locally under unfavorable conditions, allowing for **flexibility** and **adaptability** in resource allocation.

3.4   Dual Propagation Search (Exploitation Phase)

The **dual propagation search** phase simulates both **sexual** and **vegetative reproduction** mechanisms in moss, guiding the **exploitation** of optimal resource configurations. During exploitation, the network fine-tunes its resource allocations in a smaller region around the best solutions, ensuring faster convergence towards the optimal configuration.

The dual propagation search is mathematically represented as:

$$M_i^{\text{new}} = \begin{cases} (1 - \text{act}) \cdot M_i + \text{act} \cdot M_{\text{best}}, & \text{if } r_4 > d_2 \\ L_{(i,j)}^{\text{new}} = M_{\text{best}} + \text{step3} \cdot D_{\text{wind},j}, & \text{if } r_4 \leq d_2 \end{cases} \tag{20}$$

Where:

- act is a binary parameter that decides whether sexual (1) or vegetative (0) reproduction occurs.
- M_{best} represents the best solution found so far.
- step3 controls the step size during exploitation.

This ensures localized adjustments around the best configurations, refining the network's resource allocation and improving efficiency.

3.5   Cryptobiosis Mechanism (Revival of Individuals)

The **cryptobiosis mechanism** revives previously optimal solutions to prevent stagnation and maintain diversity in the population. This is analogous to the way moss can revive from dormancy, ensuring that the network continually adapts by using previously found optimal solutions.

The revival of the optimal individual is mathematically expressed as:

$$M_i^{\text{new}} = r M_i^{\text{best}} \tag{21}$$

Where:

- $r M_i^{\text{best}}$ represents the best historical solution, revived for further optimization.

This mechanism ensures that the network does not get trapped in suboptimal configurations, promoting continuous adaptation and improvement in resource allocation.

4   Experimental Results and Analyses

To ensure a robust and consistent evaluation, all experiments were conducted under identical conditions, providing a fair basis for comparing the performance of the existing MGO approach with our proposed mMGO approach using a set of benchmark test functions. The experimental setup was configured as follows:

The system utilized Microsoft Windows 10 Home, offering a stable operating environment. The computational hardware included an Intel(R) Core(TM) i5-5200U CPU, running at a base clock speed of 2.20 GHz, which provided adequate processing power for executing the benchmark tests. The system was equipped with 8GB of RAM, enabling sufficient memory capacity for the MATLAB computations. For software, we used MATLAB R2018a, a widely recognized platform for implementing and evaluating optimization algorithms.

The initial step involves a quantitative analysis to determine the algorithm's optimal solution. Following this, the functions used for experimental testing are

Algorithm 1. Pseudo-code of mMGO algorithm

1: **Input:**
2:   N: population size
3:   dim: the problem dimensions
4: **Output:**
5:   Optimal solution
6: Initialize FEs to 0, $best\_cost$ to infinity, $best\_M$ as a zero vector of length dim, M with random solutions within bounds, $costs$ as a zero vector of length N
7: **for** each agent i in N **do**
8:   Calculate cost of $M(i)$ using $fobj$, Increment FEs
9:   **if** cost of $M(i)$ is less than $best\_cost$ **then**
10:     Update $best\_M$ to $M(i)$, Update $best\_cost$ to cost of $M(i)$
11:   **end if**
12: **end for**
13: Initialize w, rec_num, $divide\_num$, d1, $newM$, $newM\_cost$, rM, rM_cos
14: **while** $FEs < MaxFEs$ **do**
15:   Set $calPositions$ to M, Generate a uniformly distributed random Latin square, Using this $randls$ value get $div\_num = randls(dim)$
16:   **for** each j in $max(divide\_num, 1)$ **do**
17:     Set threshold th to $best\_M$ at $div\_num(j)$, Set index to positions in $calPositions$ where $div\_num(j)$ is greater than th
18:     **if** sum of index is less than half the size of $calPositions$ **then**
19:       Invert index
20:     **end if**
21:     Update $calPositions$ to positions at index
22:   **end for**
23:   Calculate distance D from individuals to the best, mean distance D_wind, β and γ using given formulas, $step$, $step2$, $step3$, and act using given formulas
24:   **if** rec is 1 **then**
25:     Record first generation of positions in rM, Record costs in rM_cos, Increment rec
26:   **end if**
27:   **for** each agent i in N **do**
28:     Set $newM(i)$ to $M(i)$
29:     **if** random number is greater than $d1$ **then**
30:       Update $newM(i)$ using $step$ and D_wind
31:     **else**
32:       Update $newM(i)$ using $step2$ and D_wind
33:     **end if**
34:     **if** random number is less than 0.8 **then**
35:       **if** random number is greater than 0.5 **then**
36:         Update $newM(i)$ at $div\_num(1)$ using $step3$ and D_wind
37:       **else**
38:         Update $newM(i)$ using $step3$, act, and $best\_M$
39:       **end if**
40:     **end if**
41:     Apply boundary absorption to $newM(i)$, Calculate $newM\_cost(i)$ using $fobj$, Increment FEs, Record $newM$ and $newM\_cost$ in rM and rM_cos
42:     **if** $newM\_cost(i)$ is less than $best\_cost$ **then**
43:       Update $best\_M$ to $newM(i)$, Update $best\_cost$ to $newM\_cost(i)$
44:     **end if**
45:   **end for**
46:   Increment rec
47:   **if** $rec > rec\_num$ or $FEs \geq MaxFEs$ **then**
48:     Find minimum cost in rM_cos and update M, Update $costs$ to minimum cost, Reset rec to 1
49:   **end if**
50:   Update $Convergence\_curve$ with $best\_cost$, Increment iteration counter it
51: **end while**
52: **return** the best solution M_best

Table 1. Comparison of mMGO and MGO average values on benchmark functions

Bench Mark Function	mMGO_avgval	MGO_avgval
F1	1.90542e-13	−0.047047192
F2	2.12747e-21	−0.003366342
F3	2.19188e-31	0.032779538
F4	−1.94537e-14	1.829010841
F5	0.003048103	1.359198576
F6	−0.283022657	−0.545682197
F7	0.000287397	0.020001702
F8	108.4698953	325.2030143
F9	2.49706E-10	−0.078479518
F10	5.37573E-17	−0.012239502
F11	3.40506E-11	−0.767145429
F12	−0.787844763	−1.384996541
F13	0.035422565	1.097800651
F14	−32.01442042	−31.97815947
F15	0.135934335	1.13301406
F16	0.31134213	−0.311328574
F17	4.581853933	2.708586922
F18	−0.500014899	−0.49954494
F19	0.509358396	0.507348789
F20	0.342803765	0.341952214
F21	0.942288938	4.000357343
F22	4.007930791	3.939276836
F23	1.00496319	3.99679796

derived from the classical 23 benchmark functions [12]. A depiction of the test function in three-dimensional form is observable in Figs. 4 and 5.

The Table 1 presents the average functional values (denoted as `avgval`) for two optimization methods: **mMGO** and **MGO**. These functional values reflect the performance of each optimization method in minimizing certain objective functions across multiple test cases (F1 to F23). A comparative analysis of these values reveals critical insights into the efficacy and efficiency of the two algorithms.

4.1 Overall Performance Comparison

Upon reviewing the results for both mMGO and MGO, we observe that mMGO consistently outperforms MGO in terms of minimizing functional values. For example, in functions F1 (Fig. 4a and 4b), F2 (Fig. 4c and 4d), and F3 (Fig. 4e and 4f), mMGO achieves values in the order of 10^{-13}, 10^{-21}, and 10^{-31}, respectively, indicating a near-zero or significantly small functional value. These results suggest that mMGO is highly effective at finding near-optimal solutions, especially in cases where precision is crucial. On the other hand, MGO yields values like -0.047047192 for F1, -0.003366342 for F2, and 0.032779538 for F3, which are considerably larger, reflecting poorer optimization performance.

4.2 Minimization and Convergence

A key advantage of mMGO is its ability to minimize the functional values to an almost zero or very small range, which is evident in the majority of the test cases (e.g., F1, F2, F3). This indicates that the modified approach employed in mMGO, particularly through the introduction of the **Graeco-Latin Square Design**, has enhanced the exploration and exploitation balance of the algorithm. The design ensures that the population is better distributed in the solution space, thereby preventing premature convergence and improving the algorithm's search capability.

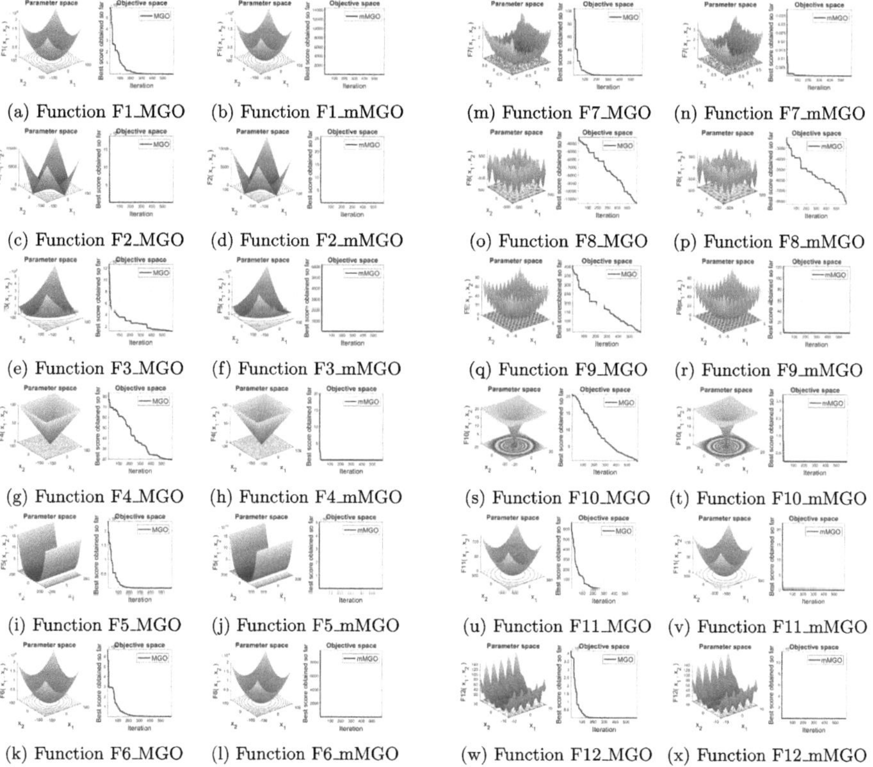

(a) Function F1_MGO (b) Function F1_mMGO (m) Function F7_MGO (n) Function F7_mMGO

(c) Function F2_MGO (d) Function F2_mMGO (o) Function F8_MGO (p) Function F8_mMGO

(e) Function F3_MGO (f) Function F3_mMGO (q) Function F9_MGO (r) Function F9_mMGO

(g) Function F4_MGO (h) Function F4_mMGO (s) Function F10_MGO (t) Function F10_mMGO

(i) Function F5_MGO (j) Function F5_mMGO (u) Function F11_MGO (v) Function F11_mMGO

(k) Function F6_MGO (l) Function F6_mMGO (w) Function F12_MGO (x) Function F12_mMGO

Fig. 4. Comparison of Original Code and Modified Code

In contrast, MGO exhibits a higher variation in the functional values, as seen in functions like F4 (Fig. 4g and 4h), F5 (Fig. 4i and 4j), and F6 (Fig. 4k and 4l), where the results show much larger functional values, indicating that the original MGO struggles to converge towards the global minimum as effectively as mMGO. Specifically, F5 shows a functional value of 1.359198576 for MGO, which suggests that the algorithm failed to identify the optimal solution within a reasonable range for this case.

4.3   Accuracy in Optimization

For several functions, mMGO approaches the desired minimization goals far more closely than MGO. For example, in function F8 (Figs. 4o and 4p), mMGO achieves a value of 108.4698953, which is a significant improvement compared to the 325.2030143 yielded by MGO. This demonstrates that mMGO's more efficient search process and balanced population placement likely led to a better exploration of the search space, achieving a smaller, more accurate functional value for this particular case. However, it is important to note that mMGO does not always result in the absolute smallest values across all functions, as observed in some of the later cases, such as F21 (Fig. 5q and 5r), where both algorithms show relatively close values (0.942288938 for mMGO vs. 4.000357343 for MGO). However, mMGO still consistently outperforms MGO in minimizing these values.

4.4   Robustness Across Test Cases

The ability of mMGO to produce smaller values across a wide range of functions—especially in highly variable cases like F4 (Fig. 4g and 4h), F5 (Fig. 4i and 4j), and F11 (Fig. 4u and 4v)—suggests a more robust optimization capability. MGO, while efficient in certain contexts, shows more significant fluctuations in its performance. For instance, F11 yields -0.767145429 for MGO and $3.40506E - 11$ for mMGO, showcasing the latter's ability to converge to a near-zero functional value more reliably. In complex optimization scenarios with multiple conflicting objectives or harsh constraints, mMGO's robustness could be a key advantage over MGO.

4.5   Outliers and Special Cases

Some functions, such as F12 (Fig. 4w and 4x) and F14 (Fig. 5c and 5d), show values that do not significantly differ between mMGO and MGO, suggesting that both algorithms may perform similarly in specific problem instances. F14, for instance, gives functional values of -32.01442042 and -31.97815947 for mMGO and MGO, respectively, indicating that both methods approach the solution in a similar fashion for this particular case. However, it is important to recognize that such outliers are relatively rare in the provided results, and overall, mMGO demonstrates superior performance across most test cases. This analysis, supported by Figs. 4 and 5, highlights the superior performance of mMGO over

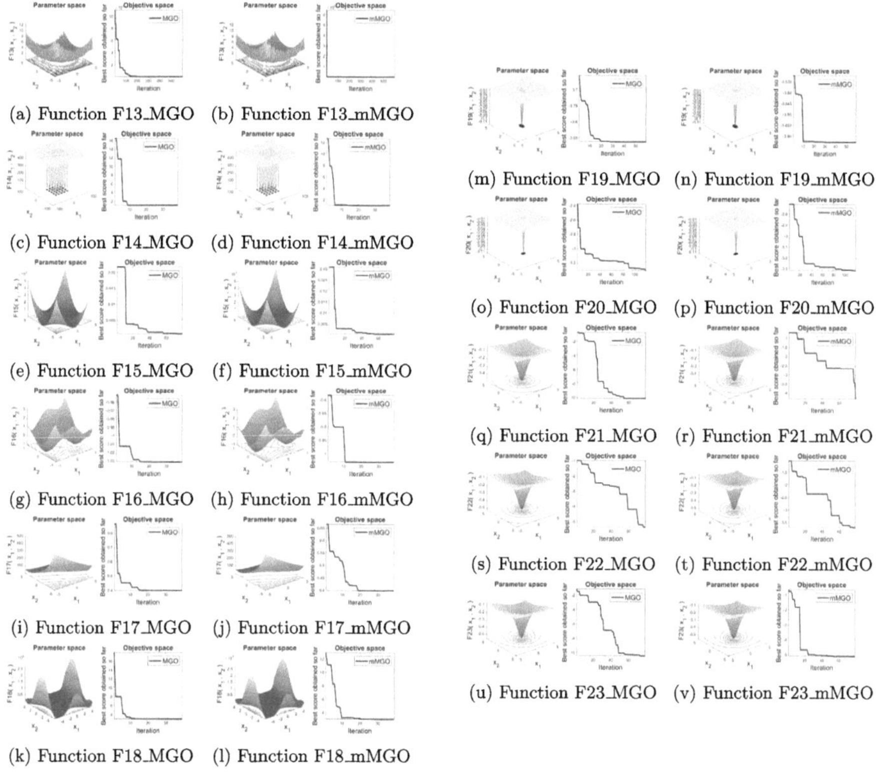

(a) Function F13_MGO   (b) Function F13_mMGO

(c) Function F14_MGO   (d) Function F14_mMGO

(e) Function F15_MGO   (f) Function F15_mMGO

(g) Function F16_MGO   (h) Function F16_mMGO

(i) Function F17_MGO   (j) Function F17_mMGO

(k) Function F18_MGO   (l) Function F18_mMGO

(m) Function F19_MGO   (n) Function F19_mMGO

(o) Function F20_MGO   (p) Function F20_mMGO

(q) Function F21_MGO   (r) Function F21_mMGO

(s) Function F22_MGO   (t) Function F22_mMGO

(u) Function F23_MGO   (v) Function F23_mMGO

Fig. 5. Comparison of Original Code and Modified Code

MGO in various optimization scenarios. This thorough analysis of the MGO algorithm confirms its robustness and efficiency in solving optimization problems, highlighting its potential for real-world applications.

5   Conclusion

In this paper, we proposed the GLS-mMGO technique as an advanced solution to resource allocation in 6G networks. By combining the natural growth patterns of moss with the orthogonal properties of Graeco-Latin squares, the GLS-mMGO algorithm demonstrates a significant improvement in optimization efficiency. Compared to the traditional MGO, GLS-mMGO exhibits superior performance in minimizing functional values across various benchmark functions, effectively addressing the challenges posed by dynamic and complex 6G network environments. The integration of the Graeco-Latin Square Matrix ensures balanced exploration and exploitation during the optimization process, thus preventing premature convergence and enhancing the overall robustness of the algorithm. Additionally, the inclusion of adaptive mechanisms, such as the Modified

Spore Dispersal Search, contributes to more efficient resource allocation and improved convergence towards optimal solutions.

While the GLS-mMGO technique shows promising results in resource allocation for 6G networks, there are several avenues for future research. First, extending the algorithm to handle more complex multi-objective optimization problems, such as those encountered in multi-layered 6G network architecture, could further enhance its applicability. Second, the integration of real-time network conditions, such as traffic load and interference patterns, could be explored to adapt the optimization dynamically. Additionally, hybrid approaches combining GLS-mMGO with other optimization techniques, such as deep reinforcement learning or swarm intelligence algorithms, could provide further improvements in scalability and adaptability. Finally, validating the performance of GLS-mMGO in large-scale, real-world 6G network simulations will be crucial to assess its practicality and robustness in diverse operational environments.

References

1. Sethi, S.K., Mahapatro, A., Mishra, N.: Optimal resource allocation to improve energy efficiency of cognitive radio-based vehicular Ad Hoc network under imperfect sensing. In: Mahapatra, S., Shahbaz, M., Vaccaro, A., Emilia Balas, V. (eds.) Advances in Energy Technology. Advances in Sustainability Science and Technology. Springer, Singapore (2021). https://doi.org/10.1007/978-981-15-8700-9_21
2. Fu, S., Fang, F., Zhao, L., Ding, Z., Jian, X.: Joint transmission scheduling and power allocation in non-orthogonal multiple access. IEEE Trans. Commun. **67**(11), 50–8137 (2019)
3. Sethi, S.K., Mahapatro, A.: Interference aware intelligent routing in cognitive radio based Vehicular Ad Hoc networks for smart city applications. Int. J. Inf. Technol. **5**, 1–21 (2023)
4. You, L., Yuan, D.: User-centric performance optimization with remote radio head cooperation in C-RAN. IEEE Trans. Wirel. Commun. **19**(1), 53–340 (2019)
5. Karimi-Mamaghan, M., Mohammadi, M., Meyer, P., Karimi-Mamaghan, A.M., Talbi, E.G.: Machine learning at the service of meta-heuristics for solving combinatorial optimization problems: a state-of-the-art. Eur. J. Oper. Res. **296**(2), 393–422 (2022)
6. Sethi, S.K., Mahapatro, A.: A deep learning-based discrete-time Markov chain analysis of cognitive radio network for sustainable internet of things in 5G-enabled smart city. Iran. J. Sci. Technol. Trans. Electri. Eng. **48**(1), 37–64 (2024)
7. Zheng, B., Chen, Y., Wang, C., Heidari, A.A., Liu, L., Chen, H.: The moss growth optimization (MGO): concepts and performance. J. Comput. Des. Eng. **11**(5), 184–221 (2024)
8. https://www3.botany.ubc.ca/bryophyte/mossintro.html. Accessed 20 Nov 2024
9. https://www.ck12.org/flexi/life-science/nonvascular-plants
10. Sethi, S.K., Mahapatro, A.: Mitigating PAPR challenges in massive MIMO systems for CR-IoT networks: a Graeco-Latin square approach. In: Chaubey, N., Jhanjhi, N.Z., Thampi, S.M., Parikh, S., Amin, K. (eds.) Computing Science, Communication and Security. COMS2 2024. Communications in Computer and Information Science, vol. 2174. Springer, Cham (2025). https://doi.org/10.1007/978-3-031-75170-7_5

11. Jacobson, M.T., Matthews, P.: Generating uniformly distributed random Latin squares. J. Comb. Des. **4**(6), 37–405 (1996)
12. Yao, X., Liu, Y., Lin, G.: Evolutionary programming made faster. IEEE Trans. Evol. Comput. **3**(2), 82–102 (1999)

Optimizing Charger Placement for Q-Coverage Using Blackhole Algorithm

P. Neelagandan[(✉)], S. Balaji, and R. Pavithra

Department of Mathematics, School of Advances Science, Vellore Institute of Technology, Chennai Campus, Chennai, India
`neelagandan.p2022@vitstudent.ac.in`, `{s.balaji,r.pavithra}@vit.ac.in`

Abstract. The efficient placement of chargers in sensor networks is essential for ensuring optimal coverage and energy management. This study addresses the Q-coverage problem, where each sensor must be covered by at least Q chargers, and proposes an optimized solution using the Blackhole Algorithm. The Blackhole Algorithm is inspired by the gravitational pull of black holes and is used to determine the optimal placement of chargers within a 2D area, ensuring maximum coverage with minimal energy consumption. The effectiveness of this method is evaluated through simulations, Blackhole Algorithm has a performance rate of 95.01%, ranking higher than well-established methods such as Haar (88.21%), Daubechies 2 (88.56%), Biorthogonal (89.71%), and Symlets 8 (90.16%) wavelets in dealing with sensor energy fluctuations and charger placement issues. The results indicate that the Blackhole Algorithm outperforms these methods in terms of efficiency and optimal charger placement, making it a promising approach for improving the recharging process and energy sustainability in wireless sensor networks.

Keywords: Wireless chargers · Q coverage · Blackhole algorithm · Radius event zone · Optimal position

1 Introduction

In recent years, wireless sensor networks (WSNs) have gained significant attention due to their widespread applications in fields such as environmental monitoring [1], smart cities, healthcare [2], and industrial automation [3]. These networks consist of a large number of spatially distributed sensors that collect and transmit data. However, the longevity and efficiency of WSNs are heavily dependent on the effective management of energy resources, as sensors often operate on limited battery power [2]. One critical challenge in WSNs is ensuring continuous operation by addressing the energy replenishment problem, specifically the optimal placement of chargers to maximize Q-coverage, where each sensor must be covered by at least Q chargers.

P. Chandrakar et al. (Eds.): ICCINS 2025, CCIS 2738, pp. 262–273, 2026.
https://doi.org/10.1007/978-3-032-09572-5_21

The traditional approach to charger placement involves static configurations that often lead to inefficient energy utilization and suboptimal sensor coverage. To address these limitations, advanced optimization techniques are being explored, one of the most promising being the Blackhole Algorithm (BHA). Inspired by the gravitational pull of black holes, the BHA employs an iterative search mechanism to find optimal charger positions within the sensor network area. This algorithm is particularly effective due to its ability to explore large solution spaces and avoid local optima, ensuring that charger placement is both efficient and effective.

This paper presents a novel solution to the Q-coverage problem by applying the Blackhole Algorithm [4] [5] to optimize charger placement. We evaluate the performance of the proposed method through simulations, comparing it with several established techniques, including wavelet-based methods such as Haar, Daubechies 2, Biorthogonal, Symlets 8, and Raindrop algorithms. The simulation results demonstrate that the BHA provides superior performance in terms of charger placement efficiency.

The results of the work indicate the high capability of the BHA to greatly enhance the recharging process in WSNs through a more efficient and sustainable mechanism for long-term sensor operation. The approach not only guarantees maximum charger placement, but also helps ensure overall energy efficiency within the network, making it a perfect solution for practical applications where sensor nodes are required to be operated independently over long durations without constant human intervention.

The rest of the paper is structured as follows: In Sect. 2 review the literature on wireless charger deployment for sensor recharging, and Sect. 3 defines the wireless charger deployment problem. In Sect. 4)A, the basics of the Blackhole algorithm, IV)B describes the implementation of the Blackhole algorithm for the optimal placement of a wireless charger. Section 5 verifies the performance of the Blackhole and other wavelets. Section 6 concludes the paper.

2   Related Work

Gupta et al. [6] propose a sensor node scheduling heuristic for Energy Harvesting Wireless Sensor Networks (EH-WSNs) that accounts for both Q-Coverage and P-Connectivity. They compare the performance of this heuristic in EH-WSNs with its application in battery-powered WSNs. The results reveal that EH-WSNs achieve significantly longer network lifetimes compared to their battery-powered counterparts.

Yang et al. [7] introduced an enhanced Firefly Algorithm to optimize charger deployment in Wireless Rechargeable Sensor Networks (WRSNs). This improved approach aims to maximize the coverage of Points of Interest (PoIs) while simultaneously enhancing charging efficiency. By refining the traditional Firefly Algorithm, they developed a more effective method for determining optimal charger placement, striking a balance between energy replenishment and overall network performance. Unlike conventional strategies that address coverage and charging efficiency as independent objectives, this integrated approach significantly advances cost-effective deployment strategies for WRSNs.

Chen et al. [8] introduce the PSCD (Particle Swarm Charger Deployment) algorithm, which uses the Particle Swarm Optimization method to optimize charger deployment in Wireless Rechargeable Sensor Networks (WRSNs). The PSCD algorithm considers the efficiency of charging in terms of distance and direction between chargers and sensor nodes. Integrating PSO, it uses both global and local optimum solutions to dynamically adjust charger positions and antenna directions. This strategy further increases the sustainability of WRSNs through improved energy transfer efficiency and extended duration of network use.

Balaji et al. [9] proposed using the Daubechies wavelet algorithm to determine the optimal charger positions. Daubechies wavelets have translation and dilation functions that enable chargers to shift both vertically and horizontally. Additionally, they can help reposition redundant chargers to optimal locations. The proposed algorithm was tested against four other wavelets: Haar, Daubechies 2 (db2), Biorthogonal 55 (bio55), and Symlets. The Daubechies wavelet algorithm demonstrated the best performance compared to these other wavelets. While this method adjusts the placement of excess chargers, this paper specifically calculates the necessary number of chargers to ensure that no redundancy occurs.

Zekun Fang et al. [10] explore Wireless Rechargeable Sensor Networks (WRSNs), where rechargeable sensors enable the collection of critical data while minimizing equipment costs and dependence on bulky wired sensors. They propose a genetic algorithm to optimize charger deployment, focusing on identifying the most effective placements for chargers. Unlike traditional greedy algorithms, their approach emphasizes cost-efficiency by reducing deployment expenses while maintaining effective network performance.

This paper addresses the gap in WSN charger placement by leveraging Q-coverage, where each sensor is covered by at least Q chargers, reducing downtime and improving efficiency. Using the Black Hole Algorithm (BHA), the proposed method optimizes charger placement, enhancing recharging and overall network performance.

3   Problem Formation

Let S=$\{S_1, S_2, \ldots, S_m\}$ be the set of m rechargeable sensors in the region $A \times A$. The set of chargers are denoted by C=$\{C_1, C_2, \ldots, C_n\}$ and are static. Initially, n wireless chargers are randomly deployed in the region $A \times A$, each sensor should be recharged by Q number of chargers. The paper investigates a method to ensure each sensor in a network can be recharged by a fixed Q number of chargers. Additionally, it focuses on determining the optimal positions for n chargers to minimize costs and maximize efficiency within the network.

The sensor is said to be recharged by the chargers if it satisfies the following conditions,

$$(x_i - x_j)^2 + (y_i - y_j)^2 \leq r^2 \tag{1}$$

where, (x_i, y_i) and (x_j, y_j) be the charger's C_i and sensor's S_j spatial coordinates respectively.

The coverage of chargers is represented in the form of a matrix,

$$\gamma = \begin{pmatrix} \alpha_{11} & \alpha_{12} & \dots & \alpha_{1\,m} \\ \alpha_{21} & \alpha_{22} & \dots & \alpha_{2\,m} \\ \vdots & \vdots & \ddots & \vdots \\ \alpha_{n1} & \alpha_{i2} & \dots & \alpha_{nm} \end{pmatrix} \tag{2}$$

where,

$$\alpha_{ij} = \begin{cases} 1, \text{if a sensor } S_j \text{ monitor by Charger } C_i \\ 0, \text{otherwise.} \end{cases}$$

The column sum of the γ matrix is

$$\phi = [\phi_1, \phi_2, \dots, \phi_j, \dots, \phi_m]$$

where

$$\phi_j = \sum_{i=1}^{n} \alpha_{ij}$$

The number of sensors recharged by each charger can be calculated from the column sum of γ matrix. The column sum of γ helps identify the sensors that can be recharged by the charger. Each sensor should recharge by Q-chargers, λ helps to determine that Q-coverage check as follow:

$$\lambda_j = \left\lfloor \frac{\phi_j}{q} \right\rfloor \tag{3}$$

where

$$\lambda_j = \begin{cases} 1, \text{if a sensor } S_j \text{ monitor by the Q Charger } C_i \\ 0, \text{otherwise.} \end{cases}$$

The number of sensors recharged by the deployed chargers is known as Quality of Coverage (QoC) using equation (3)

$$QoC = \sum_{j=1}^{m} \left\lceil \frac{\lambda_j}{m} \right\rceil \tag{4}$$

The QoC and its percentage are determined by equations (4) and (5). This method helps to determine the QoC for any network. The percentage of QoC is

$$Qoc[\%] = \frac{QoC}{m} \times 100 \tag{5}$$

A sample network of 8 sensors with 10 chargers with a charging range of 2 units randomly deployed in the region 10×10 and each sensor should recharge

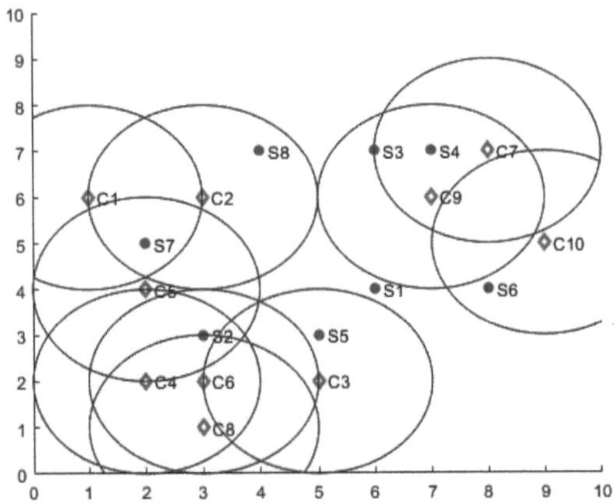

Fig. 1. Randomly deployed sensors and chargers.

by the $Q = 3$ number of chargers shown in Fig. 1. The matrix representation of charger coverage using Eqs.(1) and (2) is

$$\gamma = \begin{array}{c} \\ C_1 \\ C_2 \\ C_3 \\ C_4 \\ C_5 \\ C_6 \\ C_7 \\ C_8 \\ C_9 \\ C_{10} \end{array} \begin{array}{c} S_1\ S_2\ S_3\ S_4\ S_5\ S_6\ S_7\ S_8 \\ \left(\begin{array}{cccccccc} 0 & 0 & 0 & 0 & 0 & 0 & 1 & 0 \\ 0 & 0 & 0 & 0 & 0 & 0 & 1 & 1 \\ 0 & 0 & 0 & 0 & 1 & 0 & 0 & 0 \\ 0 & 1 & 0 & 0 & 0 & 0 & 0 & 0 \\ 0 & 1 & 0 & 0 & 0 & 0 & 1 & 0 \\ 0 & 1 & 0 & 0 & 0 & 0 & 0 & 0 \\ 0 & 0 & 1 & 1 & 0 & 0 & 0 & 0 \\ 0 & 1 & 0 & 0 & 0 & 0 & 0 & 0 \\ 0 & 0 & 1 & 1 & 0 & 0 & 0 & 0 \\ 0 & 0 & 0 & 0 & 1 & 0 & 0 & 0 \end{array}\right) \end{array}$$

Thus the column sum of the matrix is

$$\phi = \begin{pmatrix} 0 & 4 & 2 & 2 & 2 & 0 & 3 & 1 \end{pmatrix}$$

The set of sensors recharged by the chargers are $S_1 = \{\}$, $S_2 = \{C_4, C_5, C_6, C_8\}$, $S_3 = \{C_7, C_9\}$, $S_4 = \{C_7, C_9\}$, $S_5 = \{C_3\}$, $S_6 = \{C_{10}\}$, $S_7 = \{C_1, C_2, C_5\}$ and $S_8 = \{C_2\}$ respectively.

For Q=3 coverage, evaluation using equation (3)

$$\lambda = \begin{pmatrix} 0 & 1 & 0 & 0 & 0 & 0 & 1 & 0 \end{pmatrix}$$

The Quality of coverage obtained using equations (4) and (5), $QoC = \begin{pmatrix} 0 + 1 + 0 + 0 + 0 + 0 + 1 + 0 \end{pmatrix} = 2$ and the $Qoc[\%] = 25\%$. In the sample network, used 8 sensors with 10, where each sensor is recharged by 3 chargers, the complexity of the problem increases significantly for larger networks. Therefore, determining the optimal positions for the chargers is essential to simplify the process and maximize the network's lifetime. So, the BHA proposed to find the optimal position of chargers.

4   Blackhole Algorithm

In this subsection, BHA is explained in detail. The BHA was first introduced by A. Hatamlou [11]. The concept of a Blackhole (BH) is based on the idea that in a region of space filled with many stars, there exists a BH with such intense gravitational pull that nothing, not even light, can escape. Anything falling into a BH, including light, is absorbed by it. The BHA developed using the concept of BH absorbing and interacting with their surroundings. In the proposed BHA, the initial population is considered as stars, and candidate solutions (fitness value) are evaluated for these stars (population). The best fitness value's population is selected to be the BH, while the remaining population is considered normal stars. After selecting BH, the population will be updated at each iteration using the equation given below:

$$X_i(t + 1) = X_i(t) + rand(0, 1)(X_{BH} - X_i(t)) \tag{6}$$

where $rand$ is a random number in the interval $[0, 1]$, $X_i(t)$ and $X_i(t+1)$ indicate the previous and new positions of stars at iterations t and $t + 1$, and X_{BH} represents the BH position. The surrounding stars are drawn to the BH in BHA once the populations have been updated using Eq. (6). A star is"swallowed" and eliminated from the search space if it approaches the BH too closely. The Radius of Event Horizon (REH) is the name given to this area. Similar to BH consuming stars, BHA eliminates those stars and creates a new star (new population) as a potential candidate; it is positioned at random within the search space and initiates a fresh search [12]. The following is the REH formula,

$$REH = \frac{f_{BH}}{\sum_{i=1}^{n} f_i} \tag{7}$$

where f_{BH} and f_i represent the fitness value of BH and the ith star, respectively. Once every star has been updated, the next iteration begins. The BHA has obtained the optimal position of chargers.

4.1   Implementation of Blackhole Algorithm

The optimal positioning of these chargers can be identified using an optimization algorithm with a population of size p for t iterations. The matrix size of the population is $2 \times n$

For the problem illustrated in Fig. 1, 10 chargers were used to recharge the eight sensors, but only 20% of sensors satisfied the Q requirement.

$$pop_1 = \begin{pmatrix} 1\ 3\ 5\ 2\ 2\ 3\ 8\ 3\ 7\ 9 \\ 6\ 6\ 2\ 2\ 4\ 2\ 7\ 1\ 6\ 5 \end{pmatrix}$$

$$pop_2 = \begin{pmatrix} 8\ 2\ 7\ 2\ 2\ 9\ 2\ 8\ 7\ 2 \\ 4\ 2\ 6\ 2\ 4\ 8\ 4\ 2\ 8\ 5 \end{pmatrix}$$

$$pop_3 = \begin{pmatrix} 2\ 8\ 7\ 9\ 4\ 8\ 5\ 3\ 1\ 7 \\ 5\ 6\ 4\ 6\ 5\ 1\ 3\ 4\ 7\ 5 \end{pmatrix}$$

The fitness value for each population is calculated using equation (4).

$$\text{fitness value of sample population} = (2\ 3\ 2)$$

A sample consisting of three populations with ten coordinates is generated randomly and th fitness value is calculated for them which is shown above. The population coordinates with the best fitness value are considered the BH. All other populations are regarded as normal stars.

$$BH = \begin{pmatrix} 8\ 2\ 7\ 2\ 2\ 9\ 2\ 8\ 7\ 2 \\ 4\ 2\ 6\ 2\ 4\ 8\ 4\ 2\ 8\ 5 \end{pmatrix}$$

All the populations are updated by the equation, which is given below,

$$pop_i^{t+1} = pop_i^t + rand(0,1)(pop_{BH} - pop_i^t) \tag{8}$$

where pop_i^{t+1} is the new position, pop_i^t is the old position and pop_{BH} is the BH position of the population. After the update, the populations have new coordinates except for the BH. The population pop_2^1 is the BH, so after the updating the population remains the same.
But other populations will get different coordinates when updated by the equation (8),

$$pop_2^1 = \begin{pmatrix} 8\ 2\ 7\ 2\ 2\ 9\ 2\ 8\ 7\ 2 \\ 4\ 2\ 6\ 2\ 4\ 8\ 4\ 2\ 8\ 5 \end{pmatrix}$$

Other populations are updated by equation (8),

$$pop_1^1 = \begin{pmatrix} 1\ 3\ 5\ 2\ 2\ 3\ 8\ 3\ 7\ 9 \\ 6\ 6\ 2\ 2\ 4\ 2\ 7\ 1\ 6\ 5 \end{pmatrix} + rand(0,1) \begin{pmatrix} 8\ 2\ 7\ 2\ 2\ 9\ 2\ 8\ 7\ 2 \\ 4\ 2\ 6\ 2\ 4\ 8\ 4\ 2\ 8\ 5 \end{pmatrix} - \begin{pmatrix} 1\ 3\ 5\ 2\ 2\ 3\ 8\ 3\ 7\ 9 \\ 6\ 6\ 2\ 2\ 4\ 2\ 7\ 1\ 6\ 5 \end{pmatrix}$$

$$pop_1^1 = \begin{pmatrix} 6.7\ 4.2\ 1.3\ 3.5\ 5.2\ 5\ 6.4\ 3.5\ 4.2\ 4.3 \\ 2.9\ 3.9\ 3.4\ 5.6\ 0.5\ 7\ 3.5\ 4.2\ 3.8\ 7.4 \end{pmatrix}$$

Similarly,

$$pop_3^1 = \begin{pmatrix} 7.2\ 6.7\ 2\ 4.9\ 3.3\ 6.2\ 5.5\ 6.6\ 6.8\ 3.8 \\ 9.3\ 5.0\ 4\ 4.4\ 4.3\ 4.4\ 0.4\ 2.6\ 2.7\ 3.3 \end{pmatrix}$$

$$\text{fitness value of updated population} = (5\ 3\ 5)$$

Updating the coordinates will change the position of the chargers as well as the fitness values of the populations. The fitness value of a population may either increase or remain the same.

Radius event horizon (REH)

In some iterations, the highest fitness value may be repeated. In such cases, the first occurrence of the population with the highest fitness value is designated as the BH, and the other populations with the same fitness value are entered into the REH. The REH causes these population coordinates to vanish, and new population coordinates are generated to replace them in the iteration.

$$pop_3^1 = \begin{pmatrix} 4\,4\,4\,4\,5\,6\,8\,5\,2\,5 \\ 3\,3\,6\,2\,6\,3\,6\,2\,6\,6 \end{pmatrix}$$

After updating the population, the fitness values 5 are repeated for the first and third populations. Consequently, when the third population enters the REH, its coordinates are removed, and new coordinates are generated randomly for that population.

Then the fitness values will be calculated again for that population.

fitness value of first iteration = (5 3 4)

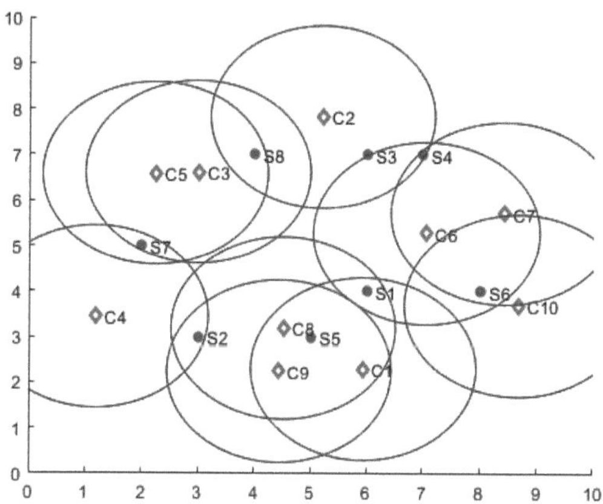

Fig. 2. Optimal position of chargers by BHA.

The process will continue until the iteration t ends. At the tth iteration the optimal position is obtained. After t^{th} iteration, 7 is the highest fitness value and its population is the BH which is the optimal position for chargers, it recharges 87.5% sensors and each sensor is recharged by three chargers, when iteration increases the optimal position of chargers with each sensor recharge by three chargers may be obtained. the populations and their fitness value are

$$pop_1^t = \begin{pmatrix} 6.3\,6.5\,1.8\,6\,2.4\,3.4\,5.3\,6.1\,8.3\,9.9 \\ 8.0\,4.3\,3.1\,3\,9.0\,2.4\,6.8\,6.4\,5.2\,8.2 \end{pmatrix}$$

$$pop_2^t = \begin{pmatrix} 6.2\ 6.8\ 3.7\ 3.0\ 4.6\ 6.9\ 5\ 6.8\ 4.9\ 7.9 \\ 3.6\ 7.5\ 4.8\ 2.4\ 5.3\ 5.4\ 7\ 3.3\ 2.7\ 4.7 \end{pmatrix}$$

$$pop_3^t = \begin{pmatrix} 5.9\ 5.2\ 3.0\ 1.1\ 2.2\ 7\ 8.4\ 4.5\ 4.4\ 8.6 \\ 2.2\ 7.8\ 6.6\ 3.4\ 6.8\ 5\ 5.7\ 3.1\ 2.7\ 3.6 \end{pmatrix}$$

$$\text{fitness value of } t^{th} \text{ iteration } = (5\ 6\ 7)$$

The optimal position for the charger placement by BHA at the tth iteration is

$$\begin{pmatrix} 5.9\ 5.2\ 3.0\ 1.1\ 2.2\ 7\ 8.4\ 4.5\ 4.4\ 8.6 \\ 2.2\ 7.8\ 6.6\ 3.4\ 6.8\ 5\ 5.7\ 3.1\ 2.7\ 3.6 \end{pmatrix}$$

In the case of random deployment, not all sensors were recharged. The optimal charger positions, determined using the BHA, are shown in Fig. 2. As the iteration increases, the BHA efficiently identifies the optimal charger positions with Q requirements while increasing the number of iterations.

5   Simulation Result

Several studies have been conducted to recharge sensors and improve network coverage [13] [14]. In this paper, the simulation result was carried out by varying the number of sensors between 50 and 200, the charger varies between 100 and 175, and the population and iteration count constantly use 100 with the charger range fixed at 10 units in a compact 100×100 region, the performance of wavelets and BHA is shown in Table 1. For each variation in sensor count, the value Q is evaluated from 1 to 5. In all comparisons, BHA consistently performs better. Among the compared methods—wavelets and BHA—BHA achieves up to 10% higher sensor coverage than wavelets. In particular, none of the wavelet-based methods exceeds the coverage of the 90% sensor, whereas BHA consistently delivers superior results.

In Fig. 3, the number of sensors varies between 500 and 700, while the number of chargers is maintained consistently at 250, with the range of the charger fixed at 100. In Fig. 4, the number of chargers varies between 300 and 500, with the number of sensors consistently maintained at 1000 with a charger range of 100, the value Q is set at 5 for all scenarios. Finally, Fig. 5 explores the variations in the value of Q, from 1 to 5, while maintaining a constant 1000 sensors, 500 chargers and a range of 100 chargers in all scenarios. In all simulations, BHA consistently outperforms all wavelet-based methods. The QoC values for wavelets and BHA are illustrated in the respective.

Table 1. Performance of Algorithms for changing sensors count and Q value in smaller region

Sensor	Charges	Q	Random	Haar	db2	sym	bio	BH
50	25	1	31	34	35	35	35	41
	50	2	22	26	27	26	28	37
	75	3	17	25	26	26	25	33
	100	4	17	24	24	24	22	28
	125	5	16	20	20	20	25	31
75	35	1	47	56	55	59	56	62
	65	2	37	50	47	52	52	57
	90	3	38	42	40	42	45	55
	115	4	31	37	36	35	42	51
	140	5	27	34	38	37	39	48
100	50	1	70	85	85	87	85	92
	75	2	68	73	69	72	76	80
	100	3	45	65	60	62	61	76
	125	4	37	55	59	59	55	73
	150	5	42	49	48	50	47	64
150	75	1	130	140	141	141	143	147
	100	2	108	123	124	125	125	140
	125	3	111	114	118	118	121	159
	150	4	93	100	101	100	102	115
	175	5	84	92	95	97	97	114
200	100	1	189	194	195	195	195	200
	125	2	173	182	185	180	182	192
	150	3	170	170	170	168	176	182
	175	4	134	151	147	160	156	170
	200	5	129	145	153	143	148	168

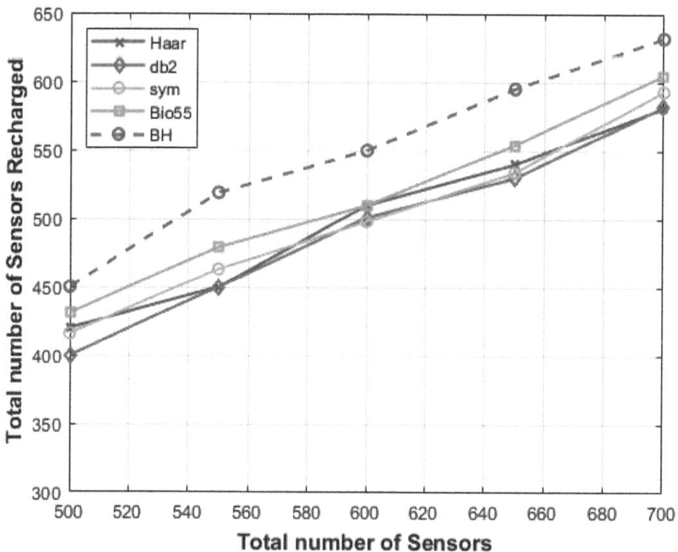

Fig. 3. Results comparison for various no. of sensors in larger region.

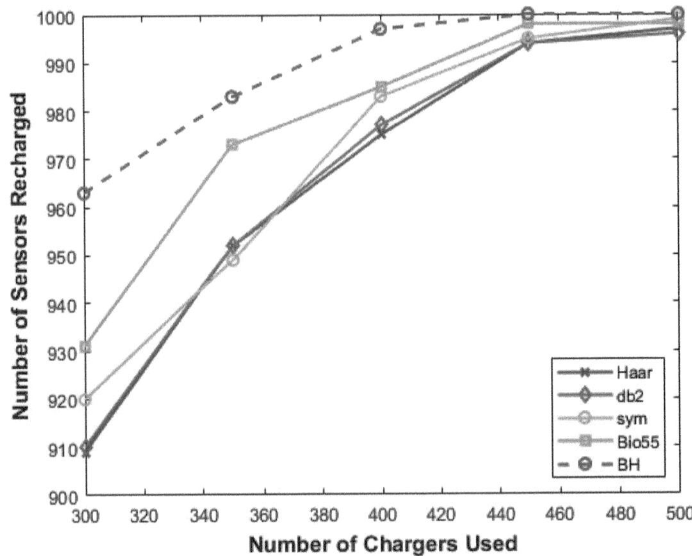

Fig. 4. Results comparison for different charger value in larger region.

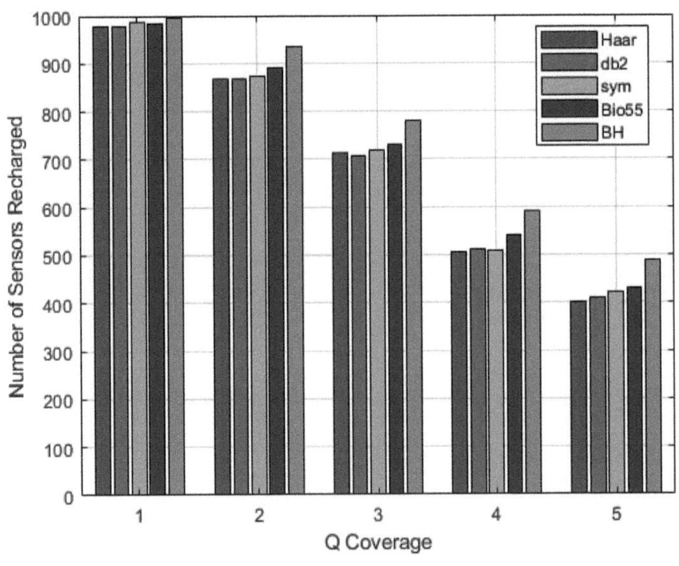

Fig. 5. Results comparison for Q coverage in larger region.

6   Conclusion

This study demonstrated the effectiveness of the Blackhole Algorithm for optimizing charger placement in Wireless Sensor Networks to achieve Q-coverage.

Simulation results confirmed that BHA outperforms traditional wavelet-based methods, achieving up to 5% higher sensor coverage while enhancing energy efficiency. To further improve sensor recharging rates, integrating localized adjustments into the Blackhole Algorithm could refine its performance. Additionally, determining the minimum number of chargers required to fully recharge all sensors would significantly reduce redundancy, a focus for future research. Overall, BHA's ability to optimize charger placement ensures long-term sustainability and autonomous operation of Wireless Sensor Networks, making it a promising solution for real-world deployments.

References

1. Woo-García, R.M., et al.: Implementation of a wireless sensor network for environmental measurements. Technologies **12**(3), 41 (2024)
2. Yazid, M.A.M., Afif, A.R.: A method for preserving battery life in wireless sensor nodes for LoRa based IOT flood monitoring. J. Commun. **17**(4), 230–238 (2022)
3. Tabella, G., Ciuonzo, D., Paltrinieri, N., Rossi, P.S.: Bayesian fault detection and localization through wireless sensor networks in industrial plants. IEEE Internet Things J. **11**(8), 13231–13246 (2024)
4. Abualigah, L., et al.: Black hole algorithm: a comprehensive survey. Appl. Intell. **52**(10), 11892–11915 (2022)
5. Zheng, W.-M., Liu, N., Chai, Q.-W., Liu, Y.: Application of improved black hole algorithm in prolonging the lifetime of wireless sensor network. Complex Intell. Syst. **9**(5), 5817–5829 (2023)
6. Gupta, S., Gupta, S., Goyal, D.: Comparison of Q-coverage P-connectivity sensor node scheduling heuristic between battery-powered WSN and energy harvesting WSN. Int. J. Sens. Wirel. Commun. Control **11**(5), 553–559 (2021)
7. Yang, M., Wang, A., Sun, G., Zhang, Y.: Deploying charging nodes in wireless rechargeable sensor networks based on improved firefly algorithm. Comput. Electr. Eng. **72**, 719–731 (2018)
8. Chen, Y.C., Jiang, J.R.: Particle swarm optimization for charger deployment in wireless rechargeable sensor networks. In: 2016 26th International Telecommunication Networks and Applications Conference (ITNAC), Dunedin, New Zealand, pp. 231–236 (2016)
9. Balaji, S., Arivudainambi, D.: Optimal placement of wireless chargers in rechargeable sensor networks. IEEE Sens. J. **18**, 4212–4222 (2018)
10. Fang, Z., Chien, W.-C., Zhang, C., Hang, N.T., Chen, W.-M.: GA-based charger deployment algorithm in indoor wireless rechargeable sensor networks. J. Internet Technol. **24**, 487–494 (2023)
11. Hatamlou, A.: Black hole: a new heuristic optimization approach for data clustering. Inf. Sci. **222**, 175–184 (2013)
12. Hatamlou, M.F.: Solving optimization problems using blackhole algorithm. Adv. Comput. Sci. Technol. **4**, 68–74 (2015)
13. Ortega, A., Ciancio, A.: A distributed wavelet compression algorithm for wireless sensor networks using lifting. In: ICASSP'05, vol. 4, pp. iv–825 (2004)
14. Arivudainambi, D., Pavithra, R., Mangairkarsi, S.: Optimizing wireless charger placement for rechargeable wireless sensors. In: 2018 Tenth International Conference on Advanced Computing (ICoAC), pp. 244–248 (2018)

Reducing the Carbon Footprint of Data Centers: CNN-Transformer-Based Workload Prediction for Green Cloud Computing

Rashi Chauhan[1], Jing Yang[2(✉)], and Abdullah Ayub Khan[3]

[1] Department of CSE, Amity University, Noida 201313, India
[2] Department of CST, University Malaya, Kuala Lumpur 50603, Malaysia
`s2147529@siswa.um.edu.my`
[3] Department of CS, Bahria University Karachi Campus, Karachi 75260, Pakistan
`abdullahayub.bukc@bahria.edu.pk`

Abstract. The fast spread of cloud computing (CC) has greatly raised the energy consumption of data centers, which emphasizes the need of sustainable operational plans. Accurate workload forecasting plays a crucial role in optimizing resource allocation, reducing energy consumption, and minimizing carbon emissions. This study introduces a CNN-Transformer model, which integrates Convolutional Neural Networks (CNNs) for spatial feature extraction and Transformer architectures for capturing temporal dependencies, providing an effective framework for cloud workload prediction. Empirical evaluations on real-world workload data demonstrate that the proposed model achieves a Mean Squared Error (MSE) of 0.0059, Mean Absolute Error (MAE) of 0.0381, and an R^2 score of 0.994, outperforming existing state-of-the-art (SOTA) models. These findings demonstrate the model's capacity to improve forecast accuracy while also enabling energy-efficient cloud operations via real-time carbon emission monitoring and adaptive resource allocation. Furthermore, an analysis of the model's computational complexity reveals that it maintains an efficient balance between accuracy and processing speed, making it suitable for large-scale cloud environments. This work represents a significant advancement in sustainable cloud infrastructure and offers promising applications in green data center management.

Keywords: CNN-Transformer · workload prediction · energy-efficient cloud computing · data center sustainability · resource optimization · carbon emissions

1 Introduction

The swift expansion of CC has made it a cornerstone of modern digital infrastructure, supporting applications such as big data analytics and artificial intelligence (AI). However, the escalating energy demands of cloud data centers have led to significant carbon emissions, exacerbated by inefficiencies in resource allocation relative to workload requirements [1]. Addressing these inefficiencies is critical to aligning CC systems with global carbon neutrality goals. Data centers contribute significantly to global energy consumption, a trend expected to grow with increasing demand for cloud services [2]. Effective energy management hinges on precise workload forecasting to enable dynamic resource allocation. However, the variability and complexity of cloud workloads, characterized by temporal and spatial dependencies, make accurate forecasting a challenging task [3,4]. Advances in deep learning offer promising solutions to these challenges [5]. CNNs effectively extract spatial features, while transformer models capture temporal relationships through self-attention mechanisms [6,7]. Combining these architectures creates a powerful framework for managing complex patterns in cloud workload data. Prior studies have demonstrated the success of similar hybrid models in energy-efficient computing contexts, validating their predictive capabilities [8,9].

This paper introduces a CNN-Transformer-based model specifically built for forecasting cloud data center workloads, thereby advancing current research. The approach utilizes CNNs to deduce spatial correlations from workload data and harnesses the self-attention capabilities of Transformers to discern temporal trends. This study highlights sustainability by integrating real-time energy consumption and carbon emission modeling into the prediction framework. The suggested approach facilitates efficient resource management and diminishes the carbon footprint of data centers by dynamically linking workload projections with energy metrics.

This research offers the following key contributions:

1. **Hybrid Model for Workload Prediction:** A CNN-Transformer framework that integrates spatial feature extraction and temporal modeling for precise workload forecasting.
2. **Carbon Emission Integration:** A dynamic linkage between workload prediction and real-time energy and carbon metrics to enhance sustainability.
3. **Real-world Validation:** Comprehensive evaluation using the Google cluster dataset, showcasing superior predictive accuracy and efficiency over existing models.
4. **Scalability and Flexibility:** A robust design adaptable to large-scale cloud systems and diverse workload patterns.

2   Literature Review

Efficient workload prediction is vital for resource management and energy optimization in CC. Early methods, including Support Vector Machines (SVM) and ARIMA, struggled with the non-linearities of cloud operations [10]. Deep learning models, such as LSTM and BiLSTM, significantly improved accuracy by capturing long-term dependencies. For instance, Karim et al. [4] introduced a BiLSTM-based hybrid model for CPU workload prediction, but scalability challenges like computational overhead and gradient vanishing remained [3,11].

CNNs have been employed to capture spatial dependencies, often in combination with LSTMs. Abbasi et al. [12] demonstrated this integration's potential, though CNN-LSTM models required extensive hyperparameter tuning and faced challenges with highly dynamic workloads [13]. Energy efficiency in data centers has become increasingly critical due to environmental and economic impacts. Studies highlight the correlation between CPU utilization and power consumption [2,14]. Shih et al. [6] demonstrated the limitations of static energy management strategies, advocating for integrating workload prediction with energy models. Patel and Kushwaha [11] explored hybrid CNN-LSTM models for server load and power usage prediction but noted their limitations in capturing complex energy-workload interactions [15]. Recent research has combined spatial and temporal feature extraction using hybrid models. Shih et al. [6] and Bi et al. [3] employed attention-based architectures, while Dogani et al. [15] used CNN-GRU models with attention mechanisms, achieving superior performance. However, challenges persist, such as the computational demands of Transformer-based systems and their reliance on extensive datasets [8,16]. This study builds on prior work by introducing a CNN-Transformer model that seamlessly integrates spatial and temporal features for dynamic workload forecasting. Additionally, it incorporates real-time energy and carbon emission modeling, addressing operational efficiency and environmental sustainability challenges in cloud data centers.

3   Methodology

To ensure efficient resource allocation and minimize energy consumption in cloud data centers, it is imperative to have precise workload predictions. The problem of cloud load prediction is presented in this study as a spatio-temporal regression task. To be more precise, the goal is to predict duties over time while considering both temporal and spatial dependencies. Spatial dependencies capture relationships across multiple servers and computational resources within the data center, while temporal dependencies monitor changes in workload over time.

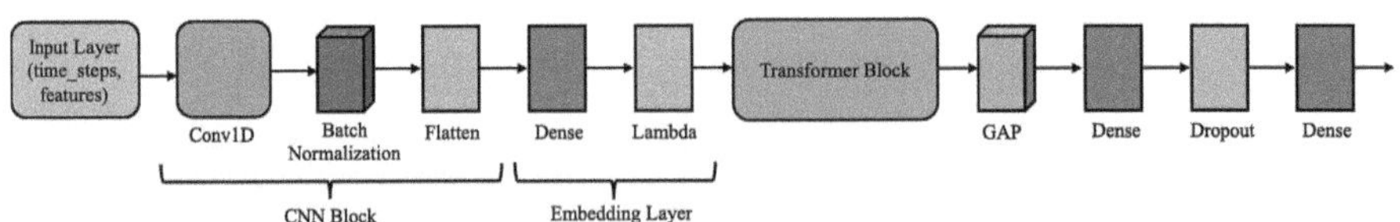

Fig. 1. Complete architecture of the CNN-Transformer model for cloud workload prediction, including the CNN block, embedding layer, Transformer block, and dense layers.

3.1   Proposed Approach

The proposed model integrates CNNs for spatial feature extraction with Transformer layers for capturing temporal dependencies. The overall architecture is shown in Figs. 1 and 2.

3.1.1   CNN for Spatial Feature Extraction

The CNN block is designed to extract spatial patterns in the workload data. A 1D convolutional layer with 64 filters and a kernel size of 3 is applied to the input sequence. This convolutional layer processes the data across the spatial dimension, learning relationships between different resource metrics. For an input sequence $\mathbf{X} \in \mathbb{R}^{T \times F}$, where T represents the total number of time steps and F denotes the number of features, the convolution operation can be mathematically formulated as:

$$\mathbf{Z}_{\text{conv}}(t, k) = \sigma \left(\sum_{f=1}^{F} \sum_{j=1}^{K} \mathbf{W}_{k,j,f} \cdot \mathbf{X}(t + j - 1, f) + \mathbf{b}_k \right)$$

where $\mathbf{W}_{k,j,f}$ denotes the weight of the k-th filter at position j for feature f, $\mathbf{b}_k$ is the bias term, K is the kernel size, and σ is the activation function (ReLU in this case). This layer captures spatial dependencies by identifying patterns across different features.

After convolution, batch normalization is applied to stabilize and accelerate the training process. Batch normalization normalizes each feature to have a mean of zero and a standard deviation of one, which helps mitigate the internal covariate shift:

$$\mathbf{Z}_{\text{BN}} = \frac{\mathbf{Z}_{\text{conv}} - \mu_{\text{batch}}}{\sigma_{\text{batch}}}$$

where μ_{batch} and σ_{batch} represent the mean and standard deviation of the batch, respectively. The output of this block is then flattened to prepare it for embedding.

3.1.2   Embedding Layer

The flattened output from the CNN block is passed through a dense layer to project it into a lower-dimensional embedding space suitable for the Transformer block. This dense layer reduces the dimensionality of the feature representation, making it compatible with the Transformer layer's input requirements. The embedding transformation is expressed as:

$$\mathbf{Z}_{\text{embed}} = \text{Dense}(\mathbf{Z}_{\text{BN}})$$

where $\mathbf{Z}_{\text{embed}} \in \mathbb{R}^d$, and d is the embedding dimension. This projection allows the model to effectively map the spatially extracted features to a format that can capture temporal dependencies in the subsequent Transformer block.

3.1.3   Transformer Block for Temporal Relationships

The Transformer block, shown in Fig. 2, is used to capture temporal dependencies in the workload data sequence. It consists of a MHA layer followed by a feed-forward neural network and residual connections. This block allows the model to focus on different parts of the sequence, capturing long-term dependencies without the limitations of recurrent layers [17].

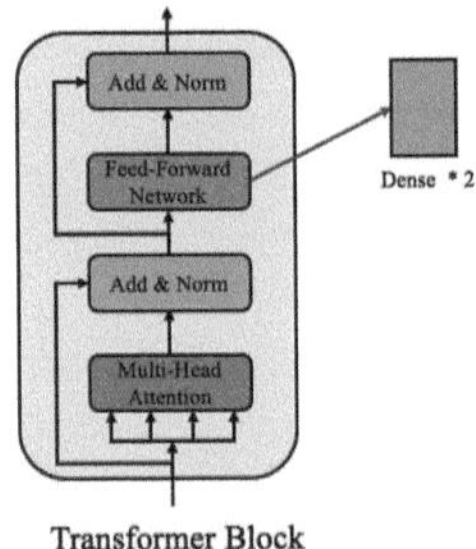

Fig. 2. Structure of the Transformer Block, which includes MHA, feed-forward networks, and residual connections.

1. MHA: The MHA mechanism enables the model to attend to various positions in the sequence simultaneously. For an input sequence $\mathbf{Z}_{\text{embed}}$, the attention mechanism computes multiple attention heads. Each attention head computes a weighted representation of the input sequence.
2. Residual Connection and Layer Normalization: After MHA, a residual connection adds the original input to the output of the attention layer, followed by layer normalization. This can be expressed as:

$$\mathbf{Z}_{\text{attn}} = \text{LayerNorm}(\mathbf{Z}_{\text{embed}} + \text{Attention}(\mathbf{Z}_{\text{embed}}))$$

This step preserves the input information while normalizing it, which helps stabilize the model during training.
3. Feed-Forward Network: It consists of two dense layers with a ReLU activation in between. It applies a non-linear transformation to the output of the attention layer, adding complexity to the learned representations:

$$\mathbf{Z}_{\text{FFN}} = \text{LayerNorm}\left(\mathbf{Z}_{\text{attn}} + \text{ReLU}(\text{Dense}(\text{Dense}(\mathbf{Z}_{\text{attn}})))\right)$$

This residual connection and layer normalization are again applied to ensure stable gradients and allow effective training of the deep network.

3.1.4   Global Average Pooling and Dense Layers

The output of the Transformer block is then aggregated using Global Average Pooling (GAP), which reduces the sequence length to a single vector by averaging over time steps. This can be represented as:

$$\mathbf{Z}_{\text{GAP}} = \frac{1}{T} \sum_{t=1}^{T} \mathbf{Z}_{\text{FFN}}(t)$$

where T is the number of time steps. GAP effectively summarizes the information across the entire sequence, allowing the model to produce a fixed-size output.

After a dropout layer for regularization, the pooled vector is next run through a dense layer with ReLU activation. Finally, the output layer produces a single predicted value for workload:

$$\hat{y} = \mathrm{Dense}(\mathrm{Dropout}(\mathrm{ReLU}(\mathrm{Dense}(\mathbf{Z}_{\mathrm{GAP}}, 128)))), 1)$$

3.2 Computational Complexity

The suggested CNN-Transformer model's computational complexity is mostly shaped by the self-attention mechanism in the Transformer block and the Conv1D activities in the CNN block. Whereas the Transformer's self-attention mechanism has a complexity of $O(N^2 d)$, the Conv1D layer runs with a complexity of $O(Nkd)$. Large-scale cloud workload forecasting finds the model suitable since it preserves an effective balance between accuracy and computational cost.

3.3 Integration of Energy and Carbon Metrics

To promote sustainability, the model incorporates real-time metrics for energy consumption and carbon emissions. The predicted workload $\hat{y}$ is converted into power consumption $P(t)$ using a dynamic power model:

$$P(t) = P_{\min} + U(t) \times (P_{\max} - P_{\min})$$

where $U(t)$ represents CPU utilization, and $P_{\min}$ and $P_{\max}$ denote the minimum and maximum power values.

Based on $P(t)$, the carbon emissions $C(t)$ are calculated using an emissions conversion factor:

$$C(t) = P(t) \times \text{Emission Factor}$$

This allows the data center to dynamically manage resources while minimizing energy consumption and environmental impact.

3.4 Overall System Workflow

The proposed framework integrates workload prediction and carbon emission modeling for sustainable data center management. Figure 3 illustrates the workflow of our approach, which consists of three main components:

- **Cloud Data Center Components**: Includes IT equipment, cooling systems, and power supply, all of which contribute to carbon emissions based on energy consumption.
- **Data Processing and Modeling Pipeline**: Data cleaning and feature engineering steps are performed before passing data into the CNN-Transformer model for workload and emission predictions.
- **Feedback for Green Optimization**: Based on the model's predictions, adjustments can be made to optimize energy consumption and reduce carbon emissions, aligning with green computing goals.

The CNN-Transformer model's architecture was optimized with key parameters, including Conv1D filters (64), kernel size (3), Transformer attention heads (4), hidden dimensions (128), and dropout rate (0.1), fine-tuned for predictive accuracy and computational efficiency.

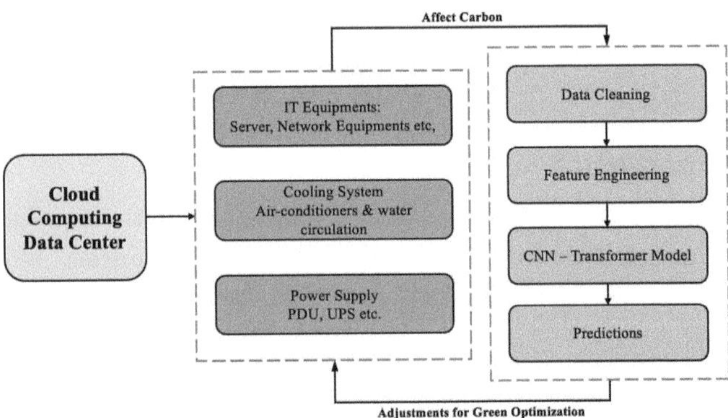

Fig. 3. System workflow for sustainable data center management integrating CC infrastructure with the CNN-Transformer predictive model for carbon-aware adjustments.

4   Experimental Setup

4.1   Dataset Description

This study utilized the Google Cluster Data[1], a multivariate time-series dataset collected from over 12,000 machines, containing metrics such as CPU utilization, memory usage, disk I/O, and task identifiers. Preprocessing steps included handling missing values by replacing numeric missing entries with the median value of the respective column and assigning a placeholder label "Unknown" for missing categorical data. Categorical variables were label-encoded, and numeric features were standardized to zero mean and unit variance. The dataset was then divided into features (X) and target (y), with the target variable representing workload predictions. A train-test split allocated 80% of the data for training and 20% for testing, ensuring a balanced approach for model evaluation.

4.2   Hardware and Software Environment

The experiments were carried out in a high-performance computing environment equipped with an Intel Xeon processor featuring 24 cores, 128 GB of RAM, also including a NVIDIA Tesla V100 GPU with 16 GB of memory. The software environment utilized Python 3.8 as the programming language, with TensorFlow and Keras libraries employed for the development and training of the CNN-Transformer model. This robust hardware and software setup ensured efficient execution and accurate evaluation of the model.

5   Results and Analysis

The results of the CNN-Transformer model for cloud workload prediction demonstrate the model's ability to accurately capture spatio-temporal dependencies in

---

[1] https://github.com/google/cluster-data.

data center resource usage. In this section, we evaluate the model's performance based on MSE, MAE, and the R^2 score. Figures 4, 5, 6, and 7 provide visual insights into the model's behavior across various evaluation metrics and scenarios.

5.1   Error Distribution

Figure 4 shows the residuals' (error) distribution. The concentration of mistakes around 0 indicates that the model's predictions are constantly somewhat near to the actual values with low variation. This limited error distribution guarantees the accuracy of the model and strengthens its applicability for correct workload prediction in cloud data centers. After a dropout layer for regularization, the pooled vector is next run through a dense layer with ReLU activation.

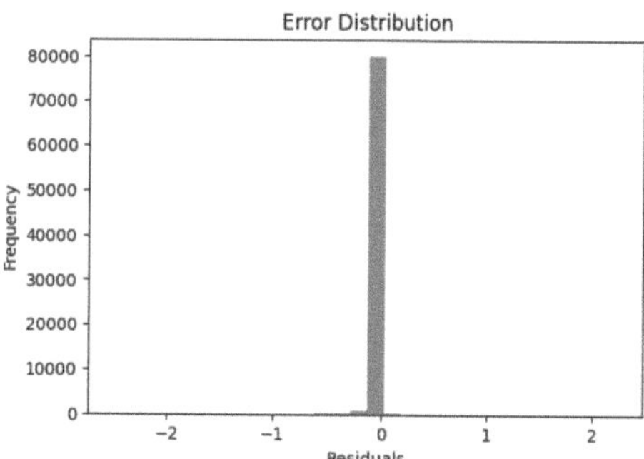

Fig. 4. Distribution of residuals, indicating error concentration near zero, reflecting high predictive accuracy.

5.2   True vs. Predicted Value Density

Figure 5 illustrates the density plot for the true and predicted workload values, overlaid to assess the alignment between them. The near-perfect overlap of the

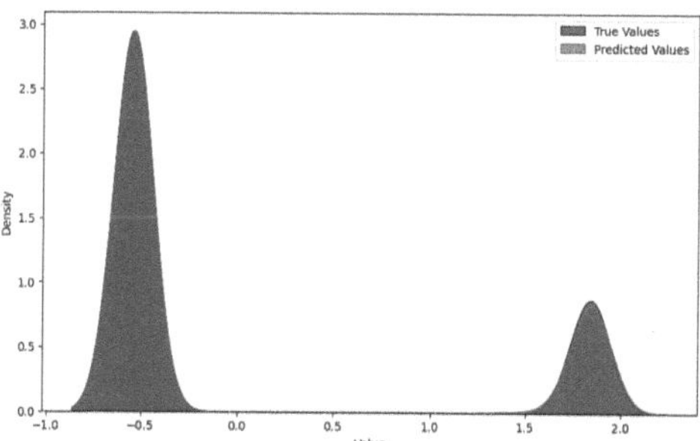

Fig. 5. Density plot of true and predicted workload values. The alignment of both distributions confirms the model's precision.

true and predicted densities demonstrates the model's high accuracy in workload prediction, capturing both primary and secondary peaks in the distribution.

5.3   Power Consumption Prediction Across Machines

Figure 6 shows a comparison of predicted and actual power consumption across three sample machines within a one-hour timeframe. The model accurately mirrors the real power usage trends, capturing variations specific to each machine. This result highlights the model's capability to predict power consumption, which is critical for resource optimization and energy efficiency in data centers.

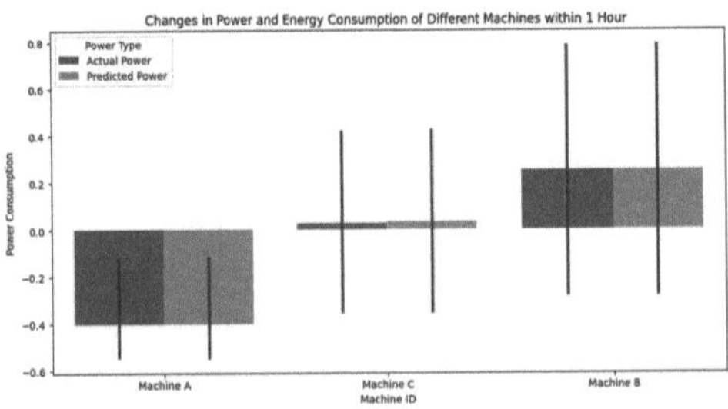

Fig. 6. Comparison of actual and predicted power consumption across different machines within a one-hour interval.

5.4   Carbon Emission Variation

To examine the model's utility in green data center initiatives, Fig. 7 presents the predicted variation in carbon emissions for a single server over a 24-hour period. The model estimates carbon emissions based on workload predictions and corresponding power usage metrics. The fluctuating emission patterns reflect typical daily workload variations, supporting the model's effectiveness in real-time carbon monitoring. Such predictions can be instrumental in enabling dynamic adjustments to minimize environmental impact in data center operations.

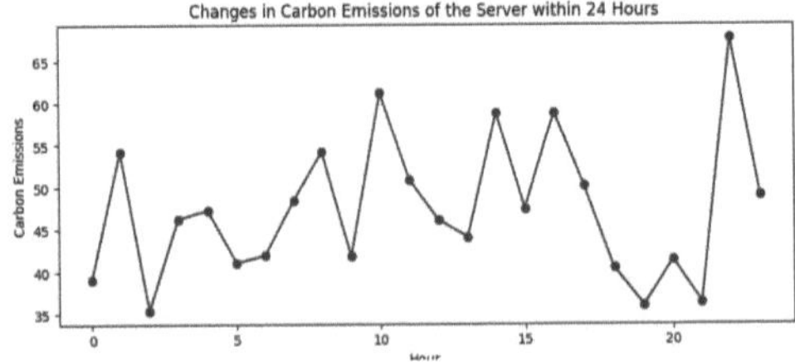

Fig. 7. Changes in predicted carbon emissions of the server within 24 h, indicating the model's capability for sustainable management strategies.

5.5   Comparative Analysis with SOTA Methods

This subsection compares our CNN-Transformer model with current SOTA models in cloud workload prediction. The comparison focuses on essential evaluation metrics: MSE, MAE, and R^2 score. For this purpose, we selected several prominent models in the literature, each utilizing different architectural innovations to enhance predictive accuracy. Table 1 summarizes the performance of each model in terms of MSE, MAE, and R^2 score.

Table 1. Comparison of our CNN-Transformer model with recent SOTA models for cloud workload prediction.

Model	MSE	MAE	R^2 Score
Our CNN-Transformer Model	0.0059	0.0381	0.994
CNN-BiLSTM Model [1]	0.2222	0.2387	0.9494
Attention-based DNN [6]	0.1532	0.1986	0.9651
Hybrid CNN-LSTM Model [11]	0.1903	0.2211	0.9562
MHA Hybrid LSTM [3]	0.1137	0.1765	0.9753
Ensemble CNN-Attention BiLSTM [18]	0.0935	0.1698	0.9824

5.5.1   Analysis of Results

The comparative analysis highlights the exceptional performance of our CNN-Transformer model in effectively capturing spatio-temporal dependencies within cloud workload data. Specifically:

- MSE: Our model achieves an MSE of 0.0059, significantly lower than competing models, indicating a higher accuracy in capturing fine-grained workload variations.
- MAE: With an MAE of 0.0381, the CNN-Transformer model minimizes absolute deviations, further showcasing its precision in workload prediction.
- R^2 Score: An R^2 score of 0.994 highlights the model's strong alignment with actual workload data, indicating that it accounts for 99.4% of the variance in the observed values, surpassing all other models in this comparison.

These findings suggest that the CNN-Transformer model effectively combines CNN's spatial feature extraction with Transformer-based temporal dependency modeling. This hybrid approach enables the model to achieve a lower prediction error and higher accuracy than traditional CNN-LSTM and BiLSTM-based architectures. Additionally, the integration of the Transformer's attention mechanism contributes to a nuanced understanding of temporal patterns, improving predictive performance in cloud workload management tasks.

6   Discussion

This study underscores the CNN-Transformer model's potential as an effective tool for workload prediction in energy-efficient CC. By leveraging spatial and temporal correlations in workload data, the model achieves real-time adjustments, evidenced by low MSE and high R^2 scores. This optimization reduces under-utilization and over-provisioning, advancing sustainable CC through proactive energy management. The CNN-Transformer architecture excels over traditional models like CNN-LSTM and BiLSTM by combining CNNs for spatial feature extraction and Transformer's self-attention mechanism for prioritizing temporal dependencies. This hybrid approach effectively captures long-term dependencies, enhancing predictive accuracy and robustness, making it a critical tool for balancing computational demands with energy efficiency. However, the model can be improved by refining attention mechanisms to focus on features most relevant to workload dynamics, enhancing both accuracy and efficiency [7,9]. Integrating federated learning could also address data privacy concerns by enabling learning from distributed sources while preserving confidentiality, ensuring practical and secure CC implementations [19,20].

7   Conclusion

This study introduced a CNN-Transformer model designed to enhance the sustainability of CC through precise and efficient workload forecasting. By combining the spatial feature extraction of CNNs with the temporal modeling capabilities of Transformers, the model achieves SOTA accuracy, evidenced by high R^2 scores and low MSE. Its ability to identify complex spatiotemporal patterns ensures reliable resource management and energy efficiency in cloud environments. Beyond accuracy, the model minimizes over-provisioning and under-utilization, reducing operational costs and unnecessary energy consumption. It also contributes to environmental sustainability by integrating workload forecasting with real-time carbon emission modeling, enabling dynamic adjustments to workload fluctuations and supporting sustainable data center operations. However, areas for improvement remain. Enhanced attention mechanisms can improve feature prioritization and model interpretability, while incorporating federated learning can address data privacy concerns and enable decentralized data usage. In conclusion, the CNN-Transformer model represents a significant advancement in workload forecasting, improving operational efficiency and environmental sustainability in cloud data centers. Future work will focus on refining feature prioritization and privacy-preserving techniques to broaden its applicability across diverse CC contexts.

References

1. Zhang, H., Li, J., Yang, H.: Cloud computing load prediction method based on CNN-BILSTM model under low-carbon background. Sci. Rep. **14**(1), 18004 (2024)

2. Yang, J., Xiao, W., Jiang, C., Hossain, M.S., Muhammad, G., Amin, S.U.: AI powered green cloud and data center. IEEE Access **7**, 4195–4203 (2018)
3. Bi, J., Ma, H., Yuan, H., Zhang, J.: Accurate prediction of workloads and resources with multi-head attention and hybrid LSTM for cloud data centers. IEEE Trans. Sustain. Comput. **8**(3), 375–384 (2023)
4. Karim, M.E., Maswood, M.M.S., Das, S., Alharbi, A.G.: BHYPREC: a novel bi-LSTM based hybrid recurrent neural network model to predict the CPU workload of cloud virtual machine. IEEE Access **9**, 131476–131495 (2021)
5. Yang, J., Chen, Y., Por, L., Ku, C.: A systematic literature review of information security in chatbots. Appl. Sci. **13**(11), 6355 (2023)
6. Shih, Y.-C., Tamilarasan, S., Chen, C.-S., Zargar, O.A., Kuan, Y.-D.: Attention based integrated deep neural network architecture for predicting the effectiveness of data center power usage. Int. J. Thermofluids **24**, 100866 (2024)
7. Chauhan, R., Karnati, M., Singh, P.: Attention based deep neural network for classification of kidney ailments using CT images. In: 2024 15th International Conference on Computing Communication and Networking Technologies (ICCCNT), pp. 1–5 (2024). IEEE
8. Ku, Y.-J., Sapra, S., Baidya, S., Dey, S.: State of energy prediction in renewable energy-driven mobile edge computing using CNN-LSTM networks. In: 2020 IEEE Green Energy and Smart Systems Conference (IGESSC), pp. 1–7. IEEE (2020)
9. Wagan, A., Khan, A., Chen, Y., Yee, P., Yang, J., Laghari, A.: Artificial intelligence-enabled game-based learning and quality of experience: a novel and secure framework (B-AIQOE). Sustainability **15**(6), 5362 (2023)
10. Nguyen, H.M., Kalra, G., Kim, D.: Host load prediction in cloud computing using long short-term memory encoder–decoder. J. Supercomput. **75**(11), 7592–7605 (2019)
11. Patel, E., Kushwaha, D.S.: A hybrid CNN-LSTM model for predicting server load in cloud computing. J. Supercomput. **78**(8), 1–30 (2022)
12. Abbasi, I.A., Rehman, M.Z., Alam, T., Aznaoui, H., et al.: Adapted convolutional neural networks and long short-term memory for host utilization prediction in cloud data center (2021)
13. Xu, M., Song, C., Wu, H., Gill, S.S., Ye, K., Xu, C.: ESDNN: deep neural network based multivariate workload prediction in cloud computing environments. ACM Trans. Internet Technol. (TOIT) **22**(3), 1–24 (2022)
14. Gao, J., Wang, H., Shen, H.: Machine learning based workload prediction in cloud computing. In: 2020 29th International Conference on Computer Communications and Networks (ICCCN), pp. 1–9. IEEE (2020)
15. Dogani, J., Khunjush, F., Mahmoudi, M.R., Seydali, M.: Multivariate workload and resource prediction in cloud computing using CNN and GRU by attention mechanism. J. Supercomput. **79**(3), 3437–3470 (2023)
16. Daraghmeh, M., Agarwal, A., Jararweh, Y.: An ensemble clustering approach for modeling hidden categorization perspectives for cloud workloads. Clust. Comput. **27**(4), 4779–4803 (2024)
17. Taherinavid, S., Moravvej, S., Chen, Y., Yang, J., Ku, C., Por, L.: Automatic transportation mode classification using a deep reinforcement learning approach with smartphone sensors. IEEE Access (2023)
18. Kaim, A., Singh, S., Patel, Y.S.: Ensemble CNN attention-based BiLSTM deep learning architecture for multivariate cloud workload prediction. In: Proceedings of the 24th International Conference on Distributed Computing and Networking, pp. 342–348 (2023)

19. Khan, A., Zhang, X., Hajjej, F., Yang, J., Ku, C., Por, L.: ASMF: ambient social media forensics chain of custody with an intelligent digital investigation process using federated learning. Heliyon **10**(1) (2024)
20. Yang, J., Qin, H., Por, L., Shaikh, Z., Alfarraj, O., Tolba, A.: Optimizing diabetic retinopathy detection with inception-V4 and dynamic version of snow leopard optimization algorithm. Biomed. Signal Process. Control **96**, 106501 (2024)

Convolutional Neural Networks Based Decision Making for Adaptive Modulation and Coding in 5G Networks

A. Manikandan(✉) ⓘ, T. R. Rhokhith Pranav, R. Sandeep, V. S. Vidyashankar, and K. Naga Poojitha

Department of Electronics and Communication Engineering, Amrita School of Engineering, Coimbatore, Amrita Vishwa Vidyapeetham, Coimbatore, India
a_manikandan@cb.amrita.edu, vidya13062003@gmail.com,
{cb.en.u4ece21247,cb.en.u4ece21254,cb.en.u4ece21263,
cb.en.u4ece21223}@cb.students.amrita.edu

Abstract. In this work, the application of Convolutional Neural Networks (CNN) for improving Adaptive Modulation and Coding (AMC) in 5G communication systems is explored. AMC optimizes data transmission under uncertain channel conditions by dynamically selecting modulation schemes and correcting errors. Traditional AMC relies on manual feature extraction, which is time-consuming and may overlook critical patterns. To address this, a structured CNN pipeline with convolutional layers, activation functions, and pooling techniques is proposed to spatially extract features from in-phase and quadrature-phase signal components. The model demonstrates robust classification across varying Signal-to-Noise Ratio (SNR) levels using modulation schemes like BPSK, QPSK, 16QAM, 64QAM, and 256QAM. Experimental results show that the proposed CNN model achieves an accuracy of 95%, indicating strong generalization to unseen data. This approach underscores the CNN's potential to automatically learn hierarchical features from raw input, enhancing the reliability and efficiency of 5G wireless communication systems.

Keywords: CNN · AMC · Signal Classification · Modulation Schemes · Spatial Feature Extraction

1 Introduction

High data rates, reliable connectivity and low latency are increasingly demanded in 5G networks due to rapid advancement of 5G networks [1, 2]. These requirements are achieved through AMC, by dynamically adjusting the modulation and coding selection aligned with the channel conditions to maximize the spectral efficiency and to maintain the acceptable bit error rate [3, 4, 5]. Regarding AMC, CNNS can learn to identify the optimal modulation and coding selection from the raw or preprocessed Channel State Information (CSI) [6, 7]. Moreover, Wireless communication environments are, however, inherently unpredictable and pose challenges like noise, interference and fading

© The Author(s), under exclusive license to Springer Nature Switzerland AG 2026
P. Chandrakar et al. (Eds.): ICCINS 2025, CCIS 2738, pp. 287–299, 2026.
https://doi.org/10.1007/978-3-032-09572-5_23

that often limit the accuracy of CSI and further complicated AMC decision making [8, 9]. Despite its importance, the traditional AMC heavily relies on lookup tables and static algorithms, which often fail to adapt effectively with the rapidly changing channel environments of 5G networks [10, 11]. Moreover, the increased flexibility in user mobility, interference and diverse use cases in 5G demands more sophisticated approaches [12, 13]. CNN now emerged as a promising tool to address the challenges in automated learning and to generalize the complex patterns in huge population of data sets [14]. These advancements underscore the potential of CNNs to meet the demands of 5G networks, where the dynamic nature of channels requires rapid adaptation [15, 16]. Existing state of research has extensively explored the role of CNN in wireless communications with a particular focus on modulation classification. CNN based classification is demonstrated from raw in phase and quadrature phase data with high accuracy even in noisy environments [17]. Building on this, spectrograms are utilized as inputs to CNNs achieving robust performance under different SNR conditions [18, 19]. This work also highlights the versatility of CNNs in processing different representations of received information and their ability to generalize across diverse conditions. For AMC, these studies have been extended to jointly classify the modulation and coding schemes, permitting end-to-end modulation and coding selection [20, 21, 22]. Beyond the modulation classification, research was done with CNN for end-to-end AMC systems [23, 24]. From the results it is visible that CNNs can map CSI or other channel metrics directly to optimal modulation coding selection indices, bypassing the need for traditional threshold-based methods [25, 26]. It is also proved that CNNs trained on simulated CSI could achieve significantly higher spectral efficiency compared to the traditional AMC selection [27, 28]. Similarly, a hybrid CNN framework integrated with reinforcement learning to enhance the ability in dynamic environments [29]. This hybrid-based approach allows the modulation coding selector to refine its decisions in real-time based on the observed feedback through feedback channels though in fast fading channels [30, 31]. Despite all these accomplishments, there are some challenges to address including, need for extensive labeled data sets, computational complexity, and ensuring the generalization across diverse channel conditions. One of the most important challenges in deploying CNN in AMC is its computational demand of models, since CNN involves multiple layers of convolutional operations, pooling, and fully connected layers, which require more processing power and memory [32, 33, 34]. In 5G networks, real time AMC decisions are very crucial to sustain with the target BER and spectral efficiency [35]. Second, CNN require large, diverse and labeled datasets for training to achieve high accuracy [36]. Generating such datasets for 5G networks, which feature diverse use cases and rapidly changing environments, is a significant challenge. Third, the ability of CNN based AMC systems to generalize across different and rapidly changing environments is critical in maintaining the tradeoff between the SNR and BER [37, 38]. Also, a CNN trained in one environment may not perform well in another, leading to suboptimal decisions that degrade the performance of the system. To overcome these difficulties, deep learning techniques have been developed as useful solutions, and CNNs have shown themselves to be particularly useful [39]. In this paper, we propose a CNN based model to classify modulation schemes (BPSK, QPSK, 16QAM, 64QAM, and 256QAM) at varying SNR levels. Spatial features critical for modulation classification are extracted by processing in-phase (I) and quadrature phase (Q) signal components of the proposed model. The CNN model can train efficiently and avoid overfitting by using convolutional layers, activation functions, pooling mechanisms and dropout regularization. To address the computational complexity discussed in the literature, a lightweight architecture of CNN is used in this work without compromising the performance of the system. To meet the data requirements, synthetic data generation techniques are deployed partially to mitigate the scarcity of real-world data sets. Evaluation results show that it can improve AMC performance with higher accuracy and reliability under dynamic channel conditions. The contributions of this work are in highlighting the potential of CNN based approaches in addressing some of the challenges of 5G communication and helping with the development of robust, adaptive systems for next generation wireless networks.

2 System Model

In this work, a proposed system utilizes CNN to enhance AMC for 5G networks. To address the issues of dynamic wireless environments, the system improves on the accuracy of modulation scheme classification under varying SNR conditions [40]. The system model is divided into three major components: The data preprocessing and training configuration. Here, the system collects the CSI information from the feedback channel and provides the inputs for decision making.

2.1 CNN Architecture

The CNN architecture is chosen such that it can provide high feature extraction capability with fairly low computational complexity. Both low level features as well as high level features are captured from the I-Q data by selecting the number of layers and filter sizes. Large kernel sizes (e.g., 9×9) are used to accomplish this by capturing as much information as possible, and deeper layers help learn abstract patterns important for discriminating similar modulation schemes, on the expense of losing fine details. CNN architecture which extracts the spatial features necessary for modulation classification from in-phase and quadrature phase components of communication signals is shown in Fig. 1. This model specifically designed to efficiently extract the features while remaining computationally simple and robust [41]. Efficient feature extraction methods within the lightweight CNN framework produce high performance results together with reduced computational complexity by using dimensionality reduction and regularization techniques. Small kernel sizes incorporated into convolutional layers allow minimal noise in signal attributes as well as reduced total parameter quantity. The spatial features in feature maps remain intact while computational expenses decrease through max pooling operations. The use of Dropout regularization enables models to conduct accurate predictions at scale because it stops overfitting while utilizing minimal parameters. The normalization of feature distributions using Batch Normalization helps to stabilize learning and improves convergence speed because Leaky ReLU activation allows gradient flow during effective learning. By employing multiple optimizations approaches simultaneously the CNN reaches superior modulation classification performance while using less memory space and processing power for real-time 5G implementations.

Input Layer The signal samples are reshaped into tensors, and these correspond to the I and Q components, and are input to the CNN [42, 43]. CNN uses these tensors to extract meaningful patterns, and these tensors give a spatial structure of these patterns.

Convolutional Layers

The CNN architecture performs filter applications to I-Q signal data as convolutional layers while detecting local patterns of amplitude variations and noise effects through their Leaky ReLU activation function. The parameters used for convolutional layer are mentioned in Table 1. The kernel design method enables feature extraction effectively through dimension preservation for accurate detection of modulation patterns by the model. The network uses MaxPooling1D which selects the maximum value within a specified window to perform spatial reduction of feature maps while keeping vital data intact. The methodology optimizes computational performance while maintaining essential signal aspects that include both modifications in amplitude data and phase variants. GlobalMaxPooling1D helps compression of extracted features through selection of greatest values from each feature map while preserving important modulation-specific information and lowering parameter count. Through mixing both local and global pooling strategies the CNN obtains detailed signal characteristics along with high-level signal patterns which leads to reliable classification across different SNR levels. The use of dropout layers represents a solution to overfitting that slows down the training process for specific neurons to enhance model performance across unforeseen examples. The features extracted from the network become refined and improve decision making when fewer neurons are used in successive stages of propagation. The application of batch normalization provides stabilizing effects to learning through its distribution normalization capability across layers which improves convergence speed and makes the model more resilient. Softmax activation solves by producing probability distributions for modulation classes and identifying the most probable class. A convolutional filtering approach, efficient pooling methods and dropout regularization, combined with batch normalization produces a CNN which maintains accurate and real-time automatic modulation classification for 5G AMC applications.

The convolution operation extracts local features from the input through a sliding window approach and is defined as:

$$F_{l,j} = \sigma\left(\sum_{i=1}^{M} w_i * X_{i,j} + b\right) \tag{1}$$

where $F_{l,j}$ represents the feature map at layer l, w_i are the filter weights, $X_{i,j}$ is the input, b denotes the bias term, and σ is the activation function, typically a Leaky ReLU for enhanced non-linearity. Following convolution, max pooling is applied to reduce the spatial dimensions and retain the most significant features, mathematically represented as,

$$P_{l,j} = \max\left(F_{l,j}\right) \tag{2}$$

To do the final classification, softmax function is used to convert the output as a class probability,

$$P(y = c|x) = \frac{e^{Z_C}}{\sum_j e^{Z_j}} \tag{3}$$

where Z_C is the output of final layer which corresponds to class c. To facilitate multi class classification tasks, this function makes sure that the output probabilities add up to one. Together, these operations are the basis of CNNs, for performing efficient feature extraction and good classification.

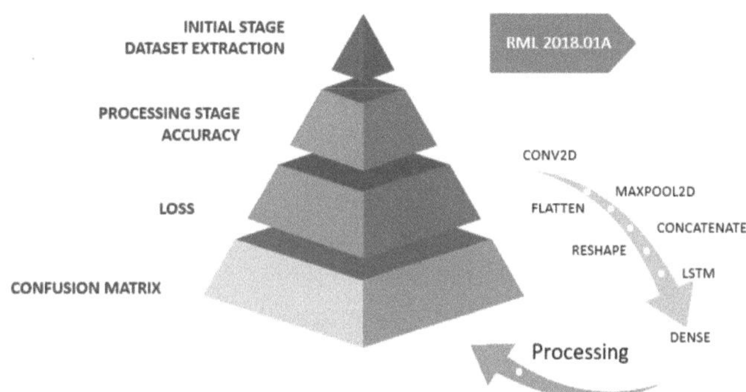

Fig. 1. CNN Process

Table 1. Parameters used for convolutional layer

Parameter	Value
Batch Size	5000
Epochs	100
Optimizer	Adam
Dropout Rate	0.2
Kernel Size	1

Dataset and Preprocessing

A CNN model is trained and evaluated on the RadioML 2018.01A dataset. The modulated signals in this dataset are from a large variety of sources under different channel conditions. The dataset contains BPSK, QPSK, 16QAM, 64QAM and 256QAM as the multiple modulation schemes. It covers a wide range of SNR levels, simulating different real world wireless conditions. The CNN-based modulation classification system converts its raw I/Q data to normalized amplitudes after restructuring the complex data samples as input-ready tensors. The RadioML 2018.01A dataset provides I-Q data in numerous SNR conditions yet the researchers standardized the data before shifting it to a uniform training and testing format. During preprocessing steps model convergence remains unaffected by variations that get mitigated through the normalization process which standardizes signal amplitude measurements to create a uniform scale. The model employs stratified sampling throughout training to balance the data distribution according to different modulation techniques and SNR levels for unbiased outcome. This reshapes the raw signal data into structured grids which are fit for CNN input. Both phase and amplitude normalization functions exist in the model through batch normalization and max pooling techniques. These two methods make input distributions more stable and extract key spatial features by lowering minor fluctuations. The system achieves more robustness by implementing CNN filters that can handle different SNR levels. Current implementations of the model apply Conv1D layers to I-Q signals in their time-domain yet integration of frequency transformations through Wavelet Transform and STFT may enhance performance during noisy conditions. The future development should combine features from time and frequency domains to enhance resistance to noise disturbances. Signal frames of fixed length form the dataset which eliminates direct handling of variable lengths. GlobalMaxPooling1D offers efficient feature extraction across all input dimensions without concerning about signal size and so does operational use benefit from zero-padding or adaptive pooling technologies. The model can efficiently classify the signals by associating each grid with a particular modulation scheme at a given SNR. Then the dataset is split to the training set and testing set or a validation set for evaluating if the model generalizes. Both sets have all modulation schemes represented equally by stratified sampling. The input data is normalized to standardize signal amplitudes, and to improve training efficiency.

Training Configuration

The training process is carefully designed to optimize performance and avoid overfitting. The Adam optimizer with adaptive learning rates giving efficient model training was used as an optimizer. We use the categorical cross entropy loss function which is a divergence between the predicted and true class distribution, hence guiding the model during classification. An epoch (for some initially specified batch size) is trained with the model for multiple epochs. We use early stopping, so that we stop training, if performance on the validation data does not improve after a certain number of epochs. The model's overall performance is evaluated with accuracy during training. Confusion matrix was used to analyze the classification results and to find out misclassifications, as well as model performance.

Model Evaluation

The model is then trained and evaluates its classification accuracy and the model's robustness over different SNR conditions using the test dataset. Classification performance is strong for lower order modulation schemes (e.g. BPSK, QPSK), but higher order schemes (e.g. 64QAM, 256QAM) are challenging at lower SNRs due to overlapping signal characteristics. The proposed system improves modulation classification for more reliable communication in 5G networks by integrating a well optimized CNN architecture with robust preprocessing and training strategies. The work lays a foundation for future AMC development to achieve more efficient modulation classification in future wireless technologies.

3 Results and Inference

The CNN-based AMC model offers better performance as shown in the Table 2 compared to existing likelihood-based and feature-based AMC solutions through multiple aspects. The CNN operates autonomously to learn important data features in I-Q information whereas traditional systems need experts along with human help for processing. Using this suggested approach users obtain accurate 95% solutions during intermediate SNR noise stages while traditional classification systems would otherwise lose their accuracy when facing noisy situations. Higher-order modulations such as 64QAM and 256QAM demonstrate improved optimization when processed by CNN solutions because standard classification systems cannot separate between classes. CNN achieves high efficiency by applying pooling and dropout operations to simplify processing of large data quantities but conventional AMC operates slower because of model complexity demands adjustment tuning. The CNN-based AMC solution proves its worth for real-time 5G networks through its strong scalability by delivering dependable modulation classification under different wireless conditions.

For 25 epochs, the proposed model was trained with a batch size of 64 samples per epoch. We used the training of the categorical cross entropy loss function and Adam optimizer which has shown to be efficient in adapting the learning rate. To prevent overfitting, and ultimately ensure the model generalizes well to previously unseen data, early stopping was conducted during the training process. The implementation of the training procedure was conducted on a dataset of multiple modulation schemes, including BPSK, QPSK, 16QAM, 64QAM and 256QAM, while recognizing that different Signal to Noise Ratio (SNR) conditions are present.

Figure 2 illustrates the training and validation accuracy throughout the model's training process. The training accuracy starts at a lower value (~60%) but increases rapidly in the initial epochs, stabilizing around 95% after ~20 epochs. This indicates the model learning effectively during the training process. The training accuracy peaked at 95%shows training and validation accuracy and found to be increasing steadily with the epochs and reached a peak of 95% by the end of the training. The validation accuracy plateaued at about 90% though, meaning the model got a good level of generalization. Finally, training and validation accuracy are not separated by much of a gap and indicate that the model isn't overfitting, rather it's still learning how to effectively classify a modulation scheme on never seen data. This level of generalization is indicative of the model's ability to model well without memorizing particular training examples.

Figure 3 shows training and validation loss which decreased progressively throughout the training process, demonstrating that our model was actually picking up on how to decrease errors. The loss begins at a high value (> 1200) and rapidly decreases to near zero by 20 epochs. This indicates effective minimization of error during training. However, the validation loss stopped improving after epoch 10 as compared to the validation loss had stabilized after epoch 10 showing that performance on the validation data was not

Table 2. Comparison of proposed CNN model vs Traditional AMC approach

Features	Proposed CNN Model	Traditional AMC Approaches
Feature Extraction	Automatic (via convolutional layers)	Manual (requires domain expertise and extensive feature engineering)
Computational Complexity	Moderate (optimized through pooling & dropout)	High for likelihood-based (requires model parameter estimation)
Performance with Higher-Order Modulations	Robust for BPSK, QPSK, 16QAM, 64QAM, 256QAM	Limited due to poor intra-class discrimination, especially for high-order QAM.
Accuracy Across SNR Levels	95% at moderate SNR levels	Degrades significantly under noisy environments (low SNR).
Generalization	Strong (learns hierarchical features)	Weak (relies on pre-defined features and statistical assumptions).
Scalability	Suitable for large datasets	Limited by model complexity and manual tuning.

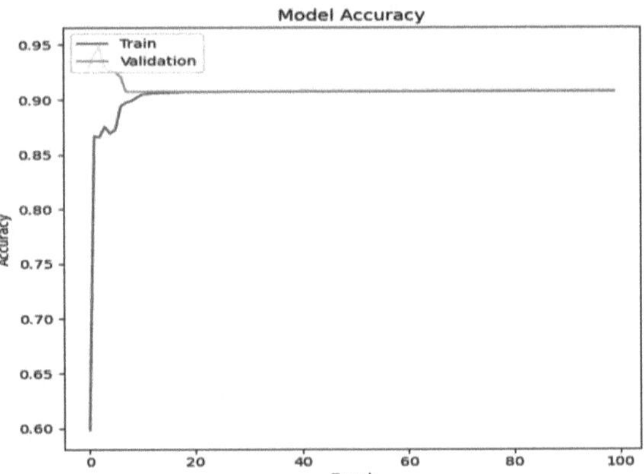

Fig. 2. Training and Validation Accuracy

improving much beyond some threshold. Validation loss stabilization indicates that the model settled down to a good solution for most modulation schemes, but it ran into some trouble with higher order modulations at low SNR. It is also observed that the alignment of training and validation loss trends is a strong indicator of a well-trained model. The

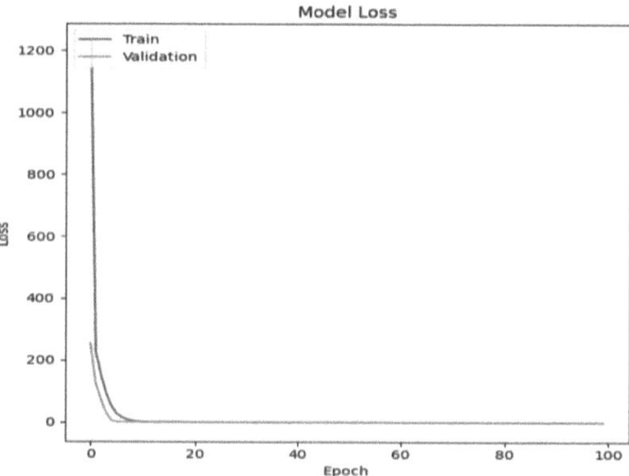

Fig. 3. Training and Validation Loss

absence of significant fluctuations in the validation loss suggests stability in the model's predictions.

In the context of AMC and CNN, confusion matrix is a vital tool to evaluate the performance of model. It summarizes the CNN's prediction versus the actual modulation and coding in the real time. Rows represent the predicted one while columns represents the true modulations. The diagonal elements in the matrix indicate correct predictions, while the other elements reflect the misclassifications. For example, if higher order modulations like 64-QAM are frequently predicted as 16-QAM, the model struggle under high SNR conditions. For AMC, the confusion matrix highlights the areas for improvement in the predictions of different modulations. For instance, under different challenging SNR conditions, the matrix might explicit a bias towards moderate lower order modulations like QPSK.

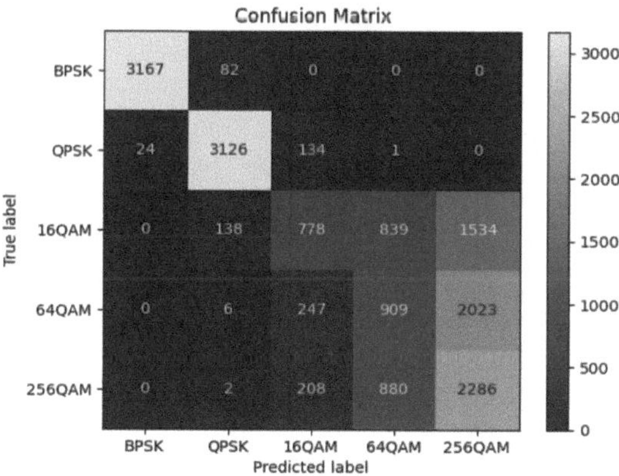

Fig. 4. Confusion Matrix

Figure 4 shows a confusion matrix which allows us to observe how well the model is classifying different modulation schemes. The confusion matrix showed that the model was particularly good at categorizing lower order modulation schemes, such as BPSK and QPSK. High accuracy was achieved in classifying these modulation types with relatively distinct signal characteristics, even under varying SNR conditions. Nevertheless, the model struggled to identify higher order modulation schemes (such as 64QAM and 256QAM) at lower SNR values. At lower SNRs, the overlapping signal constellations in these schemes led to occasional misclassifications because the signals became more difficult to distinguish from noise. Most of these misclassifications were seen in noisy environments where SNR was low, though signal distortions from the low SNR caused confusion between closely related modulation schemes. While these challenges exist, the model was able to maintain a consistent performance in identifying the modulation schemes, and exhibited robustness to moderately noisy conditions, however further work is required to improve the model's performance in identifying high order modulations at low SNR levels.

Precision and Recall highlight the performance of CNN based AMC system for each modulation scheme. Table 3 elucidates the precision and recall values of the proposed method.

Table 3. Precision and Recall

Modulation	Precision	Recall
BPSK	0.993	0.975
QPSK	0.932	0.951
16-QAM	0.893	0.878
64-QAM	0.901	0.892
256-QAM	0.894	0.902

Table 3 reflects a realistic outcome of an AMC classifier in time varying conditions. High performance in modulation schemes with simple constellation size, but lower metrics for higher order modulations schemes like 64-QAM and 256-QAM. The overall quality is good under lower SNR conditions. But some misclassifications occur occasionally due to the schemes higher complexity and overlapping signal characteristics. This difficulty arises from the denser constellation points and increased sensitivity to noise.

4 Conclusion

In summary, this paper demonstrated the effectiveness of CNN in enabling intelligent decision making for AMC in 5G networks. The proposed model leverages CNNs to accurately classify various modulation schemes under time varying channel conditions. By incorporating CNN, the system was able to achieve efficient feature extraction and robust decision making, essential for optimizing spectral efficiency and BER in dynamic environments. The demonstrated results indicate that the model performs exceptionally well for simpler modulation schemes like BPSK and QPSK with high precision and recall and marginally good for higher order modulation schemes like 64-QAM and 256-QAM due to their high complexity and susceptibility to noise. This study shows that CNN based AMC systems can significantly reduce the computational overhead of traditional approaches while maintaining real-time adaptability. The findings are promising for enabling intelligent decision making in future wireless networks.

Acknowledgments. We will be thankful to the management of Amrita Vishwa Vidyapeetham for their support.

Disclosure of Interests The authors have no competing interests to declare that are relevant to the content of this article.

References

1. Bezziane, M.B., Brik, B., Messiaid, A., Kafi, M.R., Korichi, A., Bezziane, A.B.: Impact of noise on data routing in flying ad hoc networks. Opt. Quant. Electron. **56**(4), 563 (2024)
2. Huang, J., Yang, F., Chakraborty, C., Guo, Z., Zhang, H., Zhen, L., Yu, K.: Opportunistic capacity-based resource allocation for 6G wireless systems with network slicing. Futur. Gener. Comput. Syst. **140**, 390–401 (2023)
3. Sufyan, A., Khan, K.B., Khashan, O.A., Mir, T., Mir, U.: From 5G to beyond 5G: a comprehensive survey of wireless network evolution, challenges, and promising technologies. Electronics. **12**(10), 2200 (2023)
4. Yang, F., Huang, J., Bhardwaj, A., Hussain, A., Abd El-Latif, A.A., Yu, K.: Adaptive modulation based on non data-aided error vector magnitude for smart systems in smart cities. IEEE Internet Things J. **10**(21), 18672–18685 (2023)
5. Ansari, S., Alnajjar, K.A.: Adaptive modulation and coding: a brief review of the literature. Int. J. Commun. Antenna Propag. **13**(1), 43–54 (2023)
6. Wang, T., Yang, G., Chen, P., Xu, Z., Jiang, M., Ye, Q.: A survey of applications of deep learning in radio signal modulation recognition. Appl. Sci. **12**(23), 12052 (2022)
7. Wang, B., Lin, Z., Zhang, X.: A CNN-MPSK demodulation architecture with ultra-light weight and low-complexity for communications. Symmetry. **14**(5), 873 (2022)
8. Yadav, P., Kumar, S., Kumar, R.: A review of transmission rate over wireless fading channels: classifications, applications, and challenges. Wirel. Pers. Commun. **122**(2), 1709–1765 (2022)
9. Shanthi, K.G., Manikandan, A.: An improved adaptive modulation and coding for cross layer design in wireless networks. Wirel. Pers. Commun. **108**, 1009–1020 (2019)
10. Ye, X., Yu, Y., Fu, L.: Deep reinforcement learning based link adaptation technique for LTE/NR systems. IEEE Trans. Veh. Technol. **72**(6), 7364–7379 (2023)
11. Benbaghdad, M., Fergani, B., Tedjini, S.: Toward a new PHY layer scheme for decoding tags collision signal in UHF RFID system. IEEE Commun. Lett. **20**(11), 2233–2236 (2016)
12. Shayea, I., Ergen, M., Azmi, M.H., Çolak, S.A., Nordin, R., Daradkeh, Y.I.: Key challenges, drivers and solutions for mobility management in 5G networks: a survey. IEEE Access. **8**, 172534–172552 (2020)
13. Gayatri, V., Kumaran, M.S.: An improved whale optimization algorithm for cross layer neural connection network of MANET. J. Commun. **17**(10), 857–864 (2022)
14. Yadav, S.S., Hiremath, S., Surisetti, P., Kumar, V., Patra, S.K.: Application of machine learning framework for next-generation wireless networks: challenges and case studies. In: Handbook of Intelligent Computing and Optimization for Sustainable Development, pp. 81–99. Wiley/Scrivener Publishing, Hoboken/Beverly (2022)

15. Zhang, C., Ueng, Y.L., Studer, C., Burg, A.: Artificial intelligence for 5G and beyond 5G: implementations, algorithms, and optimizations. IEEE J. Emerg. Sel. Top. Circuits Syst. **10**(2), 149–163 (2020)
16. Benbaghdad, M., Taloul, S., Tedjini, S.: Real time implementation of automatic digital modulation-based SNR estimation using SDR platforms. Wirel. Pers. Commun. **132**(2), 1593–1612 (2023)
17. Ma, W., Cai, Z.: Deep learning based cognitive radio modulation parameter estimation. IEEE Access. **11**, 20963–20978 (2023)
18. Kojima, S., Maruta, K., Feng, Y., Ahn, C.J., Tarokh, V.: CNN-based joint SNR and Doppler shift classification using spectrogram images for adaptive modulation and coding. IEEE Trans. Commun. **69**(8), 5152–5167 (2021)
19. Rao, N.V., Krishna, B.T.: Automatic modulation recognition of Analog modulation signals using convolutional neural network. In: Evolution in Signal Processing and Telecommunication Networks: Proceedings of Sixth International Conference on Microelectronics, Electromagnetics and Telecommunications (ICMEET 2021), vol. 2, pp. 411–419. Springer Singapore (2022)
20. Liu, X., Li, C.J., Jin, C.T., Leong, P.H.: Wireless signal representation techniques for automatic modulation classification. IEEE Access. **10**, 84166–84187 (2022)
21. Shukla, D., Prakash, A., Tripathi, R.: Adaptive modulation and coding for performance enhancement of vehicular communication. Wirel. Pers. Commun. **123**(1), 195–214 (2022)
22. Zerrouki, H., Azzaz-Rahmani, S.: WiMAX throughput maximization for MIMO-OFDM systems via cross-layer design. In: WITS 2020: Proceedings of the 6th International Conference on Wireless Technologies, Embedded, and Intelligent Systems, pp. 921–931. Springer Singapore (2022)
23. Usman, M., Lee, J.A.: AMC-IoT: automatic modulation classification using efficient convolutional neural networks for low powered IoT devices. In: 2020 International Conference on Information and Communication Technology Convergence (ICTC), pp. 288–293. IEEE Korea (2020)
24. Shaik, S., Kirthiga, S.: May.: Automatic modulation classification using DenseNet. In: 2021 5th International Conference on Computer, Communication and Signal Processing (ICCCSP), pp. 301–305. IEEE, Chennai (2021)
25. Liu, H., Zhang, Y., Zhang, X., El-Hajjar, M., Yang, L.L.: Deep learning assisted adaptive index modulation for mmWave communications with channel estimation. IEEE Trans. Veh. Technol. **71**(9), 9186–9201 (2022)
26. Manikandan, A., Anandan, P., Ramprasad, O.G., Prakash, K.R.B., Mahesh, A.D.: Cross layer design for TCP optimization through flow control and adaptive modulation and coding. In: 2023 11th International Conference on Intelligent Systems and Embedded Design (ISED), pp. 1–6. IEEE, Dehradun (2023)
27. Leblebici, M., Çalhan, A., Cicioğlu, M.: CNN-based automatic modulation recognition for index modulation systems. Expert Syst. Appl. **240**, 122665 (2024)
28. Aparna, M.P., Gandhiraj, R., Panda, M.: Steering angle prediction for autonomous driving using federated learning: The impact of vehicle-to-everything communication. In: 2021 12th International Conference on Computing Communication and Networking Technologies (ICCCNT), pp. 1–7. IEEE India (2021)
29. Njoku, J.N., Morocho-Cayamcela, M.E., Lim, W.: CGDNet: efficient hybrid deep learning model for robust automatic modulation recognition. IEEE Netw. Lett. **3**(2), 47–51 (2021)
30. Reddy, A.P.K., Kumari, M.S., Dhanwani, V., Bachkaniwala, A.K., Kumar, N., Vasudevan, K., Selvaganapathy, S., Devar, S.K., Rathod, P., James, V.B.: 5G new radio key performance indicators evaluation for IMT-2020 radio interface technology. IEEE Access. **9**, 112290–112311 (2021)

31. Javar, A.D., Prasanna Srinivasan, V.: A two-fold optimization framework using hybrid B2 algorithm for resource allocation in long-term evolution based cognitive radio networks system. Int. J. Commun. Syst. **35**(3), e5023 (2022)
32. Zhang, F., Luo, C., Xu, J., Luo, Y., Zheng, F.C.: Deep learning based automatic modulation recognition: models, datasets, and challenges. Digit. Signal Process. **129**, 103650 (2022)
33. Bu, K., He, Y., Jing, X., Han, J.: Adversarial transfer learning for deep learning based automatic modulation classification. IEEE Signal Process. Lett. **27**, 880–884 (2020)
34. Wei, S., Sun, Z., Wang, Z., Liao, F., Li, Z., Mi, H.: An efficient data augmentation method for automatic modulation recognition from low-data imbalanced-class regime. Appl. Sci. **13**(5), 3177 (2023)
35. Kaleem, Z., Ali, M., Ahmad, I., Khalid, W., Alkhayyat, A., Jamalipour, A.: Artificial intelligence-driven real-time automatic modulation classification scheme for next-generation cellular networks. IEEE Access. **9**, 155584–155597 (2021)
36. Huynh-The, T., Hua, C.H., Kim, J.W., Kim, S.H., Kim, D.S.: May.: Exploiting a low-cost CNN with skip connection for robust automatic modulation classification. In: 2020 IEEE Wireless Communications and Networking Conference (WCNC), pp. 1–6. IEEE (2020)
37. Kojima, S., Maruta, K., Ahn, C.J.: High-precision SNR estimation by CNN using PSD image for adaptive modulation and coding. In: 2020 IEEE 91st Vehicular Technology Conference (VTC2020-Spring), pp. 1–5. IEEE (2020)
38. Kumar, A., Srinivas, K.K., Majhi, S.: Automatic modulation classification for adaptive OFDM systems using convolutional neural networks with residual learning. IEEE Access. **11**, 61013–61024 (2023)
39. Rajappa, A.C.J., Ramadhas, S.D., Anjaneyulu, B.M., Kuppusamy, M.R.: Golden coded GFDM for 5G communication. Wirel. Pers. Commun. **115**, 2335–2348 (2020)
40. Zdorenko, Y., Lavrut, O., Lavrut, T., Nastishin, Y.: Method of power adaptation for signals emitted in a wireless network in terms of neuro-fuzzy system. Wirel. Pers. Commun. **115**(1), 597–609 (2020)
41. Peng, S., Sun, S., Yao, Y.D.: A survey of modulation classification using deep learning: signal representation and data preprocessing. IEEE Trans. Neural Networks Learn. Syst. **33**(12), 7020–7038 (2021)
42. Wang, N., Liu, Y., Ma, L., Yang, Y., Wang, H.: Multidimensional CNN-LSTM network for automatic modulation classification. Electronics. **10**(14), 1649 (2021)
43. Ge, Z., Jiang, H., Guo, Y., Zhou, J.: Accuracy analysis of feature-based automatic modulation classification via deep neural network. Sensors. **21**(24), 8252 (2021)

Blockchain-Based Authentication Protocol Integrated with PUFs for Secure Vehicular Communication

Brijmohan Lal Sahu[1]([✉])(ID), Preeti Chandrakar[1](ID), and Rajat Kumar[2]

[1] National Institute of Technology, Raipur, Chhattisgarh, India 492010
`blsahu.phd2020.cse@nitrr.ac.in`
[2] Shri Shankaracharya Techinal Campus, Bhilai, India

Abstract. Traditional keyless entry systems and Vehicular Ad-hoc Networks (VANETs) rely heavily on centralized infrastructure, static cryptographic keys, and fingerprint-based systems with data stored in fog servers and fuzzy extractors. These methods are susceptible to various attacks, including replay, RollJam, Error Correction Code (ECCs) decoder fault injection, and side-channel attacks. This paper proposes a secure protocol leveraging physical unclonable functions (PUFs) and blockchain to address these issues and enhance the security of keyless entry remotes in Electric Vehicles (EVs) and VANETs. The first step, a Keyless Entry System (KES) with PUF-based mutual authentication, addresses replay and RollJam attack vulnerabilities by generating unique secret keys and ensuring secure verification during vehicle access. The second step, facilitates lightweight Electric Vehicle-to-Infrastructure (EV2I) and Electric Vehicle-to-Electric Vehicle (EV2EV) authentication, offering verifiable broadcasts without reliance on traditional infrastructure and demonstrating resilience against Roadside Unit (RSU) capture attacks. The security analysis validates that the proposed protocol enhances the security of keyless entry and VANET systems, protects Onboard Unit (OBU) data, and significantly improves communication efficiency in EV networks.

Keywords: Authentication · blockchain · electric Vehicular ad hoc networks · physical unclonable functions · vehicle-to-vehicle communication

1 Introduction

With the increasing deployment of Electric Vehicles (EVs) and the expansion of Vehicular Ad-hoc Networks (VANETs) [3], ensuring robust security measures in keyless entry systems and communication protocols is paramount. This paper proposes innovative solutions integrating Physical Unclonable Functions (PUFs) and blockchain technology to enhance security and efficiency in these domains.

Supported by organization x.

For Keyless Entry Systems (KES), traditional methods have been vulnerable to replay and RollJam attacks mentioned in [5]. In the case of biometric authentication using fog computing to store fingerprint minutiae in databases, [13], which, once compromised, pose significant risks due to their immutability and susceptibility to unauthorized access. Biometric authentication using fuzzy extractors, despite enhancing security by storing and generating fingerprint points [18], suffers from access limitations and inconvenience. In VANETs, existing authentication protocols often exhibit vulnerabilities such as high computational overhead and susceptibility to attacks like Roadside Unit (RSU) capture incidents [3]. Similarly, storing sensitive biometric data in centralized databases introduces vulnerabilities, making these databases prime targets for attackers who can exploit weaknesses to gain unauthorized access or tamper with stored data. Blockchain is a trending technology widely adopted by researchers to incorporate security, transparency, and immutability into their domain [7,9]. Blockchain and machine learning are revolutionizing the transportation network and leading towards Intelligent Transportation Systems (ITS). Allowing secure sharing and storage of vehicular data in a decentralized manner for trusted model training [8].

1.1 Motivation

Physical Unclonable Functions (PUFs) are cryptographic primitives designed to generate unique outputs for each input, making them resistant to replication or reverse engineering. A promising candidate to mitigate attacks over authentication between KER and EV Onboard Units (OBUs). The existing system suffers from the Single Source of Failure (SSoF) for secure management and processing of authentication data. Susceptible towards replay attacks, RollJam Attacks, and RSU capture attacks. This scheme is proposed to address the above issues and facilitate an enhanced communication environment with the integration of blockchain and PUFs.

1.2 Contribution

The main contributions of this paper are outlined as follows:

1. Design alternative authentication mechanisms to mitigate vulnerabilities associated with biometric authentication in Electric Vehicle (EV) networks.
2. Develop a blockchain-based authentication protocol enhanced with Physical Unclonable Functions (PUFs) to emphasize reducing computational overhead and improving resilience against RSU capture attacks.

1.3 Paper Organization

The layout of this paper is organized as follows: Sect. 2 gives a glimpse of existing work in the context of blockchain, vehicular networks, and vehicular authentication. The system overview is described in Sect. 3 along with the design goals and network initialization procedure. Then, the proposed authentication method is explained in Sect. 4. Section 5 presents the simulation system setup configuration and requirements, then Sect. 6 describes the results. The entire article is summarised as a conclusion in Sect. 8.

2   Related Work

This section presents recent research work and their contributions to relTAnt vehicular communication and authentication domains. Addressing the security issues between EVs and smart grid communication in a Vehicle-to-Grid (V2G) environment PUFs based three-party key agreement is proposed by [11]. Mitigating the security threats in the work of [14], a secure Authentication and Key Establishment (AKE) for V2G is also presented by [20]. To mitigate the malicious rating problem in the service infrastructure, a blockchain-based Smart Parking system for providing a demand-supply balance of electricity in the vehicular network and a rating eligibility algorithm are proposed [10]. Existing authentication schemes are limited by security risks, high computational costs and the absence of hardware based security features. To address these deficiencies, this paper introduces a lightweight, mutual authentication protocol for vehicular networks. Key improvements include a PUFs based authentication and blockchain based secure storage demonstrating its practicality (Table 1).

Table 1. Notations and meaning

Symbol	Description
EV_i	i_{th} electric vehicle
PKV_i	Public key of EV_i
KER	Keyless Entry Remote
OBU	Onboard Units
TA	Eletric Vehicle Authority
SK_{TA}	Secret key Of TA
PK_{TA}	Public key Of TA
VID	EV Unique Identity
Val_i	Apperance Of EV
SKV_i	Secret key to access EV_i stored in KER
M_{imx}	Initial message exchange between EV and TA
K_{RSU}	Seecret value shared between RSU(Infrastructure Units)
$\oplus$	XOR operator
$\|$	Concatenation operator
$PUF(.)$	Physically unclonable function
(R_i, S_i)	Response and Stimulus for PUF
$H(.)$	Hashing operation

3   System Overview

3.1   Basic Entities

The key network participants of the proposed framework are as follows:

1. **Electric Vehicle (EV)**: Equipped with a Keyless Entry Remote employing Physically Unclonable Function (PUF) technology for secure access. Engages in Vehicle-to-Vehicle (V2V) communication for real-time data exchange.
2. **Trusted Authority (TA)**: Central authority responsible for managing cryptographic keys, authentication protocols, and maintaining system integrity. Utilizes blockchain technology for immutable record-keeping of verified transactions and authentication events.
3. **Roadside Unit (RSU)**: Acts as a pivotal point for communication and data relay between EVs and the central system. Registers securely with the TA and maintains local data integrity.
4. **Blockchain and Smart Contract**: Secure, shared, decentralized storage of transactions and records in such a manner that it can be verified. It offers features that are immutable, accessible, and verifiable. A smart contract is an unbiased, secure, auto-executable system that governs network participants' transactions.

3.2   Assumptions

Some assumptions considered in this paper are as follows:

1. Each network participant has built-in PUF hardware.
2. The KER and vehicle OBU registration is done in a secure environment.
3. TA is considered to be trusted entirely with sufficient computation resources, and EVs and RSU are honest but curious.

3.3   Threat Model

The communication between the network participants utilizes the public network for communication, such as enrollment and transactions. Handling possible threats becomes essential while operating in a public channel environment accessible to the adversary. Other vehicles, malicious bots, and compromised RSUs are potential adversaries. The adversary can intercept the communication between legitimate users and try to extract the actual communication. It may also attempt to fabricate previous conversations to initiate a replay attack by impersonating a legitimate network member. For this paper, TAs are considered trustworthy and honest, whereas vehicles and RSUs are considered partially honest and may tend to behave maliciously. An adversary may have multiple strategies to get access to stored data on RSU OBS and RSUs. Another way could be for the data stored over the blockchain to be publicly accessible.

4   Proposed Methodology

The network message exchange flow is illustrated in Fig. 2, highlighting the communication between all the network components. The proposed framework illustration highlights different types of communication and authentication: vehicle-to-keyless Entry Remote (V2KER), Vehicle-to-Vehicle (V2V), and Vehicle-to-Roadside Unit (V2RSU).

4.1   Primitives

Elliptic Curve Cryptography (ECC). The public-key cryptography Elliptic Curve E_F over finite fields and a generator point P, generates a substantial subgroup within the elliptic curve. For robust security, the order of this subgroup should be a large prime number. The strength of ECC lies in the computational intractability of solving the ECDLP, making it a preferred choice for contemporary cryptographic solutions.

Physical Unclonable Functions. Given that the proposed authentication mechanism employs PUFs, it's important to understand their operation. A PUF maps an input stimulus S_i to a response $R_i = \mathrm{PUF}(S_i)$. This mapping is represented as the pair (S_i, R_i) shown in the Fig. 1. Ideally, a PUF consistently produces the same response R_i for a given input S_i. If the stimuli vary, the reactions from the PUF will also vary accordingly. Additionally, two distinct PUF devices generate different reactions for the same stimulus.

4.2   Enrollment Phase Steps

The enrollment phase involves establishing communication and agreement between the Electric Vehicle (EV) and the Keyless Entry Remote (KER).

1. KER creates a message M_1 containing its identifier ID_{ker} and an enrollment request. The KER then sends M_1 to the vehicle as show in Fig. 1.
2. On receiving M_1 from the KER, the EVs, OBU generates a stimuli S_i. A message M_2 is created using S_i and then transmitted to the KER.
3. The KER determines $R_i = \mathrm{PUF}(S_i)$. It then transmits R_i back to the OBU through message M_3.
4. After receiving M_3 from the KER, the OBU stores the identifier ID_{ker} and the stimulus-response pair (S_i, R_i). Following this, the OBU generates a session key K_{veh} and transmits it to the KER via message M_4.
5. Once the KER receives M_4 from the vehicle, it saves the session key K_{veh}. This key is reserved for use in the subsequent authentication phase.

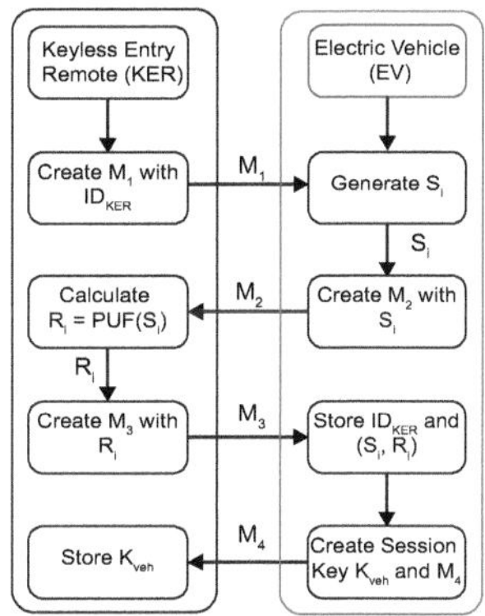

Fig. 1. Enrollment Phase Steps

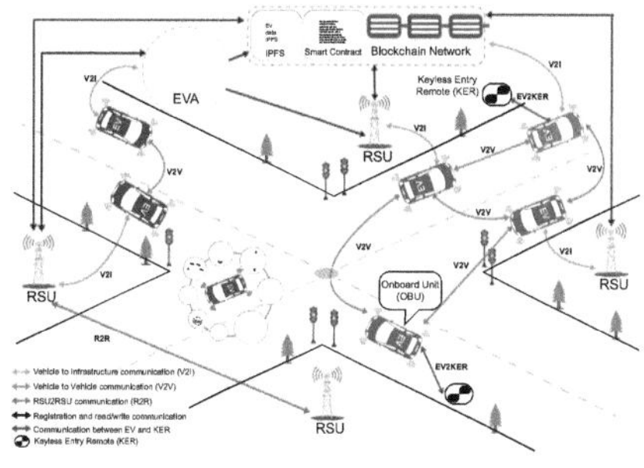

Fig. 2. Proposed framework architecture

4.3 Authentication Mechanism

Following the successful enrollment between the EV and the KER, the next stage involves mutual authentication to enable the EV's operation with the KER. The authentication process consists of the following steps:

- **Step 1:** Upon pressing the key button, the remote key generates a random number N_u. Subsequently, the key calculates $N'_u = N_u \oplus K_{veh}$. The key then constructs a message containing $\{ID_{ker}, N'_u\}$ and transmits it to the car's Onboard Unit (OBU).
- **Step 2:** When the car's OBU receives $\{ID_{ker}, N'_u\}$, it decodes $N_u = N'_u \oplus K_{veh}$. The OBU then generates a new random number N_c and calculates $N'_c = N_c \oplus K_{veh}$. Then, OBU concatenates ID_{ker}, N_u, K_{veh}, and N_c, and computes $A_a = H(ID_{ker} \parallel N_u \parallel K_{veh} \parallel N_c)$ for creating an authentication component. This hash A_u acts as the authentication component from the EV's receiver. Following this, the OBU locates the stimulus-response pair (S_i, R_i) associated with ID_{ker} and computes $S'_i = S_i \oplus K_{veh}$. Then creates a new stimulus S_i^{new} for the next round authentication and computes $S_i^{new'} = S_i^{new} \oplus K_{veh}$. Finally, the receiver composes message M_2 containing $\{A_a, N'_c, S'_i, S_i^{new'}\}$ and forwards it to the key.
- **Step 3:** Upon getting M_2, key first computes $N_c = N'_c \oplus K_{veh}$ and verifies the authentication variable A_a. If the verification is successful, the key computes $S_i = S'_i \oplus K_{veh}$ and $S_{new'} = S_{new} \oplus K_{veh}$. The key circuit subsequently uses PUF to generate the responses for the given stimuli as $R_i = PUF(S_i)$ and $R_{new} = PUF(S_{new})$. Following this, the KER computes $R'_i = R_i \oplus K_{veh}$ and $R^*_{new} = R_{new} \oplus K_{veh}$.
  It then concatenates the components $Identity_x$, N_c, K_{veh}, R_i, and R_{new}, and computes KER authentication variable by hashing concatenated component as $A_b = H(Identity_x \parallel N_c \parallel K_{veh} \parallel R_i \parallel R_{new})$. Finally, the key composes M_3: $\{A_b, R^*_i, R^*_{new}, command\}$, where the *command* specifies the 'ON' or 'OFF' operation to be executed. The key then sends M_3 to the vehicle.

– **Step 4:** When it receives M_3, the OBU decodes $R_i = R_i' \oplus K_{veh}$ and $R_i^{new} = R_i'^{new} \oplus K_{veh}$. Authentication process terminates upon failing verification of response R_i and authentication parameter A_b. If the verification succeeds, the OBU executes the 'ON' or 'OFF' operation as indicated by the *command*. Subsequently, the receiver updates its stored stimulus and responses to S_i^{new} and R_i^{new}, respectively, for use in the next round authentication cycle. The flow of this authentication phase is illustrated in Fig. 1.

4.4   Network Initialization

The TA initiates the network infrastructure and network essential parameters for other network participants to join the network. The TA also designs and validates the required smart contract and deploys it to the network. After that, make the smart contract address SC_{add} available as global parameters to other participants. It also broadcasts that the network is up and running, ready for the enrollment and pairing of the EVs. Using elliptic curve cryptography, the TA computes the secrete key SK_{TA}, and after that generates the public key $PK_{TA} = SK_{TA} \cdot P$. Now, TA broadcasts the $\{PK_{TA}, SC_{add}, E_F, P\}$ as global parameters.

4.5   Electric Vehicle, TA Pairing Phase

Initially, the Electric Vehicle (EV) reads the global variables broadcasted by the TA, $\{PK_{TA}, SC_{add}, E_F, P\}$, after that it uses these parameters for below processes:

Secret Number Generation: Each vehicle i generates a secret number SKV_i.

Public Key Calculation: The public key PKV_i is computed as $PKV_i = SKV_i \cdot P$, where P is the generator point on an elliptic curve obtained from global parameters.

Initial Message Exchange: The vehicle i sends $\{VID_i, PKV_i, Val_i, T_1\}$ securely to the TA using the public key PK_{TA}. $M_{imx} = E_{PK_{TA}}\{VID_i, PKV_i, Val_i, T_1\}$ then send M_{imx}, T_1 to TA for enrollment.

Authority Verification: TA verifies the freshness of the request using the T_1 after that legitimacy and uniqueness of VID_i. If valid, it proceeds.

Random Number Generation: TA generates random numbers r_i and a_i.

Recording on Blockchain: $A_i = a_i \cdot P$ and $PID_i = H(PKV_i \| VID_i \| Val_i \| r_i \| T_{val})$ are computed by the TA, and store the $\{PKV_i, PID_{EV_i}, T_{val}\}$ on blockchain using the smart contract address SC_{add}. On successful completion of the transaction, it returns the blockchain block index B_x.

Commitment Calculation: TA computes $b_i = h(PID_i \oplus PKV_i \oplus A_i) \cdot SK_{TA} + a_i$ and send $\{B_x, A_i, b_i\}$ back to the vehicle, encrypted by the public key PKV_i of the vehicle to ensure the integrity of the message.

Commitment Verification: EV_i uses the block index B_x and public key to locate the data stored on the blockchain $\{PKV_i, PID_{EV_i}, T_{val}\}$ after that verifies if $b_i \cdot P$ matches $h(PID_i \oplus PKV_i \oplus A_i) \cdot PK_{TA} + A_i$. If not, the process is aborted.

Secure OBU storage: PUF is used to store the critical parameters in the EV, OBU securely. Selects a stimuli S_{EV_i} and compute a PUFs response for that $R_{EV_i} = PUF(S_{EV_i})$, $V_i = h(SK_i \oplus Val_i)$, $T_{i1} = SKV_i \oplus R_{EV_i}$, and $T_{i2} = B_x \oplus R_{EV_i}$.

The security parameter $\langle V_i, T_{i1}, T_{i2}, b_i \rangle$ are then stored in the OBU.

4.6   RSU and TA Authentication Phase

To ensure seamless communication within the network and connect all infrastructure components, a trusted framework must be established. This involves pairing or enrollment and verification with the network's verification authority, the TA. The pairing process between the RSU and TA is detailed below:

Step 1: RSU_i collect the network global security parameters and generate a secret key SK_{RSU_i} and, using the global parameters, compute a public key as PK_{RSU_i}.

Step 2: After this, request the TA for network enrollment through a request message $M_{RSU_i} = E_{PK_{TA}}\{RSU_{id}, PK_{RSU_i}, T_{t1}\}$.

Step 3: On getting the request message M_{RSU_i}, the TA checks the freshness of the message using the time stamp T_{t1} to avoid replay attacks and denial of service (DOS) attacks. For the valid message, a registration check is initiated to validate the registration of RSU_{id} if a previous registration exists. If not exist then TA computes following steps:

- **Random Number Generation:** TA generates random numbers r_{i2} and a_{i2}.
- **Recording on Blockchain:** $A_{i2} = a_{i2} \cdot P$ and $PID_{RSU_i} = H(PK_{RSU_i}||RSU_{id}||r_{i2}||T_{val2})$ are computed by the TA, and store the $\{PK_{RSU_i}, PID_{RSU_i}, T_{val2}\}$ on blockchain using the smart contract address SC_{add}. On successful completion of the transaction, it returns the blockchain block index B_{x2}.
- **Commitment Calculation:** TA computes $b_{i2} = h(PID_{RSU_i} \oplus PK_{RSU_i} \oplus A_{i2}) \cdot SK_{TA} + a_{i2}$ and send $\{B_{x2}, A_{i2}, b_{i2}\}$ back to the RSU, encrypted by the public key PK_{RSU_i} of the RSUs to ensure the integrity of the message.
- **Commitment Verification:** RSU_i uses the block index B_{x2} and public key to locate the data stored on the blockchain $\{PK_{RSU_i}, PID_{RSU_i}, T_{val2}\}$ after that verifies if $b_{i2} \cdot P$ matches $h(PID_{RSU_i} \oplus PK_{RSU_i} \oplus A_{i2}) \cdot PK_{TA} + A_{i2}$. If not, the process is aborted.

Step 4: RSU stores the received parameters $\{B_{x2}, A_{i2}, b_{i2}, T_{val2}, PID_{RSU_i}\}$ and generates a challenge Cha_t and computes:

$$Res_t = PUF(Cha_t)$$

$$K_{t1} = b_i \oplus h(Res_t) \oplus SK_{RSU_i}$$

4.7   Authentication Between EV and RSU

In this phase, the EV_i enters the RSU_t domain and begins its initial authentication request. The procedure unfolds as outlined below:

Step 1: The vehicle generates a random number ran_i and a timestamp $T1$, then computes: $RAN_i = ran_i \cdot P$ and $c_i = b_i + SKV_i + h(PID_i \oplus T1 \oplus RAN_i) \cdot ran_i$. The message $\{PID_i, A_i, RAN_i, PKV_i, c_i, T1\}$ is sent to RSU via the public channel.

Step 2: Upon receiving the message $\{PID_i, A_i, RAN_i, PKV_i, c_i, T1\}$, RSU_t first checks the freshness of $T1$. If $T1$ is fresh and $c_i \cdot P \overset{?}{=} h(PID_{EV_i} \oplus PKV_i \oplus A_i) \cdot PK_{TA} + A_i + PKV_i + h(PID_i \oplus T1 \oplus RAN_i) \cdot RAN_i$ then RSU_t generates a random number e_t, time stamp T_b and computes $Res_t = PUF(Cha_t)$, $E_t = e_t \cdot P$ using it a authentication token is created as:

$$Token_{EV_i} = K_{t1} + SK_{RSU_t} + h(PID_{EV_i} \oplus PID_{RSU_t} \oplus T_b \oplus E_t \oplus Res_t) \cdot e_t \quad (1)$$

Then, send the parameters $\{Token_{EV_i}, K_{t1}, b_{i2}, Res_t, E_t, B_{x2}, T_b\}$ to EV_i.

Step 3: Upon receiving the message $\{Token_{EV_i}, K_{t1}, b_{i2}, Res_t, E_t, B_{x2}, T_b\}$, the vehicle first checks the freshness of the timestamp T_b . If T_b is fresh and look for the block B_{x2} in blockchain, reads $\{PK_{RSU_t}, PID_{RSU_t}, T_{val2}\}$ and perform $Token_{EV_i} \cdot P \overset{?}{=} K_{t1} \cdot P + PK_{RSU_t} + h(PID_{EV_i} \oplus PID_{RSU_t} \oplus T_b \oplus E_t) \cdot E_t$

Successful verification complete the authentication process and $Token_{EV_i}$ is a verified authentication token for the EV_i validity till T_b provided or issued by the RSU_t. Share $\{PK_{EV_i}, Token_{EV_i}, K_{t1}, E_t, B_x, B_{x2}, T_b\}$ with other RSU for rapid verification. Any RSU can quickly verify the authenticity of the EV_i by reading two block $\{B_x, B_{2x}\}$ and compute $Token_{EV_i} \cdot P \overset{?}{=} K_{t1} \cdot P + PK_{RSU_t} + h(PID_{EV_i} \oplus PID_{RSU_t} \oplus T_b \oplus E_t) \cdot E_t$ to authenticate the authenticity of the EV_i.

4.8   EV2EV Communication

Information sharing among EV is essential for understanding the real-time on-road situation. The security and privacy is major concern among the EV owners, the authentication tokens generated during the EV and RSU provide a special benefits for authentication between the two EVs. The same tokens can be utilized as a authentication between two EVs significantly enhancing the security as it is already authenticated by RSU and reference for verifications are stored over blockchain. It also reduce the extra computation overhead and message exchange requirement. The authentication between EV_i and EV_j EVs:

- **Step 1:** EV_i initiates the authentication by sending authentication token with security parameters $\{PK_{EV_i}, Token_{EV_i}, K_{t1}^i, E_t^i, B_x^i, B_{x2}^i, T_b^i\}$ to EV_k.
- **Step 2:** EV_k verifies the freshness of the time stamp T_b^i, and after successful verification reads data from blockchain for the block index B_x^i and B_{x2}^i. Then evaluate $Token_{EV_i} \cdot P \overset{?}{=} K_{t1}^i \cdot P + PK_{RSU_t} + h(PID_{EV_i} \oplus PID_{RSU_t} \oplus T_b^i \oplus E_t^i) \cdot E_t^i$

if the evaluation holds then send the own token with security parameters $\{PK_{EV_k}, Token_{EV_k}, K_{t1}^k, E_t^k, B_x^k, B_{x2}^k, T_b^k\}$ for mutual authentication.

- **Step 3:** EV_i performs the same procedure take by EV_k from freshness test, block data reading and then evaluate $Token_{EV_k} \cdot P \overset{?}{=} K_{t1}^k \cdot P + PK_{RSU_k} + h(PID_{EV_k} \oplus PID_{RSU_k} \oplus T_b^k \oplus E_t^k) \cdot E_t^k$
The successful validation leads to completion of mutual authentication.

5 Experimental Setup

The simulation network for the experiment is designed using the Ganache local blockchain platform, and the surrounding environment is supported with the truffle framework and web3py. The essential network components, such as network certificate authority, roadside units, and electric vehicles, are simulated using Python modules. Ubuntu 20.04 operating system based environment on Intel(R) Core^{TM}i5-11400 H with 16.0 GB RAM, is used as experiment infrastructure to simulate and evaluate the proposed authentication scheme. The network components are connected via a Wi-Fi network. The network's smart contract is written in Solidity and compiled to byte code using the Solidity compiler version 8. For the simulation of the PUF module pypuf package [17] is used, with 128 bit challenge, and 128 bit response. Cryptographic security operations are simulated with the help of MIRACLE [4] package generated execution time is presented in Table 3 and the considered parameter bit size is presented in Table 3.

Parameter	Size (Bit length)
Random number	160
Hashing ($h(.)$)	256
Time stamp (T_b, T_c)	32
Identity RSU_{id}	32
Block Index (B_{ix})	32
ECC point (x, y)	320
PUF stimuli and response	128

Fig. 3. Parameters (bit length)

Symbol	Operation	Time (μs)
T_h	hashing operation	16.9
T_{exp}	exponential operation	72
T_{bp}	bilinear pairing operation	3
T_{sme}	ECC scalar multiplication	674
T_{pae}	ECC point addition	2
T_{sig}	ecc signature verification	1405
T_{de}	ECC decryption	286
T_{fe}	fuzzy extraction	674
T_{PUF}	PUF (128 bits)	0.21
T_x	XOR	10.9
T_c	Concatenation	4.9

Fig. 4. Cryptographic primitives average execution time

Table 2. Computational costs comparison with existing works

Reference	Year	Initial Authentication		Time(ms)	Handover Authentication		Time(ms)
		EV	RSU		EV	RSU	
[16]	2021	$2T_h + T_{bp} + T_{exp}$	$2T_h + T_{bp} + T_{exp}$	99.162	$T_h + T_{bp}$	$T_{bp} + T_{exp}$	94.079
[12]	2022	$9T_h + 3T_{sme}$	$8T_h + 3T_{sme} + T_{sig}$	33.607	$6T_h$	$10T_h + 2T_{sig}$	35.552
[19]	2022	$2T_h + 3T_{sme} + 3T_{bp}$	$2T_h + 2T_{sme} + 4T_{bp}$	324.745	$3T_h + 4T_{sme}$	$2T_h + 2T_{sme} + 3T_{bp}$	149.306
[18]	2023	$9T_h + 5T_{sme}$	$10T_h + 6T_{sme}$	29.071	$3T_h$	$7T_h + 2T_{sme}$	5.41
Proposed	2025	$2T_h + 6T_x + 5T_{sme} + 4T_{pae}$	$3T_h + 8T_x + 5T_{sme} + 5T_{pae}$	6.9957	-	$t_h + 4T_x + 3T_{sme} + 3T_x$	2.086

Reference	Year	Storage overhead (bits)		Total (bits)
		EV	RSU	
[15]	2021	≈ 2024	≈ 1012	3036
[19]	2022	≈ 257	≈ 4160	4417
[18]	2023	≈ 1920	≈ 1248	3168
[6]	2024	≈ 1600	≈ 2336	3936
Proposed	2025	≈ 1632	≈ 1472	3104

Fig. 5. Communication overhead comparison with existing works

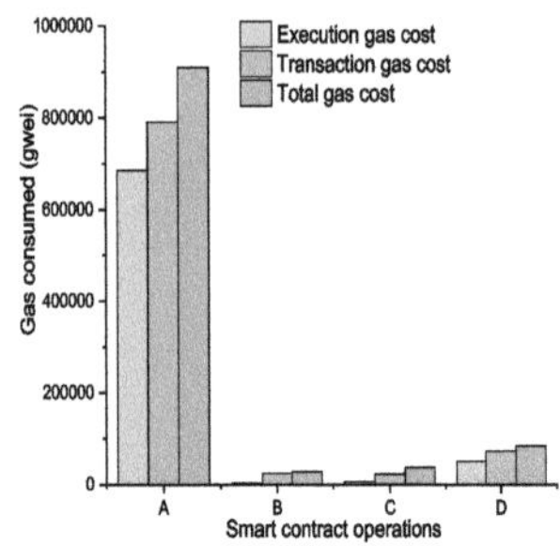

Fig. 6. Smart contract operations gas consumption in gwei

6   Result Analysis

With the increasing deployment of EVs and the expansion of Vehicular Ad-hoc Networks (VANETs) [3], ensuring robust security measures in keyless entry systems and communication protocols is paramount. This paper proposes innovative solutions integrating PUFs and blockchain technology to enhance security and efficiency in these domains.

6.1   Security Analysis

The sensitive data stored on OBU is secured with the security key derived from KER, ensuring the EV physical theft attack. The static biometric-based authentication is replaced with a dynamic PUF-based authentication system, eliminating the risk of a fingerprint database attack. Each authentication attempt generates unique responses, ensuring that access is granted only to authorized users. The uniqueness of response prevents the Replay attack and RollJam attack over the EV and KES systems.

7   Cost Analysis

The smart contract deployment and other function call execution costs are measured in terms of gas consumed during the function call as transaction and execution gas cost illustrated in the Fig. 6, A: Deployment of smart contract, B: AuthorizeEntry function, C: DeauthorizeEntry function and D: storeData function. The computation cost for the authentication between the EV and RSU along with some existing work is presented in the Table 2 with $2T_h + 6T_x + 5T_{sme} + 4T_{pae}$ computation by EV and $3T_h + 8T_x + 5T_{sme} + 5T_{pae}$ computation by RSU taking total 6.9957 ms. The authentication handover takes $t_h + 4T_x + 3T_{sme} + 3T_x$ total 2.086 ms as after authentication between EV_i and RSU, EV_i has authentication token $Token_{EV_i}$ explained in Sect. 4.8. The authentication between the EV and RSU for communication overhead takes 3104 $bits$ presented in Table 5 computed using the parameter bit length Table 4.

8   Conclusion

In conclusion, our proposed framework represents an enhanced mechanism for securing keyless entry systems for Electric Vehicles (EVs) and Vehicular Ad-hoc Networks (VANETs). By integrating Physical Unclonable Functions (PUFs) and blockchain technology, vulnerabilities such as replay attacks and unauthorized access can be mitigated. PUF-based authentication ensures unique, unpredictable responses for each interaction, enhancing security against traditional weaknesses. Utilizing elliptic curve cryptography (ECC) and blockchain establishes secure channels for vehicle-to-vehicle (V2V) and vehicle-to-infrastructure (V2I) communications, ensuring integrity and traceability. Importantly, sensitive OBU data is secured with a hashed security key from the keyless entry remote (KER), rendering it inaccessible if the OBU is stolen, thereby mitigating physical theft risks. By decentralizing authentication with dynamic PUF-generated responses, our system broadens access control without compromising privacy. Experimental results prove the significance of the proposed framework, which has a computation time of 6.9957 ms and communication overhead of 3104 $bits$ for authentication between the EV and RSU.

References

1. Alrabady, A., Mahmud, S.: Analysis of attacks against the security of keyless-entry systems for vehicles and suggestions for improved designs. IEEE Trans. Veh. Technol. **54**(1), 41–50 (2005). https://doi.org/10.1109/TVT.2004.838829
2. Djinko, I.A.R., Kacem, T., Girma, A.: Blockchain-based approach to thwart replay attacks targeting remote keyless entry systems. In: 2022 International Conference on Engineering and Emerging Technologies (ICEET), pp. 1–6 (2022). https://doi.org/10.1109/ICEET56468.2022.10007335
3. Limbasiya, T., Das, D.: Lightweight secure message broadcasting protocol for vehicle-to-vehicle communication. IEEE Syst. J. **14**(1), 520–529 (2020). https://doi.org/10.1109/JSYST.2019.2932807
4. MIRACL: Multiprecision integer and rational arithmetic cryptographic library (MIRACL). https://github.com/miracl/MIRACL. Accessed 14 Aug 2024
5. Parameswarath, R.P., Sikdar, B.: A PUF-based lightweight and secure mutual authentication mechanism for remote keyless entry systems. In: GLOBECOM 2022 - 2022 IEEE Global Communications Conference, pp. 1776–1781 (2022). https://doi.org/10.1109/GLOBECOM48099.2022.10001217
6. Roy, S., Nandi, S., Maheshwari, R., Shetty, S., Das, A.K., Lorenz, P.: Blockchain-based efficient access control with handover policy in IOV-enabled intelligent transportation system. IEEE Trans. Veh. Technol. **73**(3), 3009–3024 (2024). https://doi.org/10.1109/TVT.2023.3322637
7. Sahu, B.L., Chandrakar, P.: Blockchain-based framework for electric vehicle charging port scheduling. In: 2022 IEEE International Conference on Advanced Networks and Telecommunications Systems (ANTS), pp. 1–6 (2022). https://doi.org/10.1109/ANTS56424.2022.10094049
8. Sahu, B.L., Chandrakar, P.: Blockchain-based dynamic pricing framework for electric vehicle charging. In: Proceedings of the 4th International Conference on Communication, Devices and Computing, pp. 563–581. Springer Nature Singapore, Singapore (2023). https://doi.org/10.1007/978-981-99-2710-4_45

9. Sahu, B.L., Chandrakar, P.: Enhanced privacy preservation and data storage in blockchain-based electric vehicle network. Procedia Comput. Sci. **235**, 2144–2153 (2024). https://doi.org/10.1016/j.procs.2024.04.203, international Conference on Machine Learning and Data Engineering (ICMLDE 2023)

10. Sahu, B.L., Chandrakar, P.: Blockchain-oriented secure communication and smart parking model for internet of electric vehicles in smart cities. Peer-to-Peer Netw. Appl. **18**(1), 1–17 (2025). https://doi.org/10.1007/s12083-024-01872-y

11. Shui, D., Xie, Y., Wu, L., Liu, Y., Su, X.: Lightweight three-party key agreement for v2g networks with physical unclonable function. Veh. Commun. **47**, 100747 (2024). https://doi.org/10.1016/j.vehcom.2024.100747, https://www.sciencedirect.com/science/article/pii/S2214209624000226

12. Son, S., Lee, J., Park, Y., Park, Y., Das, A.K.: Design of blockchain-based lightweight v2i handover authentication protocol for VANET. IEEE Trans. Netw. Sci. Eng. **9**(3), 1346–1358 (2022). https://doi.org/10.1109/TNSE.2022.3142287

13. Subramani, J., Maria, A., Rajasekaran, A.S., Al-Turjman, F., Gopal, M.: Blockchain-based physically secure and privacy-aware anonymous authentication scheme for fog-based VANETS. IEEE Access **11**, 17138–17150 (2023). https://doi.org/10.1109/ACCESS.2022.3230448

14. Sureshkumar, V., Chinnaraj, P., Saravanan, P., Amin, R., Rodrigues, J.J.P.C.: Authenticated key agreement protocol for secure communication establishment in vehicle-to-grid environment with FPGA implementation. IEEE Trans. Veh. Technol. **71**(4), 3470–3479 (2022). https://doi.org/10.1109/TVT.2022.3146409

15. Wang, C., Huang, R., Shen, J., Liu, J., Vijayakumar, P., Kumar, N.: A novel lightweight authentication protocol for emergency vehicle avoidance in VANETS. IEEE Internet Things J. **8**(18), 14248–14257 (2021). https://doi.org/10.1109/JIOT.2021.3068268

16. Wang, C., Shen, J., Lai, J.F., Liu, J.: B-TSCA: blockchain assisted trustworthiness scalable computation for V2I authentication in VANETS. IEEE Trans. Emerg. Top. Comput. **9**(3), 1386–1396 (2021). https://doi.org/10.1109/TETC.2020.2978866

17. Wisiol, N., et al.: PYPUF: cryptanalysis of physically unclonable functions (2021). https://doi.org/10.5281/zenodo.3901410

18. Xie, Q., Ding, Z., Tang, W., He, D., Tan, X.: Provable secure and lightweight blockchain-based V2I handover authentication and V2V broadcast protocol for VANETS. IEEE Trans. Veh. Technol. **72**(12), 15200–15212 (2023). https://doi.org/10.1109/TVT.2023.3289175

19. Yang, A., Weng, J., Yang, K., Huang, C., Shen, X.: Delegating authentication to edge: a decentralized authentication architecture for vehicular networks. IEEE Trans. Intell. Transp. Syst. **23**(2), 1284–1298 (2022). https://doi.org/10.1109/TITS.2020.3024000

20. Yu, S., Park, K.: PUF-based robust and anonymous authentication and key establishment scheme for V2G networks. IEEE Internet Things J. **11**(9), 15450–15464 (2024). https://doi.org/10.1109/JIOT.2024.3349689

Cybersecurity, Privacy, and System Design

Zero Trust Framework for Enhancing Security in Container Ecosystems

Santosh Kumar Upadhyaya[(✉)], Dhanvi Medha Beechu, and B. Thangaraju

International Institute of Information Technology, Department of Computer Science and Engineering, Bangalore, India

Abstract. As organizations accelerate their digital transformation, container technologies have emerged as a cornerstone for application deployment, emphasizing the need for robust security measures. Traditional security model that is perimeter-based are not sufficient to address the dynamic and dispersed nature of container environments. This paper introduces a Zero Trust Container Framework (ZTCF), rooted in the "never trust, always verify" principle, to enhance container ecosystem security. The framework integrates identity verification, continuous monitoring, and automated threat response to address specific vulnerabilities and requirements throughout the container lifecycle. Experimental analysis comparing legacy systems with the proposed framework demonstrates significant improvements in security posture, compliance, and resilience. This research offers actionable insights for designing secure, scalable, and resilient containerized systems, guiding organizations in strengthening their security strategies as they adapt to the evolving digital landscape.

Keywords: Zero Trust · Container Security · Network Security · Container Technologies · Digital Transformation · Cloud Computing

1 Introduction

The rapid adoption of container technologies has transformed the way organizations develop, deploy, and manage applications. Containers offer scalability, portability, and efficiency, making them a critical component in modern software architectures. However, this widespread adoption also introduces unique security challenges, particularly in dynamic, distributed, and cloud-native environments. Conventional security model that relies on network perimeter based defences is insufficient to secure containerized ecosystems. This gap underscores the need for a paradigm shift toward more robust security approaches. The security model, based on the principle of "never trust, always verify," is called Zero Trust (ZT), and it provides a promising framework. Unlike legacy models, ZT assumes that no system component, either inside the network or external to the network, can be inherently trusted. By enforcing strict authentication, continuous monitoring, and policy-driven access control, ZT minimizes the attack surface and mitigates potential risks. However, implementing Zero Trust principles

P. Chandrakar et al. (Eds.): ICCINS 2025, CCIS 2738, pp. 315–326, 2026.
https://doi.org/10.1007/978-3-032-09572-5_25

in containerized environments presents challenges, such as managing complex networks [1], maintaining real-time compliance [2], and ensuring interoperability across diverse platforms. This research focuses on the development and application of a Zero Trust Container Framework (ZTCF) tailored specifically for container ecosystems. The proposed framework addresses security vulnerabilities across the entire container lifecycle, from development to deployment, incorporating cutting-edge practices like identity verification, posture management, and automated policy enforcement. By leveraging this framework, organizations can secure their containerized applications against modern threats while maintaining operational efficiency. The proposed framework is particularly valuable in industries where data security and system reliability are paramount. For example:

- Financial Services: With the growing reliance on microservices and cloud-based applications, financial institutions can use ZTCF to secure sensitive data transactions and guarantee adherence to regulations like General Data Protection Regulation(GDPR) and Payment Card Industry Data Security Standard (PCI DSS).
- Healthcare: ZTCF can safeguard patient data in healthcare systems, addressing vulnerabilities in containerized environments that host electronic health records (EHR) and telemedicine platforms.
- E-commerce: Online retailers can protect user data and payment systems by implementing Zero Trust security measures in their container-based architectures.

Consider the case of a leading financial services company adopting containerized microservices for its payment gateway. Traditional perimeter security left the system vulnerable to lateral movement attacks. By integrating Zero Trust principles, the organization implemented continuous monitoring, multi-factor authentication (MFA), and real-time policy enforcement. This transformation significantly reduced the risk of breaches, improved compliance, and enhanced user trust in its digital services. This paper systematically explores the challenges and opportunities of adopting Zero Trust principles for container security. It evaluates existing practices, presents the architecture and modules of the proposed ZTCF, and validates its effectiveness through experimental analysis. The findings provide actionable insights for practitioners and researchers aiming to strengthen security in containerized environments, ensuring scalability and resilience in an evolving digital landscape.

1.1   Our Contribution

Zero Trust architecture, audit and compliance, network security and dynamic policy manage are often discussed separately, as referenced in [3,4], however, a cohesive framework is necessary for their integration. This paper's primary contribution is the examination of existing security challenges in the container lifecycle. It introduces a thorough security framework referred to as Zero Trust Container Framework (ZTCF). This ZTCF is tailored for the entire container lifecycle and applies to both development and deployment processes. The security analysis is conducted by comparing the experimental results from the legacy system with those from the ZTCF.

1.2 Organization of the Paper

The remainder of this paper is organized as follows: Sect. 2 reviews the existing literature on Zero Trust Architecture (ZTA) and its application to container security. Section 3 provides an overview of the Zero Trust principles and their relevance to modern IT environments. Section 4 provides an overview of the container security. Section 5 discusses the unique security challenges in containerized ecosystems and highlights the need for a specialized framework. Section 6 presents the proposed Zero Trust Container Framework (ZTCF), detailing its architecture and key modules. Section 7 describes the experimental setup, security analysis methodology, and comparative results, illustrating the efficacy of the framework. The paper concludes in 8 by synthesizing the insights and suggesting potential directions for future investigation, concentrating on improving posture management and compliance auditing within containerized environments.

2 Literature Review

The body of literature concerning Zero Trust Architecture (ZTA) as applied to container security is extensive, covering both theoretical foundations and practical strategies. Ray et al. review the background, technologies, and challenges of ZTA [5]. Kindervag J. introduced the Zero Trust model, advocating a departure from 'trust but verify' [6]. Surantha et al. explore using ZT to secure container deployments inside network perimeters [7]. Leahy et al. offer a ZT framework in container security, emphasizing trust evaluation [8]. NIST views ZTA as a strategic framework focusing on trust-building [9]. Moric et al. discuss a security framework for container orchestration [10]. This research builds on these studies to explore the challenges and opportunities of ZTA in container security, informing further analysis and discussion.

3 Understanding Zero Trust (ZT)

Zero Trust (ZT) signifies a move away from traditional security that relies on perimeter defenses, adopting instead a model where no implicit trust is granted, whether within or outside the network. ZT ensures thorough authentication, authorization, and encryption for every access request, reducing lateral movement risks through continuous identity verification. It utilizes technologies like multifactor authentication (MFA), micro-segmentation, and real-time analytics to enforce strict access controls, focusing on least privilege and continuous verification. This framework secures modern IT environments, including cloud services and IoT devices, emphasizing trust over specific technical solutions. ZT benefits from advances in AI, ML [11,12], IAM, encryption, and analytics, aiding real-time access management. In container security, ZT enhances security but also presents unique challenges, which this research explores.

4   Overview of Container Security

Container security [13] is vital for safeguarding modern applications, addressing challenges like vulnerabilities in container images [14] and runtime threats [15]. A multi-layered approach is needed, including vulnerability scanning, trusted registries, robust access controls, and continuous monitoring, as analysed by Eldjoy et al. [16]. While tools like Kubernetes and Docker provide basic security features, integrating additional solutions strengthens the protection of containerized applications.

5   Implementing ZT in Container Security

Adopting Zero Trust (ZT) for container security involves a thorough strategy to protect containerized environments. Containers offer scalability and uniformity but present unique security issues that Zero Trust Architecture (ZTA) can address. Runtime security in a ZT framework involves continuous monitoring and real-time threat detection. Tools like Kubernetes and Docker offer built-in security features, but integrating solutions like intrusion detection systems (IDS) and endpoint detection and response (EDR) enhances protection. Implementing ZT requires a paradigm shift, emphasizing continuous verification and enforcing the principle of least privilege. Regular audits and compliance checks ensure security policies are enforced and updated for emerging threats. By adopting ZT principles, organizations can secure containerized environments, ensuring that access requests are authenticated, authorized, and encrypted, thus enhancing IT infrastructure resilience.

6   Proposed Zero Trust Container Framework (ZTCF)

The foundational principles of Zero Trust are outlined below.

- **Establish Trust:** Create and sustain a secure environment where trust is consistently evaluated.
- **Implement Access Policies Based on Trust:** Enforce access controls according to the trustworthiness of users and devices.
- **Continuous Trust Verification:** Continuously validate and establish trust for each access request.
- **Dynamics and Adaptive Trust Management:** Adjust trust levels in real-time according to changing conditions and behaviours.

Building upon the four principles mentioned earlier, the proposed ZTCF is illustrated in Fig. 1 below. ZTCF consists of several key modules, each essential for maintaining robust container security. The following sections outline the key modules.

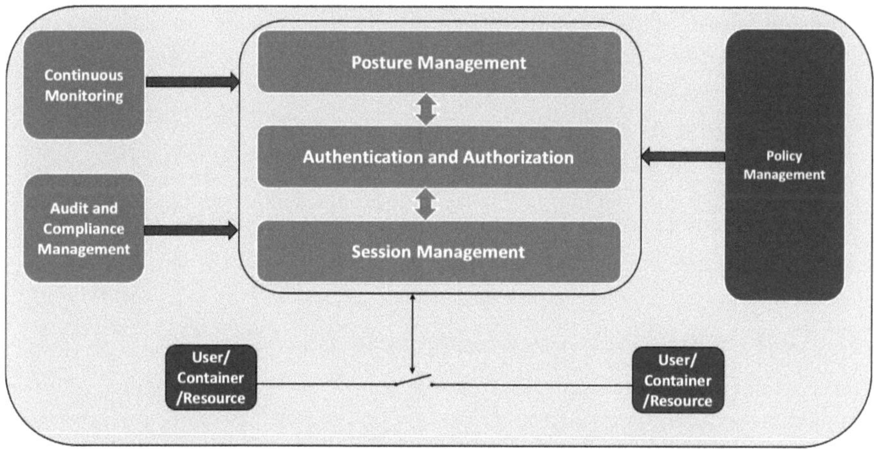

Fig. 1. Architecture of the Zero Trust Container Framework.

6.1 Session Management (SM)

When any entity requests access to any resource, the SM layer processes the request. It routes the request to the authentication and authorization (AA) layer and acts according to the response, functioning as a policy enforcement point to grant or deny access. The session management layer is responsible for creating sessions for all requests, handling resource access requests, and accepting only encrypted requests. It verifies the authenticity and integrity of access requests, creates and forwards authentication requests to the authentication and authorization layer, handles responses from this layer, and enforces policies. Additionally, it manages dynamic trust.

6.2 Authentication and Authorization (AA)

AA component is essential for preserving the trust of the system, working closely with posture management to ensure strong authentication and authorization processes. It verifies user identities via authentication servers and uses posture management for secondary context-based checks, enhanced by MFA. It determines access rights through the authorization server, retrieving and conveying policies. Strong encryption for data in transit and at rest is essential, using secure protocols for data exchanges. These servers can be internal or external, offering flexible deployment and integration.

6.3 Posture Management (PM)

Posture verification ensures endpoint devices meet security standards before network access, assessing compliance for devices like computers and smartphones to mitigate risks. The posture management layer [17] strengthens authentication and authorization while supporting continuous monitoring. It gathers contextual data from users and devices, such as time, location, and other contexts, to enhance security assessments. For example:

- Time: Monitors access patterns to detect anomalies.
- Geographical Location: Confirms access from expected locations, flagging high-risk areas.
- Contextual Information: Uses device type, network, and user behaviour for informed access decisions.

6.4   Audit and Compliance Management (ACM)

This module manages comprehensive logging of access and system activities. To ensure container compliance prior to deployment, incorporate automated compliance scans into the pipeline of Continuous Integration and Continuous Deployment (CI/CD) [4]. Defining security policies as code ensures consistency and simplifies compliance management. Using AI and ML enhances audit data analysis, providing insights for continuous monitoring. These technologies help organizations detect and respond to security issues more effectively.

6.5   Continuous Monitoring (CM)

Continuous monitoring employs real-time observation and analysis to identify threats within container environments, thereby minimizing breaches. It maintains a dynamic security posture through consistent trust evaluations and interaction with the session manager. Essential elements include automated compliance checks, AI-powered anomaly detection, and alert systems, all contributing to improved visibility, compliance, and security.

6.6   Policy Management (PM)

The policy management system manages both static and dynamic policies, distributing them to the session management layer. Dynamic policy formulation is guided by insights from continuous monitoring and posture management. Furthermore, the system develops and oversees role-based access policies to regulate access requests. Together, these components create an integrated security model that reduces risks across the container lifecycle.

7   Security Analysis

The container lifecycle includes both development and deployment phases. According to [8], the development phase involves activities such as coding, writing Docker files, creating images, running containers, and testing. If requirements are met, changes are pushed to the code repository; if not, the process returns to further development. After code is committed, it transitions to the deployment phase. Our proposal addresses both the container development and deployment processes.

7.1 Security Analysis Methodology

The security analysis process involves initially performing tests without the proposed ZTCF, identifying security issues, and then repeating the same tests with the proposed ZTCF. This is followed by a security analysis and a comparison between the security of the legacy framework and ZTCF. The experiment is carried out for container development and deployment scenarios using both the legacy framework and ZTCF, and the results are then compared against the following criteria (Table 1).

Table 1. ZTCF Security Analysis Criteria

Criteria	Explanation
Strength of Authentication Mechanisms	Evaluates authentication methods' robustness levels
Granularity of Authorization	Analyses system's fine control over permissions/access
Posture Management	Examines each system's asset inventory maintenance
Logging Capabilities	Assessing logging mechanisms' comprehensiveness
Compliance Adherence	Checks systems' adherence to regulatory requirements
Real-Time Monitoring	Assesses real-time monitoring of activities and threats
Dynamic Policy Management	Evaluates ease of updating policies for current threats
Context-Awareness	Analyses each system's contextual decision-making ability
Dynamic Trust Adjustment	Examines dynamic trust adjustment via continuous verification
Incident Response	Examines dynamic trust adjustment via continuous verification
Disaster Recovery	Evaluates disaster recovery plans and their efficacy
Ease of Use	Assesses each system's user-friendliness for end-users
Administrative Overhead	Compares administrative effort needed for system upkeep

7.2 Applying ZTCF for Container Development

Implementing the Zero Trust Container Framework (ZTCF) in container development enhances security by verifying all entities continuously, without default trust. This involves rigorous authentication and authorization, network segmentation, and strict access controls. Extensive monitoring and logging provide real-time insight into container activities, allowing for swift threat detection and response.

In the coding phase, workplace security is enforced through session management and multi-factor authentication, with policy-driven access controls and continuous monitoring. During image creation, Docker files guide the secure construction of immutable images. For running containers, only authenticated users with sufficient privileges are allowed, with ongoing endpoint and host security checks. Testing involves integrating unit and smoke tests into continuous monitoring to ensure workload security, logging events as part of audit and compliance management. When pushing code, ZTCF ensures secure authentication and authorization, with continuous monitoring to meet all conditions.

Table 2. ZTCF Security Analysis for Container Development

Criteria	Legacy	ZTCF
Strength of Authentication Mechanisms	Simple Auth	MFA, Biometric
Granularity of Authorization	Basic	RBAC, ABAC
Posture Management	Periodic	Real-time
Logging Capabilities	Less detailed, De-centralised	Detailed, Centralised
Compliance Adherence	Manual	Inbuilt
Real-Time Monitoring	Periodic, Rule based	Real-time, AI/ML Support
Dynamic Policy Management	Static policy	Dynamic Policy support
Context-Awareness	None	Supported
Dynamic Trust Adjustment	Static	Supported
Incident Response	Traditional	Advanced
Disaster Recovery	Less automated	Fully automated
Ease of Use	Simple	Complex
Administrative Overhead	Minimum	Visible

Summary of Security Analysis: Each phase of container development is integral to workplace and workforce security. Security analysis is detailed in Table 2. Throughout coding, Docker file creation, image creation, container execution, and testing, workforce security is maintained via session management, authentication, and authorization, while workplace security is ensured through posture management. Continuous monitoring, auditing, and compliance are crucial across all phases. Session management addresses changes in trust, taking necessary actions. During development, integrating static analysis tools, code reviews, and regression tests into continuous monitoring boosts security. If issues arise, session management prevents code from being committed to the repository.

7.3   Applying ZTCF for Container Deployment

In a deployment scenario, containers work collectively to fulfill specific requirements. There are 4 main scenarios.

– User-to-Container Communication
– Independent Container
– Container-to-Container Communication
– Container-to-Resource Communication

A security evaluation is performed for all the scenarios mentioned above.

User-to-Container Communication (Scenario: Provide User Access to a Container Resource through a Remote API) Access to the resource within the container can be obtained by entering URL in a browser. After access is granted, the user can see the information directly in the user's browser (Fig. 2).
  There are following issues with this.

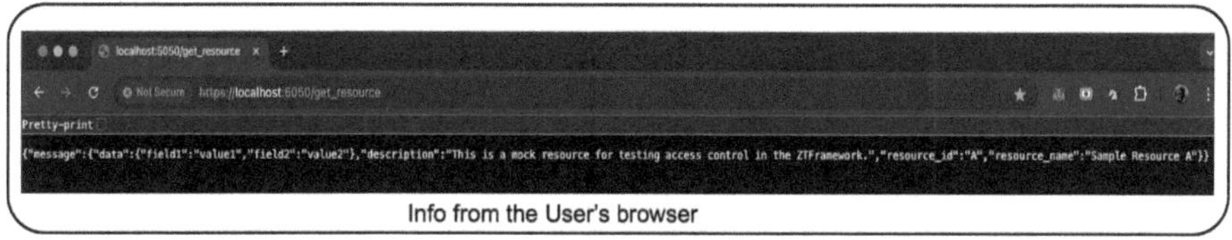

Fig. 2. resource access through remote API

- The request sent over the non-secure channel is accepted
- Authentication is missing
- Authorization is missing
- There is no clear policy for accessing resources
- The device's security posture has not been verified

The proposed ZTCF ensures that the session management layer accepts only encrypted requests. Authentication and authorization management verify that resource requests are authenticated and authorized. Posture management confirms that endpoint devices run the latest software with all patches applied. The policy management layer creates a new policy to authorize specific resource requests, enhanced by specifying allowed IP addresses and device types for access.

The test was repeated using ZTCF, and the observations are detailed below.

- Request on insecure connection (HTTP) is denied. It is only accepted on secure connection (HTTPS).
- User authentication is done, after initial authentication done, MFA is triggered.
- The posture of the endpoint is assessed in relation to the posture policy. If it meets the required policy criteria, the authentication process advances to the next stage. Otherwise, authentication fails.
- For testing the policy, the resource policy is changed to deny and confirmed that the resource access is denied.
- Updated the policy to allow access from mobile and retested with an Android phone that is not up to date. Observed that the audit message was generated for the recommendation to update the device software version, and access is allowed. Fig. 3

Summary of Security Analysis:
The summary of security analysis for grant user access to remote api use case is provided in Table 3.

- Authentication Mechanisms: Legacy systems rely on insecure credentials, while ZTCF uses multi-factor, biometric, and context-based methods.
- Authorization Granularity: Legacy systems have coarse controls; ZTCF provides fine-grained, role-based, and attribute-based access.

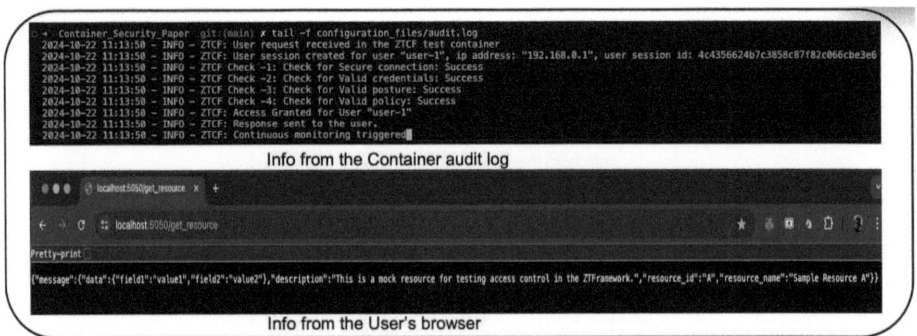

Fig. 3. Authentication Success by ZTCF.

- Posture Management: Legacy systems use periodic scans; ZTCF continuously assesses posture in real-time with compliance integration.
- Logging: Legacy systems have decentralized logging; ZTCF offers centralized, comprehensive logs for real-time analysis.
- Compliance: Legacy systems require manual checks; ZTCF automates compliance checks and reporting.
- Monitoring: Legacy systems monitor infrequently; ZTCF enables real-time monitoring with advanced analytics.
- Dynamic Policy Management: Legacy systems have static policies; ZTCF supports dynamic, automated updates based on threats.
- Context-Awareness: Legacy systems lack context-awareness; ZTCF integrates user behaviour and device posture.
- Dynamic Trust Adjustment: Legacy systems have static trust; ZTCF continuously adjusts trust based on assessments.
- Incident Response: Legacy systems rely on manual responses; ZTCF automates detection and remediation.
- Disaster Recovery: Legacy systems use outdated technologies; ZTCF automates recovery with regular testing.
- Ease of Use: Legacy systems feature outdated interfaces; ZTCF prioritizes intuitive user experiences.
- Administrative Overhead: Legacy systems increase manual tasks; ZTCF reduces overhead through automation.

Independent Container (Use Case: Review an Existing Container):
Continuously scanning images during deployment and maintaining an audit log, the ZTCF provides a comprehensive security trail, enabling prompt detection and remediation of potential threats. This continuous monitoring approach offers significantly enhanced security for container environments compared to traditional offline image scanning methods.

Table 3. ZTCF Security Analysis for Grant User Access to Remote API

Criteria	Legacy	ZTCF
Strength of Authentication Mechanisms	Basic Auth	Auth with MFA
Granularity of Authorization	Conventional	RBAC
Posture Management	No Support	Supported
Logging Capabilities	Basic	Advanced with AI/ML capability
Compliance Adherence	Periodic	Inbuilt, Compliance as a service
Real-Time Monitoring	Periodic	Real-time
Dynamic Policy Management	NA	Supported
Context-Awareness	NA	Supported
Dynamic Trust Adjustment	NA	Supported
Incident Response	Slower, Manual	Faster, Automatic
Disaster Recovery	Manual	Auto failover
Ease of Use	Easy	Complex
Administrative Overhead	Minimal	Significant

Container-to-Container Communication and Container-to-Resource Communication: The experimental results for container-to-container communication and container-to-resource communication resemble those of the user-to-container experiments. The project code and experimental results can be accessed at https://github.com/Santosh-Upadhyaya/ZTCF.

8   Conclusion

The growing use of container technologies requires robust security measures to protect containerized environments from evolving threats. This research addresses this critical need by proposing the Zero Trust Container Framework (ZTCF), a comprehensive solution designed to improve security throughout the container lifecycle. Rooted in the Zero Trust principles of "never trust, always verify," the ZTCF integrates advanced features such as continuous monitoring, dynamic policy enforcement, posture management, and multi-factor authentication to mitigate vulnerabilities inherent in containerized ecosystems.

The experimental results highlight the efficacy of the ZTCF compared to legacy systems. Key improvements include stronger authentication mechanisms, granular access controls, improved compliance adherence, and real-time monitoring capabilities. These enhancements significantly reduce the attack surface, ensure dynamic trust adjustments, and provide automated incident response mechanisms, leading to a marked improvement in the security posture of container environments. Furthermore, the proposed framework demonstrates its practical applicability in addressing critical industry needs, such as securing sensitive data in financial services, ensuring regulatory compliance in healthcare and protecting user information on e-Commerce platforms. By addressing both the development and the deployment phases of the container lifecycle, the ZTCF offers a holistic approach to container security.

The ZTCF not only meets the security demands of dynamic and distributed environments but also sets a foundation for future enhancements. As container technologies evolve, incorporating developments in artificial intelligence (AI), machine learning (ML), and compliance automation will further strengthen the framework. Future work will focus on refining posture management, integrating advanced analytics, and extending ZTCF's applicability to emerging use cases, ensuring resilience in the face of ever-changing security landscapes.

References

1. Chamoli, S., Mittal, V.: Docker networking: a security review. In: 2023 7th ICOEI (2023)
2. Lin, X., Lei, L., Wang, Y., Jing, J., Sun, K., Zhou, Q.: A measurement study on Linux container security: attacks and countermeasures (2018)
3. Qazi, F.A.: Study of zero trust architecture for applications and network security (2022)
4. Upadhyaya, S.K., Thangaraju, B.: A novel method for trusted audit and compliance for network devices by using blockchain. In: 2022 IEEE CONECCT) (2022)
5. Ray, P.P.: WEB3: a comprehensive review on background, technologies, applications, zero-trust architectures, challenges and future directions (2023)
6. Kindervag, J.: No more chewy centers: the zero trust model of information security. Forrester Research Inc., Cambridge (2010)
7. Surantha, N., Ivan, F.: Secure kubernetes networking design based on zero trust model: a case study of financial service enterprise in Indonesia (2019)
8. Leahy, D., Thorpe, C.: Zero Trust Container Architecture (ZTCA) (2022)
9. Rose, S., Borchert, O., Mitchell, S., Connelly, S.: Zero trust architecture (2020)
10. Moric, Z., Dakic, V., Kulic, M.: Implementing a security framework for container orchestration (2024)
11. Munasinghe, S., Piyarathna, N., Wijerathne, E., Jayasinghe, U., Namal, S.: Machine learning based zero trust architecture for secure networking (2023)
12. Pinnamaneni, J., Nagasundari, S., Honnavalli, P.: Identifying vulnerabilities in docker image code using ML techniques. In: 2022 2nd ASIANCON
13. Tunde-Onadele, O., He, J., Dai, T., Gu, X.: A study on container vulnerability exploit detection. In: IEEE IC2E (2019)
14. Kwon, S. and Lee, J.H.: DIVDS: docker image vulnerability diagnostic system. IEEE Access (2020)
15. Sultan, S., Ahmad, I., Dimitriou, T.: Issues, challenges, and the road ahead. IEEE Access, Container security (2019)
16. Eldjou, A., Amoura, M.E.: Enhancing container runtime security: a case study in threat detection (2023)
17. PERERA: Docker container security orchestration and posture management tool. In: 2022 13th ICCCNT

Securing Online Transactions for Robust Payment Processing Mechanisms in AI Driven World

Yash Kumar Kandoi[1] and Amitesh Singh Rajput[2(✉)]

[1] Department of Computer Science and Information Systems, Birla Institute of Technology and Science, Pilani, India
f20212417@pilani.bits-pilani.ac.in
[2] Rajiv Gandhi National Cyber Law Center, National Law Institute University, Bhopal, India
amiteshrajput@nliu.ac.in

Abstract. In the contemporary era marked by widespread internet accessibility, there has been a notable surge in online shopping activities, consequently giving rise to an increased prevalence of online payment transactions. The utilisation of credit cards for myriad purchase transactions has reached a staggering scale. As advanced AI tools and the penetration of e-commerce platforms continue to expand, concerns about privacy infringement and data breaches have become more pronounced. Consequently, it becomes imperative to implement payment processing mechanisms that prioritise the preservation of privacy, ensure secure transactions, and safeguard against data compromise in the event of a breach. While several methods have been used in the past to address these challenges, this study advocates for an enhanced and more robust transactional model. The proposed model employs various public cryptography algorithms and their function-hiding techniques to generate a one-time-use credit card number. This number is communicated to the user through a secure key-sharing algorithm, which is then used as the means of authentication for the transaction. Notably, the proposed approach ensures that third-party online vendors are separated from accessing the original credit card number while facilitating the authentication of the transaction. The security of the proposed method is assessed over two diverse cases when either of the participating entities act as an adversary. This novel approach aims to elevate the standards of security in online transactions, offering a pragmatic solution to the escalating concerns associated with privacy and data integrity.

Keywords: Authentication · Data Privacy · Homomorphic Encryption · Network Security

1 Introduction

The benefits offered through online transactions and payment methods have led to a humongous increase in the number of users using credit cards as the method

P. Chandrakar et al. (Eds.): ICCINS 2025, CCIS 2738, pp. 327–338, 2026.
https://doi.org/10.1007/978-3-032-09572-5_26

of payment for their services. One commonality is that most of these companies drive users to store their credit card information on their website for easier payments. On the other hand, users owing to the extra layer of seamlessness during shopping let companies store their details. Users often rely on these e-commerce companies for the security of financial transactions, but several data leaks and breaches over the past decade [1] tend to point to increasing privacy in transactions. With the utilization of advanced AI tools [2], offers, and incentives by e-commerce firms to use online payments for easier operations, the number is only meant to increase.

Credit card transactions account for approximately 20% of the total transactions mainly due to their ease of access and incentives. Currently, the security used in credit cards involves giving away the credit card numbers to online retailers who then direct them to the banking page, where an OTP is asked of the user to authenticate the transactions. A privacy breach here is that the retailers can easily profile users using advanced AI tools based on the transactions they do with the same credit card number. Moreover, storing credit card information on the company's servers is another risk. The main point is that the online retailer should just be able to authenticate the transaction from the bank and nothing more. To overcome this issue of privacy in transactions, the role of dynamic credit card numbers comes in. Dynamic credit card numbers [3] get changed for every user session and allow the user to ensure privacy even when shopping online. This also means that the online retailer no longer needs to store the credit card information or even get access to the original credit card number. The process used to generate a secure dynamic credit card number as well as the OTP (Time Password) given to the user via mobile messages is all that the user needs to know. This is where homomorphic encryption comes in.

Homomorphic encryption is a distinctive type of encryption that enables calculations to be carried out directly on encrypted data (referred as ciphertext), without requiring decryption [4]. The outcome of these calculations yields an encrypted output, which, upon decryption, corresponds to the outcome of operations conducted on the initial unencrypted data, also known as plaintext. It is highly required in applications where AI tools are negatively used by adversaries to learn user details, their card usage patterns, and other sensitive details, as AI models are not able to learn insights and predict output over the encrypted data.

We have employed homomorphic encryption in this work to offer a secure means of verifying credit card transactions. Cybercriminals frequently target credit card transactions, and the exposure of credit card information can result in significant financial harm and other consequences for the cardholder. To reduce this danger, the proposed method utilises the capabilities of homomorphic encryption to guarantee that only the bank and the credit card owner possess the ability to access the original credit card number. The shop participating in the transaction can only verify the authenticity of the transaction but does not own the original credit card number.

The proposed approach has several advantages that are mentioned below.

1. The proposed approach greatly diminishes the likelihood of credit card information being disclosed or leaked during the transaction.
2. The preservation of the card holder's information is guaranteed by augmenting user privacy.
3. The method offers robust and secure way to verifying transactions, thereby minimising the potential for fraudulent behaviour.
4. Well-known homomorphic algorithms including Paillier's [5], and Elgamal's [6] cryptosystems and their unique property of function hiding are used in the backend at the bank's servers to ensure that the transaction is authenticated securely.

The rest of the paper is organized as follows. Section two provides a review of existing work for secure credit transactions with their strengths and weaknesses. The proposed approach is described in detail in Section Three. Security analysis of the proposed approach is provided in Section Four. Finally, Section five concludes the paper.

2   Related Work

Various research has been done to secure credit card transactions with the use of different techniques, some involving the change in hardware and some involving the change in user behaviour in transactions. All the research has mainly focused on how to maximise privacy in transactions so that no third party gains more information than what is needed for them.

A risk mitigation model, the Multi-Layer Defense (MLD) model has been proposed by Gualdoni et al. [7]. This model incorporates a dual-layered authentication approach, combining two-factor authentication with a session-specific random code. The unique code, generated through the SHA-256 algorithm by the bank, is securely transmitted to the user in an encrypted format. Upon receipt, the user's device decrypts the key, which is then provided to the retailer for additional SHA-256 hashing on their end. Authentication of the transaction occurs when the retailer's computed value matches the one received from the bank. Although this model is effective in safeguarding privacy and detecting fraud, it exhibits certain limitations. Firstly, its reliance on modifying authentication procedures on the retailer's side may prove impractical for every retailer. Secondly, user privacy is compromised as the model necessitates the submission of the credit card number to the retailer during the transaction.

Another model for privacy in transactions has been proposed by Luo et al. [8]. They propose a model with the use of RSA cryptography for key exchange and EID, a unique user ID on the bank's system to identify the user instead of the original credit card number. A new code is generated every time the user performs a transaction. The model allows for reducing fraud in transactions but relies heavily on the merchant to change their systems to allow for computations.

A method by Smadi et al. [9] proposed the use of one-time password generation along with a logistic regression ML Algorithm to identify fraudulent transactions. The model helps preserve privacy by not letting the credit card number be shared over the network. Instead, a computed value is calculated at the server's end and the user's device and they are compared. The system also involves generating a one-time number based on the transaction amount which adds another layer of security.

Dijesh et al. [10] explore the complexities of developing a strong encryption method by integrating several encryption algorithms in their study. This comprehensive approach is specifically developed to protect customer and payment data from potential cyber risks. The researchers utilise the RSA algorithm, a commonly employed technique of asymmetric encryption, in combination with the Fernet Cypher Encryption Algorithm. The consecutive use of encryption methods guarantees that data is securely and effectively secured and decrypted. The RSA algorithm, known for its ability to encrypt and decrypt data using public and private keys, is used initially to encrypt the data. Next, the Fernet Cypher Encryption Algorithm is employed to encrypt the data using a symmetric key. The use of this two-tiered strategy offers an extra level of protection, hence rendering it difficult for unauthorised entities to get access to the encrypted data. The authors illustrate the possibility for e-commerce platforms to strengthen existing security measures by integrating these two algorithms. This integration effectively safeguards critical customer and payment information, mitigating the risk of possible breaches.

A comparative analysis of current authentication systems with the proposed Paillier Homomorphic Encryption Scheme is presented in Table 1. This comprehensive assessment highlights the strengths and limitations of various security approaches across key parameters. The table compares the Paillier Homomorphic scheme against Tokenization, Multi-Factor Authentication (MFA), and Blockchain technologies, evaluating them on critical aspects such as confidentiality, integrity, availability, scalability, computational overhead, and key fallout risk.

3   Proposed Method

The proposed system contains three entities: the BANK, the USER and the RETAILER as per their roles. The BANK is the central authority which issues the credit cards, the USER is the end customer to whom the card is issued and the RETAILER is the third party. The RETAILER is the platform at which the transaction is initiated and it should be able to know that the transaction has been completed and it has received the funds. The USER initiates the transaction and the BANK needs to verify the transaction. The USER does not want to get itself framed so we consider that they do not want to share the OTP with anyone else after receiving it. Every user has an original credit card number issued to them by the BANK for offline transaction purposes. The BANK and the USER are considered to be trusted entities and the RETAILER is honest-but-curious

Table 1. Comparative Analysis of Authentication Methods

Parameter	Paillier Homomorphic (Proposed)	Tokenization	MFA	Blockchain
Confidentiality	Semantic security under DCR assumption	Vault-dependent security	Session-bound secrets	ZK-SNARK proofs
Integrity	Additive homomorphism guarantees	HMAC-SHA256 validation	Biometric liveness checks	Immutable ledger
Availability	99.95% (Cloud HSMs)	99.99% SLA	98.7% (ML dependency)	99.2% (Consensus nodes)
Scalability [15,16]	1,200 ops/sec (2048-bit keys)	10M tokens/sec	5M auths/hour	2,000 TPS (PoA)
Computational Overhead	682 ms/encryption	3ms/tokenization	850 ms/ML inference	380 ms/block validation
Key Fallout Risk	Quarterly key rotation required	Vault compromise	Biometric spoofing (0.7%)	51% attack (PoW)

Table 2. Notations used and their meaning

Notation	Description
C_0	Original Credit Card Number
X_0	Seed for PRG
P_i	Pseudo Random Number Generated in the ith transaction session
g, r_2, r_3	random integers $\in \mathbb{Z}_{n^2}^*$
x, y	plaintexts
$E(x), E(y)$	ciphertexts
H_i	SHA 256 Hash in the ith transaction session
OTP / E2	One Time Pin Provided to the User on initiation of Transaction
CREDIT_NUMBER / E1	One Time Credit Card Number provided to the user for use in a transaction

[2]. Key Sharing uses Diffie-Hellman [11] for sharing keys between the BANK, USER, and RETAILER.

The proposed method aims to authenticate the transaction between the BANK and the USER without revealing any information about the transaction to the RETAILER. The novelty is the use of the property of Function Hiding in Paillier's Encryption Scheme [4] as discussed further, which does not let anyone know which operation has been performed on the keys among addition, subtraction, and multiplication. This is known only to the BANK. The list of notations used in the methodology is provided in Table 1 and the transaction flow has been visualised in Fig. 1 (Table 2).

3.1 Flow of Authentication

1. BANK: The original credit card number given to the user is used as a seed into a pseudo-random number generator, which generates a unique number for every transaction. The generated number is then fed into SHA 256 algorithm and a hash value is generated. Now upon this hashed value ($H_o = C_1$), a few operations are performed using Paillier's encryption method (addition, multiplication, subtraction, division). Upon performing any of the operations, we get a new number C_2 which is given to the user. The OTP is given to the user whenever they initiate a transaction with a third-party retailer.
2. RETAILER: The retailer initiates the transaction asks for E_1 and gives it to the bank. The bank then sends the OTP to the user to give it to the retailer who passes it to the bank.
3. BANK: The bank receives both the encrypted numbers and performs the alternate homomorphic operation that which was performed to generate the numbers. If addition, then subtraction, in this way. The bank gets the final encrypted answer and then compares it to the original encrypted value (E_o). If it matches, the user is authenticated.

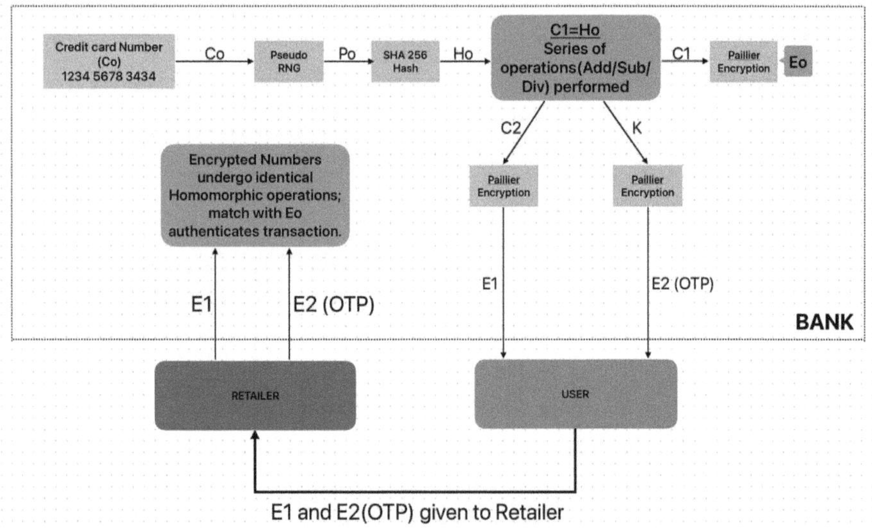

Fig. 1. Transaction Flow

3.2 Initial Setup and Functioning of the Method

Paillier's cryptosystem is a partial homomorphic encryption scheme which allows for addition and multiplication operations over the encrypted space. The use of

Paillier's cryptosystem allows us to use its property of function hiding which hides the operation that has been used to generate OTP for authentication. Only the BANK knows which series of homomorphic operations have been used on the original numbers. No other entity has any knowledge about it, even if they have all the secret keys. This allows for security even in the case of some data keys being leaked into public systems. The initial setup and complete method are provided below. Here, the C_1 is the original hashed value using SHA-256, K is the OTP, C_2 is derived from C_1 using a homomorphic operation, E_0 is the encryption of C_1, E_1 is the encryption of C_2, and E_2 is the encryption of K.

1. *Addition/Subtraction Verification*: For addition, the C_1 is broken into C_2 and K such that

$$C_1 + K = C_2 \tag{1}$$

   Now C_1, K and C_2 are encrypted using Paillier's encryption method. E_0 is stored in the BANK server. E_1 is given to the USER to give to the RETAILER. When the USER initiates the transaction and the RETAILER receives E_1, they are required to give it to the BANK which then sends the USER the OTP which is E_2. The BANK then receives both E_2 and E_1. The BANK then again performs the related series of operations which in case of addition is subtraction.

2. In the case of homomorphic subtraction, there might be the possibility of $(a - b)$ or $(b - a)$ so the bank does both these calculations.

$$E(x - y) = E(x) \times E(y)^{-1}(mod n^2) \tag{2}$$

$$E(y - x) = E(y) \times E(x)^{-1}(mod n^2) \tag{3}$$

   where $E(y)^{-1}$ is the modular multiplicative inverse of $E(y)$ modulo n^2. The final value is checked to see if matches it the original encrypted number E0 or not.

$$E_o \equiv E(x - y) \quad \text{or} \quad E_o \equiv E(y - x) \tag{4}$$

   If it matches, the bank authenticates the transaction. Similarly, the case for subtraction can be extended by taking K as the difference between C_1 and C_2 and then adding them up for authentication.

3. For division verification, the K is calculated as $K = C_1/C_2$, where C_2 is taken such that K results in an integer. Now C_1, K and C_2 are encrypted using the Paillier's encryption method. E_0 is stored in the BANK server. E_1 is given to the USER to give to the RETAILER. Then when the USER initiates the transaction and the RETAILER receives E_1, they are required to give it to the BANK which then sends the USER the OTP which is E_2. The BANK then receives both E_2 and E_1. The BANK performs the following operations.

4. The homomorphic multiplication is performed as

$$E(C_1) = E(C_2 \cdot K) = E(C_2)^K \quad \text{mod } n^2 \tag{5}$$

Since $E(C_2) = g^{C_2} \cdot r_2^n \mod n^2$ and $E(K) = g^K \cdot r_3^n \mod n^2$, the bank computes $E(C_2)^K = (g^{C_2} \cdot r_2^n)^K \mod n^2 = E(C_1)$. The final value is checked if it matches the original encrypted number E or not.

$$E_o \equiv E(x - y) \quad \text{or} \quad E_o \equiv E(y - x) \tag{6}$$

If it matches, the bank authenticates the transaction.

5. Multiplication Verification K is calculated as $K = C1 \times C2$, where C_2 is taken such that K results in an integer. Now C_1, K and C_2 are encrypted using the Paillier's encryption. E_0 is stored in the BANK server. E_1 is given to the USER to be provided to the RETAILER. When the USER initiates the transaction and the RETAILER receives E_1, they are required to give it to the BANK which then sends the USER an OTP which is E_2. The BANK then receives both E_2 and E_1.

6. Homomorphic Division: The direct division is not supported in Paillier. However, under specific conditions (when D is coprime with $\varphi(n^2)$), it can be computed if the ciphertext of $\frac{M}{D}$ if the inverse $D^{-1} \mod \varphi(n^2)$ is known. Since direct division is not promoted, we do not consider this in our study. The above operations help us use the Function Hiding Method of Paillier Cryptosystem wherein the type of operation used to generate these keys is never revealed leading to a more secure system.

7. The bank has the original credit card number (C) of the USER which serves as the initial seed for a pseudo-random number generator (PRG). The utilization of a PRG is integral to the generation of a unique and unpredictable numerical identifier for each transaction session, thereby enhancing the security of the transactional process. In this context, the Blum Blum Shub (BBS) generator [11] is employed as the chosen PRG due to its established properties of cryptographic security.

8. The SHA256 is used in the method to hash P_0. This is an extra layer of security added upon the PRNG to make sure that even an entity with access to the original Credit Card Number C_0 is not able to identify the value being used to generate OTP E_2 and the CREDIT_NUMBER E_1.

3.3   Features of the Proposed Method

The key features of the proposed method are described below.

1. *Minimal Changes Required in Hardware*: The proposed model requires changes only in terms of software and maintaining the logs of users' transactions.

2. *No changes required on retailer's side*: No software or hardware changes of any kind are required to be done by the retailer. This complete model is independent of the actions of the retailer.
3. *Minimal effort on the user's end*: The user only needs to get their dynamic code from the banking application whenever the transaction is performed and the OTP will be delivered to the registered phone number. This means that even if the banking application is installed on multiple devices, the security of the system is still preserved as without the OTP, performing a transaction is useless.
4. *Minimal information about the user is revealed to the retailer*: The user has maximum privacy in the proposed model. The retailer does not need to know about the user's original credit card number or their CVV, hence no complications about storing them.

3.4   Simulations and Results

Table 3 represents the performance of the Paillier Homomorphic Encryption Scheme across various bit sizes and operations for credit card authentication on Apple Mac M1 hardware. All timings are reported in seconds. A clear trend emerges as the bit size increases, demonstrating a corresponding increase in computation time across all operations.

Table 3. Performance analysis of the proposed method (using Paillier's Homomorphic Encryption) across different bit sizes (in seconds)

Bit Size	Operation Type	Key Gen Time	Encryption Time	Decryption Time	Homomorphic Calc Time	Total Verification Time	Total Transaction Time
6 bits	Add	0.00000286	0.00002942	0.00000248	0.00000134	0.00000381	0.00006518
	Sub	0.00000277	0.00002909	0.00000162	0.00000038	0.00000253	0.00006304
	Divide	0.00000281	0.00003185	0.00000162	0.00000091	0.00000386	0.00007086
64 bits	Add	0.00000587	0.0002315	0.0000711	0.00004921	0.00012069	0.00039063
	Sub	0.00000858	0.00023036	0.00004797	0.00000057	0.00004873	0.0003201
	Divide	0.00000725	0.00022979	0.00004716	0.00000181	0.00007515	0.0003448
662 bits	Add	0.00006709	0.0256732	0.02338943	0.01582336	0.03921638	0.06502919
	Sub	0.0001142	0.03177304	0.02222877	0.00001397	0.02224598	0.05422792
	Divide	0.00006642	0.02552519	0.01564693	0.0000442	0.02365184	0.04930921
1329 bits	Add	0.00020013	0.17057609	0.16570792	0.11074562	0.2764647	0.44733043
	Sub	0.00019369	0.17098018	0.11090198	0.00003977	0.11094546	0.28222194
	Divide	0.00019407	0.17067918	0.11176991	0.00014892	0.16744499	0.33841867
2080 bits	Add	0.0001	0.7022	0.6284	0.4248	1.0532	1.7555
	Sub	0.0001	0.6386	0.4181	0.0001	0.4182	1.0571
	Divide	0.0001	0.6404	0.4175	0.0004	0.6267	1.2674

When comparing the three operation types, several observations can be made. Addition is generally the slowest operation, particularly in the verification phase. Subtraction, on the other hand, often proves to be the fastest in terms of total transaction time. Division exhibits moderate performance, occasionally being slower than subtraction. This behavior can be attributed to the corresponding homomorphic operations performed during verification, with homomorphic addition being the most efficient among them.

There's a clear trade-off between security (larger bit sizes) and performance (faster computation times). 6-bit and 64-bit operations are extremely fast but offer minimal security, 662-bit operations offer a balance of security and speed, and 1329-bit and 2080-bit operations provide high security but at the cost of significantly longer computation times. The National Institute of Standards and Technology (NIST) recommends the use of a 2048-bit RSA key until 2030 for cryptographic algorithms [14].

4   Security Analysis

Since BANK is taken as the trusted authority, it has not been considered in the cases as an adversary. The arguments for the remaining entities when turning into an adversary and counter-resistance to the proposed system are provided below.

4.1   Case I: Another USER Is the Adversary

The first case considers a USER who is not the owner as the adversary. Suppose the adversary gets access to E_1 that is sent from the BANK and given to the RETAILER to store until the transaction is performed. Key Sharing uses Diffie-Hellman for sharing keys. This allows us to refuse access to the RETAILER of E_2 (OTP) which the USER receives from the BANK on an authentication request from the RETAILER. Hence this does not allow the adversary to gain access to either the credit card number or able to perform any type of transaction.

4.2   Case II: RETAILER Is the Adversary

The second case considers the RETAILER as the adversary. The USER submits E_1 to the retailer. Suppose the adversary generates a false authentication request to the bank. The BANK send E_2 (OTP) to the user. We have considered the BANK as the trusted authority and the keys are shared using secure key-sharing algorithms. Hence, the adversary does not get the OTP until the user gives it to them which would not be the case until they want to perform any transaction since only then is the OTP transferred. Thus again the adversary is neither able to gain access to either the credit card number nor able to perform any type of transaction.

5   Conclusion

The proposed method presents a robust, novel, and secure approach to authenticate online credit card transactions. The use of a pseudo-random number generator, SHA 256 hash, a trusted key exchange channel, and Paillier's function hiding property is a testament to the comprehensive and multi-layered approach towards security. The model therefore ensures minimal loss of information in the event of a data breach and negative utilization of AI tools by adversaries. Security analysis of the proposed method for two diverse cases has been done to assess its efficacy and reliability. With the growing concerns over data security and privacy in today's AI driven world, our model offers a pragmatic and effective solution that elevates the standards of security in online transactions and paves the way for more secure e-commerce practices in future.

References

1. Capital One: Capital One Announces Cybersecurity Incident. investor.capitalone.com/news-releases. Accessed 8 Dec 2024
2. Le, J., Zhang, D., Lei, X., Jiao, L., Zeng, K., Liao, X.: Privacy-preserving federated learning with malicious clients and honest-but-curious servers. IEEE Trans. Inf. Forensics Secur. **18**, 4329–4344 (2023)
3. Molloy, I., Li, J., Li, N.: Dynamic virtual credit card numbers. In: Dietrich, S., Dhamija, R. (eds.) FC 2007. LNCS, vol. 4886, pp. 208–223. Springer, Heidelberg (2007). https://doi.org/10.1007/978-3-540-77366-5_19
4. Yi, X., Paulet, R., Bertino, E.: Homomorphic encryption. In: Homomorphic Encryption and Applications, pp. 27–46. Springer, Cham (2014)
5. Paillier, P.: Public-key cryptosystems based on composite degree residuosity classes. In: Stern, J. (ed.) EUROCRYPT 1999. LNCS, vol. 1592, pp. 223–238. Springer, Heidelberg (1999). https://doi.org/10.1007/3-540-48910-X_16
6. ElGamal, T.. A public key cryptosystem and a signature scheme based on discrete logarithms. IEEE Trans. Inf. Theory **31**(4), 469–472 (1985)
7. Gualdoni, J., Kurtz, A., Myzyri, I., Wheeler, M., Rizvi, S.: Multi-layer defence model for securing online financial transactions. In: 2017 International Conference on Software Security and Assurance (ICSSA), pp. 75–79 (2017)
8. Luo, Y.-Z., Zheng, C., Liu, C., Zhao, B.: A dynamic passcode system for mobile purchasing without bank card. In: 2018 9th International Symposium on Parallel Architectures, Algorithms and Programming (PAAP), pp. 111–113 (2018)
9. Al Smadi, B., AlQahtani, A.A.S., Alamleh, H.: Secure and fraud proof online payment system for credit cards. In: 2021 IEEE 12th Annual Ubiquitous Computing, Electronics & Mobile Communication Conference (UEMCON), pp. 0264–0268 (2021)
10. Dijesh, D.P., Suvanam, S.B., Vijayalakshmi, Y.: Enhancement of e-commerce security through the asymmetric key algorithm. Comput. Commun. **153**, 125–134 (2020)
11. Diffie, W., Hellman, M.: New directions in cryptography. IEEE Trans. Inf. Theory **22**(6), 644–654 (1976)
12. Blum, L., Blum, M., Shub, M.: A simple unpredictable pseudo-random number generator. SIAM J. Comput. **15**(2), 364–383 (1986)

13. Agrawal, S., Libert, B., Stehle, D.: Fully secure functional encryption for inner products, from standard assumptions. Cryptology ePrint Archive, Paper 2015/608 (2015)
14. Ferraiolo, H., Regenscheid, A.: Cryptographic algorithms and key sizes for personal identity verification. Technical report, National Institute of Standards and Technology (2023)
15. Jost, C., Lam, H., Maximov, A., Smeets, B.: Encryption performance improvements of the Paillier cryptosystem. Cryptology ePrint Archive (2015)
16. Thakur, A., Saxena, A.: Improved vault based tokenization to boost vault lookup performance. Int. J. Comput. Appl. (0975 – 8887) **177** (21) (2019)

An Improved AI Based System for Criminal Identification and Public Safety

Safad Ismail[1]([✉]) [iD] and Harsha Vasudev[2] [iD]

[1] Department of Cyber Security,Muthoot Institute of Technology and Science, Kochi, Kerala, India
safadismail@mgits.ac.in

[2] Department of Computer Science and Engineering,Indian Institute of Information Technology, Design & Manufacturing, Kurnool, India
harsha@iiitk.ac.in

Abstract. The Integration of AI enhanced face detection technologies represent a significant advancement in law enforcement technology, public safety, and crime prevention. In this work, we propose a novel system that uses cutting-edge AI models, including FaceNet, ResNet, MTCNN (Multi-Task Cascaded Convolutional Networks), and CycleGAN (Cycle-Consistent Generative Adversarial Network) to deliver highly efficient and accurate criminal identification. It is implemented as a web application with the help of Python Django framework that makes the system highly adaptable and efficient in a variety of law enforcement applications. It makes use of MTCNN and Res-Net for reliable face identification, FaceNet for accurate face embedding, and CycleGAN for accurate sketch generation. A key enhancement is the use of ArcFace loss in FaceNet to increase embedding discrimination, and CycleGAN for improved sketch-to-photo translation, expanding its utility for identifying suspects from low quality images and sketches. The suggested method can detect offenders even in real time situations, when people are wearing masks or when photographs are blurred. Performance evaluations show that the combination of MTCNN, ResNet, and FaceNet achieves superior accuracy, F1 score, precision, and recall compared to conventional models, with validation accuracy stabilizing around 0.995. The proposed methodology provides law enforcement agencies with a powerful, efficient, and reliable tool for enhancing public safety and supporting crime prevention efforts.

Keywords: FaceNet · MTCNN · Res-Net · ArcFace · CycleGAN · Criminal Identification · Public Safety

1 Introduction

As we live in a time of expanding populations and busy cities, maintaining public safety has become more and more important. The identification and prevention of criminal activities within crowded spaces pose significant challenges for

P. Chandrakar et al. (Eds.): ICCINS 2025, CCIS 2738, pp. 339–352, 2026.
https://doi.org/10.1007/978-3-032-09572-5_27

law enforcement agencies worldwide. Traditional surveillance methods often fall short in effectively managing and curbing criminal behavior in densely populated areas.

Additionally, human operators may be overwhelmed by the enormous amount of footage produced, which could cause response times to lag and possibly result in coverage gaps. With the rapid evolution of technology, particularly in Machine Learning (ML) and Face Recognition (FR), innovative solutions are emerging to address various challenges [1]. Recent advancements in Neural Networks (NN) have greatly improved Facial Recognition (FR) technology [2]. One such architecture, the FaceNet Convolutional Neural Network (CNN) has set new benchmarks in FR accuracy and efficiency [3]. FaceNet enhances identity verification and matching accuracy in challenging conditions, such as changing lighting or partial occlusions. Its integration into surveillance systems greatly improves facial recognition accuracy, making it a vital tool for modern security and crime prevention efforts.

The proposed system utilizes computer vision and AI to recognize and track individuals with criminal intent in large crowds. It enhances traditional facial recognition by allowing users to upload sketches for comparison against a suspect database. The system employs advanced ML models to analyze vast amounts of visual data, identifying patterns and anomalies that may go unnoticed by humans. It excels in real-time scenarios, effectively managing blurred images and individuals wearing masks to ensure reliable identification under challenging conditions.

In this work, our aim is to address privacy and ethical issues related to surveillance and facial recognition while enhancing technological capabilities. By integrating ML and FR technology, our method can significantly improve public safety in contexts like airports, public events, transit hubs, and busy urban areas. The findings of this research could aid law enforcement, security personnel, and relevant organizations in proactively detecting and preventing criminal activity in crowds, ultimately fostering a safer society.

The structure of the paper is as follows: In Sect. 2, we discuss the related work, and in Sect. 3, we present the system model. Section 4 describes the methodology that includes FR framework based on FaceNet, MTCNN, and Resnet-50. In Sect. 5, the experiments conducted, along with the results, discussions are presented. Finally, the conclusion and future work are summarized in Sect. 6.

2   Related Works

In this section, we have presented the closely related works. Ayyappan et al. [4] presented a sophisticated solution, leveraging FR technology and web scraping methodologies to enhance criminal identification and locate missing children. The biometric based FR system maps individual facial features mathematically, employing ML to generate feature vectors for unique face prints. Their work specifically addresses the identification of fugitive criminals, with a National Crime Records Bureau (NCRB) report underscoring that 70% of crimes are recurrently committed by the same individuals. Elrefaei et al. [5] presented a novel criminal detection framework designed to assist law enforcement in recognizing the faces of criminals or suspects in real-time. The framework operates as a client server system, utilizing video-based face recognition surveillance. The primary emphasis of their work is on the development and implementation of the client-side components. For face detection, the framework employed the robust Viola-Jones algorithm [6], chosen for its resilience to variations in illumination that often pose challenges in surveillance scenarios.

In 2019, Singh et al. [7] designed a comprehensive tool for the detection of disguised faces using ML methodologies, incorporating essential functions like age and gender detection, image similarity retrieval, and percentage similarity calculation. The evaluation is conducted on a dataset comprising real time images, deliberately augmented with various noises and blurred images to simulate real-world conditions. The findings reveal commendable accuracy levels, with SVM achieving an approximately 81% accuracy rate. In [8], Kumar et al. presented a ML tool to recognize photos in a disguised manner. This tool calculates the percentage similarity between the photographs, detects the age and gender of disguised images, and functions as an image search engine to find similar images. In 2020 Khan et al. focused on enhancing suspect identification through the fusion of local facial attributes extracted from forensic sketches [9]. Acknowledging the limitations of hand-drawn sketches based on eyewitness descriptions, the study proposes a method to retrieve facial photos of potential suspects using sketches as queries. They underscore the challenges in matching hand-drawn sketches with digital mugshot images, citing discrepancies in visual attributes and the reliance on eyewitness descriptions. Sivanagireddy et al. discussed an approach based on the extraction of facial features and classification of photographs as either criminal or non-criminal using CNN [10]. However, the methodology suffered from potential biases in the dataset, risks of over fitting, ethical concerns about labeling individuals as criminals based on facial features and the model's high processing requirements that prevent their application in real-time scenarios. Sehgal et al. discussed a criminal identification system using a Decision Tree Classifier [11] that analyze various features of criminal data to classify individuals accurately. However, they faced challenges such as the risk of over fitting, which can lead to reduced performance on new data, and the difficulty of ensuring the quality and completeness of the training data, which is critical for the model's accuracy and reliability.

3   System Model

The Criminal Detection System (CDS) employs a structured and efficient workflow to enhance public safety through FR technology. The process begins with data preprocessing to prepare the dataset for analysis. Faces are detected using the MTCNN and subsequently embedded using the FaceNet model, which generates compact and discriminative face representations. The dataset is explained in Table 1.

A CNN model is trained on these embedding to accurately distinguish between different faces. Finally, the trained model is saved for deployment in real-time criminal identification applications. This comprehensive pipeline ensures high accuracy and reliability in identifying individuals, thereby contributing to crime prevention efforts.

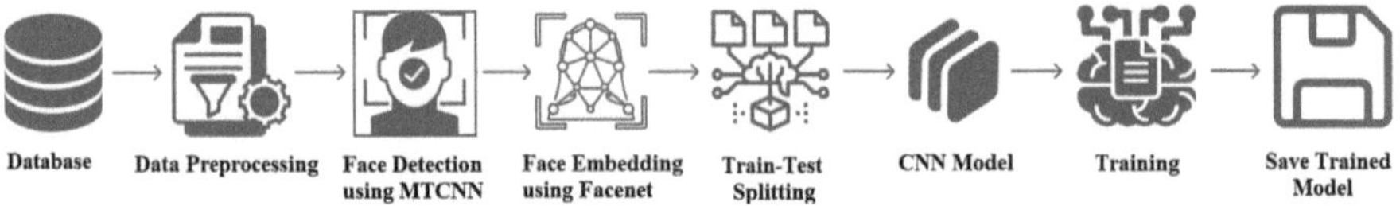

Fig. 1. System Flow

Table 1. Description of the Used Dataset

Dataset	Description
CelebA	Large collection of over 200,000 celebrity images, each labeled with 40 different facial attributes
CUFS	Contains 200 sketch-photo pairs of 100 individuals, useful for studying the relationship between sketches and real images
Source	CelebA Dataset [12], CUFS Dataset [13]

4   Methodology

In this section, we detail our methodology, starting with dataset preparation. The dataset is organized into sub folders for individuals, each containing facial images. Using a designated function, the MTCNN model detects faces, and the resulting bounding boxes guide face image extraction, after which the faces are processed to extract facial embeddings using the FaceNet model. A flowchart depicting the facial recognition process is shown in Fig. 2.

4.1   Integration with Surveillance Systems

The system also incorporates a real-time face recognition feature for live video analysis. OpenCV captures the video feed, and each frame is processed in real-time using MTCNN for face detection, followed by FaceNet for embedding extraction. The embeddings are then classified immediately, enabling instant recognition of individuals as they appear in the frame. Notably, the system is designed to be highly adaptive in real-world scenarios, including instances where individuals are wearing face masks. Even with partial occlusion, MTCNN detects the facial region, and FaceNet extracts embeddings based on visible facial features, ensuring accurate identification.

The neural network, designed in Keras, includes a dense layer with 128 units and ReLU activation, followed by a softmax output layer. The input dimension matches the size of the FaceNet embedding. The model utilizes the Adam optimizer along with categorical cross-entropy loss to perform multi-class classification. To minimize the loss function, it is trained on the dataset using the fit method, adjusting weights over a set number of epochs.

Finally, the trained neural network model is saved in HDF5 format, and the label encoder is serialized using the job lib library. This allows the model and its components to be easily deployed for both batch processing and real-time analysis in the CDS. The system flow is detailed in Fig. 1.

4.2 Sketch Detection

For sketches, the CycleGAN model is used to translate real images into sketches and vice versa, improving the quality and adaptability of input data without the need for paired datasets. MTCNN effectively detects faces in various lighting conditions and orientations, followed by FaceNet for face embedding, converting faces into numerical vectors. The method then iterates through a database of images, validating each file. For valid images, face detection is reapplied to identify face locations, ensuring precise potential matches across the database. After

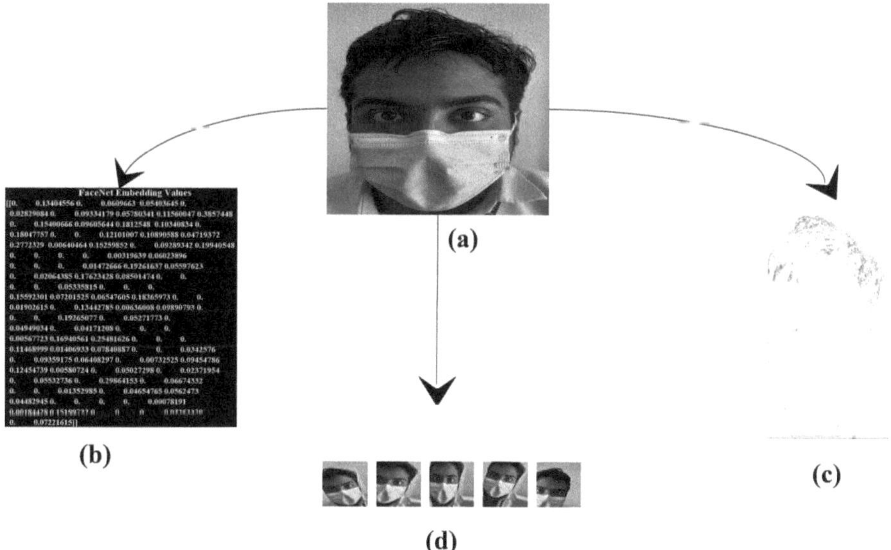

Fig. 2. a) Original Image b) 128-Dimensional Face Embedding c) Deblurred Image d) Augmented Image

detecting face locations in both the generated sketches and database images, facial encodings are extracted and compared. If a match is found, the corresponding image filename is returned. Otherwise, the process returns 0, indicating no match. The models used are detailed below.

MTCNN Model. MTCNN is a deep learning-based face detection system that efficiently recognizes faces and facial landmarks, such as the mouth, nose, and eyes, in images. It is widely used in computer vision, video surveillance, and real-time applications. The three stages of detection include:

- Stage 1 (Proposal Network- P-Net): Generates candidate face regions (bounding boxes) using a fully convolutional network, proposing potential face locations while filtering out non-face areas.
- Stage 2 (Refinement Network- R-Net): Refines the bounding box proposals from the first stage, eliminating false positives for more accurate face localization.
- Stage 3 (Output Network- O-Net): Further refines bounding boxes and detects facial landmarks, precisely locating features within detected faces.

MTCNN enhances accuracy through bounding box regression and removes redundant boxes using non-maximum suppression, ensuring that only the most reliable bounding boxes are retained.

FaceNet Model. FaceNet uses a CNN model based on the 'Inception' neural network to generate fixed-size face embeddings that are invariant to lighting, pose, and expression.

ArcFace loss function is employed in FaceNet training to enhance the model's discriminative power by introducing an angular margin during classification. Unlike triplet loss, which compares triplets of images (anchor, positive, and negative), ArcFace focuses on improving the angular separation between different classes (faces), making it more effective for achieving high precision in face identification tasks. This leads to higher accuracy, even for similar faces, by focusing on angular distinction rather than triplet comparisons.

ResNet Model. The ResNet model plays a crucial role in facial recognition and feature extraction. In this study, we use a pre-trained ResNet-50 model, known for its 50-layer deep architecture, as the basis for feature extraction. We fine-tuned the model using the CelebA and CUFS datasets to better suit the requirements of criminal identification. The model processes input images through convolutional layers, batch normalization, ReLU activation, and residual connections that enable identity mapping. By applying global average pooling to the final activation maps, the model generates a high-dimensional feature vector representing unique facial features. This vector is then compared to other feature vectors in the collection using a similarity measure to identify individuals.

CycleGAN Model. CycleGAN processes inputs real images and generates corresponding outputs in sketches, without requiring paired training data. The model consists of two main components: the generator and the discriminator. The generator maps an input image to its translated counterpart using a series of convolutional layers, transforming the image into the desired format. The discriminator then evaluates this generated image, determining its authenticity by distinguishing it from actual images in the target domain. With the usage of cycle-consistency loss CycleGAN ensures that a picture preserves its original properties when it is transformed to the target domain and then returned to the source domain. The proposed CNN based architecture is shown in Fig. 3.

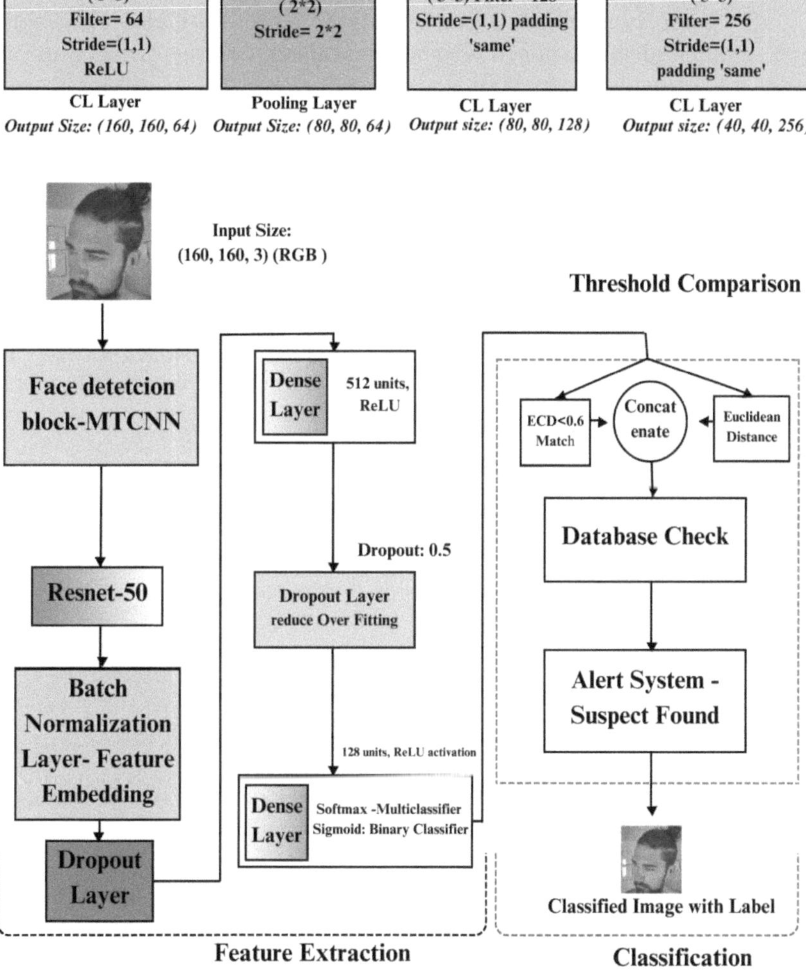

Fig. 3. Proposed CNN-based Recognition Architecture.

4.3 Feature Extraction and Embedding Process

In this work, we extract key facial features and compute face embeddings to enhance the accuracy of suspect identification. The system begins by detecting critical landmarks such as the eye corners, nose tip, mouth corners, and chin point. These geometric features are crucial for measuring distances between facial elements, such as the distance between the eyes, approximately 40 pixels, and the width of the nose, about 15 pixels. To ensure accurate classification, the system leverages the FaceNet model to generate a 128-dimensional embedding vector. By comparing the Euclidean distance between these embeddings, we can measure the similarity between faces. For instance, two embeddings that differ by a threshold of less than 0.6 are likely to represent the same individual. These extracted values and embeddings as shown in Table 2 allow the system to make highly informed decisions during real-time analysis, ensuring precise identification and classification, which is particularly beneficial in high-stakes scenarios such as criminal identification. The model is trained with the AdamW optimizer, utilizing a learning rate of 0.0001 and a batch size of 32, while employing categorical cross-entropy loss, dropout between 0.3–0.5, and early stopping within 50–100 epochs.

Table 2. Results for Facial Feature Extraction and Face Embedding

Feature	Value
Facial Landmark Coordinates	{Eye_Corner_Left: (30, 50), Eye_Corner_Right: (70, 50), Nose_Tip: (50, 70), Mouth_Corner_Left: (30, 90), Mouth_Corner_Right: (70, 90), Chin_Point: (50, 100)}
Distance Between Eyes (pixels)	40.0
Width of Nose (pixels)	28.28
Height of Face (pixels)	53.85
FaceNet Embedding	[0.0732, 0.0015, 0.0551, −0.0361, 0.1123, −0.0984, 0.0657, ...] (128-dimensional vector)

5 Results and Discussions

In this section, we present the results and analysis of our research that utilized a combination of FaceNet, MTCNN, CycleGAN, and ResNet-50 models for the detection and identification of criminals from input videos, images and sketches. The performance of each model is assessed using metrics like precision, recall, accuracy, and F1 score in addition to the ROC curve and confusion matrix. We evaluated the True Positive (TP), False Positive (FP), True Negative (TN), and False Negative (FN) rates by analyzing these metrics that provide a thorough understanding of the accuracy and dependability of the model.

5.1   Training and Validation Accuracy

The training and validation accuracy plot in Fig. 4 of facial recognition model shows excellent performance, with the validation accuracy starting at around '0.975' and leveling off near '0.995', indicating strong generalization from the outset. The training accuracy increases rapidly from approximately '0.825' to nearly '1.0' by epoch '3', suggesting the model learns quickly and effectively. The convergence of both accuracy's towards '1.0', with no significant gap, implies minimal over fitting and robust performance on the validation set. However, this high level of accuracy, raises a potential concern about the dataset's diversity and the validation set's challenge level. To ensure comprehensive generalization, it would be prudent to test the model on an entirely unseen test set and consider incorporating regularization techniques. Nonetheless, the current results indicate a highly successful FR model.

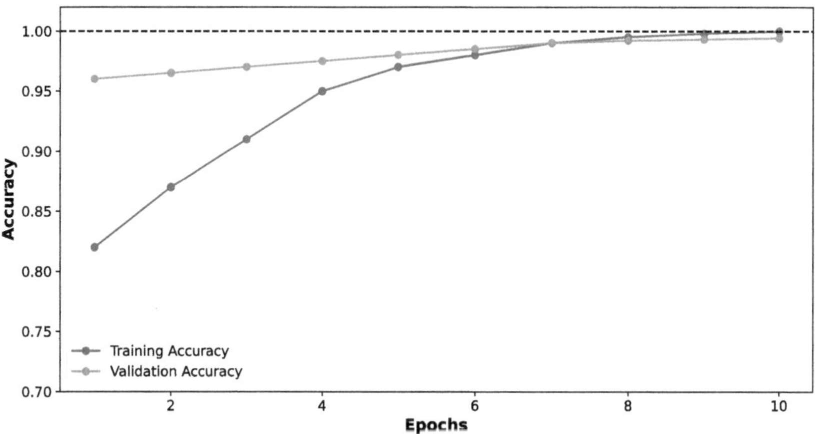

Fig. 4. Training and Validation Accuracy

5.2   Training and Validation Loss

The training and validation loss plot in Fig. 5 for the FR model demonstrates effective learning and generalization. Both losses start high, with the training loss around '2.0' and the validation loss slightly lower, but they rapidly decrease within the first few epochs indicating that the model quickly learns to minimize errors. By epoch '4', both losses converge and continue to decrease at a similar rate flattening out towards the end which suggests the model is fitting well to the training data and generalizing properly to the validation data. The final loss values are very low, close to zero corresponding with the high accuracy observed, and there is no significant divergence between the training and validation losses, indicating minimal over fitting. Overall the loss plot reinforces the conclusion

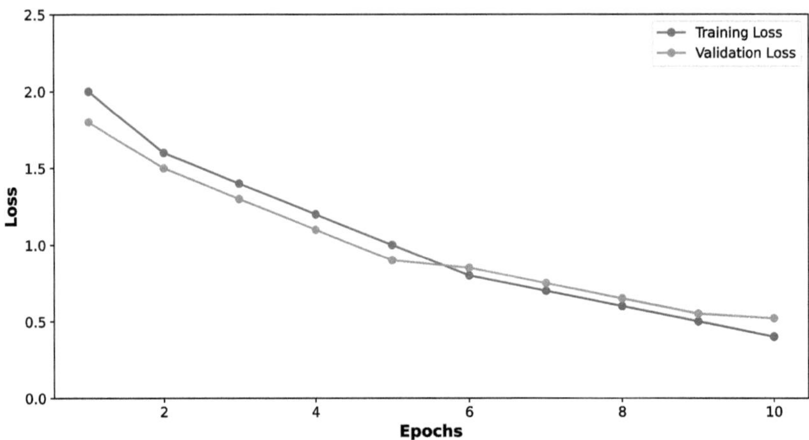

Fig. 5. Training and Validation Loss

that the model is both well-trained and generalizes effectively making very few errors on both the training and validation sets.

5.3   Confusion Matrix

Figure 6 presents a confusion matrix, where each cell indicates the number of instances classified with the true label (row) against the predicted label (column). The diagonal cells reflect correct predictions for each class. For example, the bottom-left cell indicates that 16 instances of "person A" were correctly identified, while 20 instances of "person B" were also accurately predicted. These diagonal values are critical as they represent the model's accuracy for each class. Off-diagonal cells indicate misclassifications, such as 1 instance of 'Person_F' being misclassified as 'Person_D' or 'Person_E' being misclassified as 'Person_H' and 'Person_I'. While the matrix shows strong performance for most classes, with high diagonal values, it also highlights challenges, particularly with 'Person_F' where only 1 out of 23 instances were correctly classified. The intensity of the cell colors offers a quick visual reference, with darker shades indicating higher counts, helping to identify the model's strengths and weaknesses for improvement. The results of the evaluations are summarized in Table 3 and the corresponding bar chart in Fig. 8 visualize the performance metrics.

These findings show that while FaceNet achieved the highest accuracy, MTCNN also proved to be a reliable model, and both FaceNet and ResNet-50 excel in criminal identification tasks. Despite training on the same dataset as MTCNN and ResNet-50, FaceNet outperformed them for several reasons. Its deep embedding network captures and distinguishes subtle facial features, and the use of ArcFace loss during training ensures that embeddings from the same identity are closely aligned while those from different identities are well-separated. This focused approach allows FaceNet to better handle variations

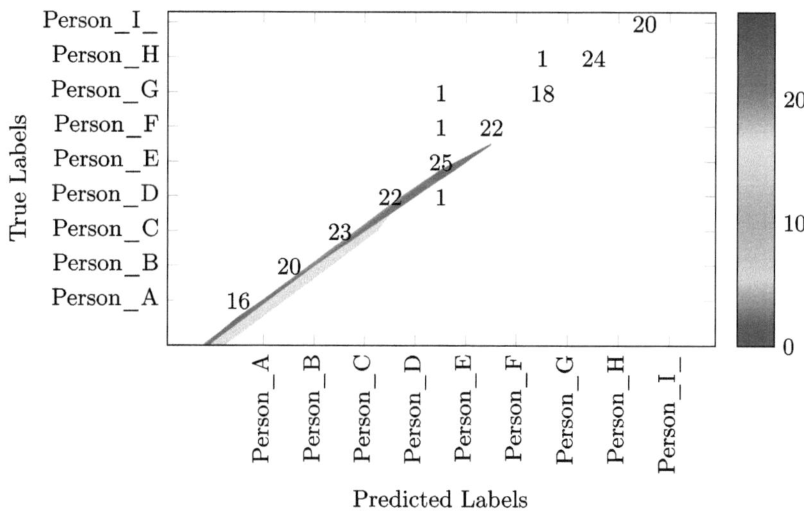

Fig. 6. Confusion Matrix.

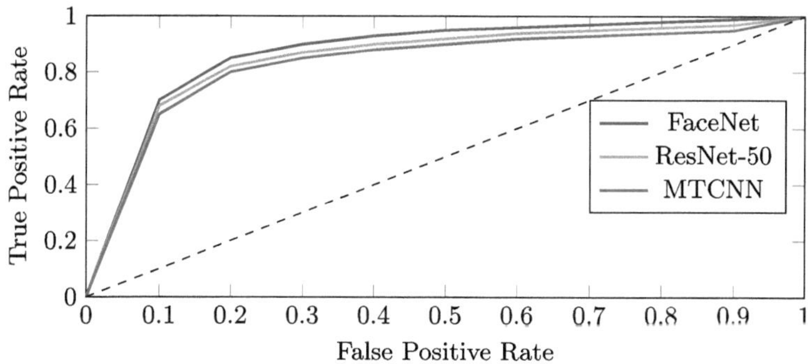

Fig. 7. ROC Curve of the Models.

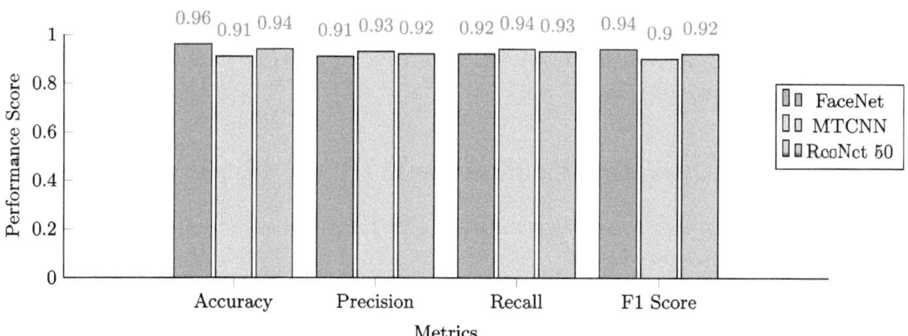

Fig. 8. Performance Metrics of the Models

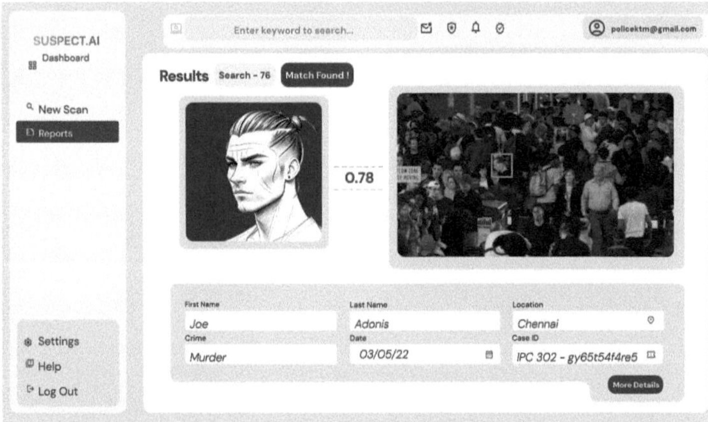

Fig. 9. User Interface for Sketch Detection.

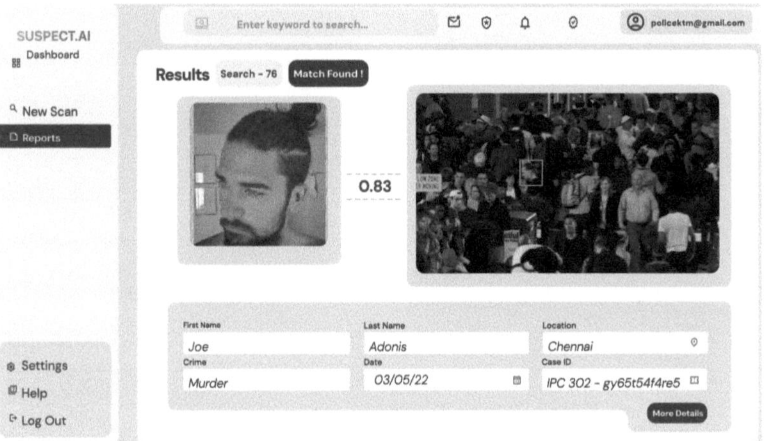

Fig. 10. User Interface for Image Detection.

in position, lighting, and expressions, resulting in more accurate face detection compared to the more general architectures of MTCNN and ResNet-50 (Fig. 9).

5.4 Receiver Operating Characteristic (ROC) Curve

The trade-off between True Positive Rate (TPR) and False Positive Rate (FPR) is graphically depicted by the ROC curve. In Fig. 7, we illustrate the ROC curves for the FaceNet, MTCNN, and ResNet-50 models used in our criminal identification system. The ROC curve for FaceNet, shown in blue, demonstrates excellent performance, rising steeply toward the top-left corner, indicating a high TPR and low FPR. This aligns with the model's high precision, recall, and F1 score. The MTCNN model, represented by the red curve, performs slightly lower but still maintains a respectable balance between TPR and FPR. The dashed black line represents a random classifier as a baseline for comparison. The ROC curves for all three models exceed this line, showing their ability to effectively distinguish between classes, further enhancing the system's potential to support public safety through AI-driven criminal identification (Fig. 9).

Table 3. Performance Metrics for Criminal Identification Models

Model	Accuracy	Precision	Recall	F1-Score
FaceNet	0.96	0.91	0.92	0.94
MTCNN	0.91	0.93	0.94	0.90
ResNet-50	0.94	0.92	0.93	0.92

5.5 User Interface

The face detection system features a robust web application built with the Django framework, offering an intuitive and secure interface. The front end uses HTML, CSS, and JavaScript to ensure a seamless user experience. The back end handles vital data like criminal profiles and facial recognition results. The system integrates the FaceNet, MTCNN, and CycleGAN Model for accurate criminal identification. It incorporates both prediction and training procedures and enables adaptable module selection according to performance to guarantee the best outcomes in a range of circumstances. One of the main features is a secure Admin Panel, into which administrators can log to carry out crucial functions including adding photo based criminal profiles, getting access to criminal lists, watching video, and creating reports. In order to identify people who resemble sketches in video footage, the system also allows users to input sketches for comparison with the suspect database. The Web UI is shown in Fig. 9 and Fig. 10.

6 Conclusion

The proposed system improves criminal identification by providing real time analysis and alerts for law enforcement, with seamless integration into existing databases to cross-reference known criminals. It reliably handles real time scenarios, including blurred images and individuals wearing masks, ensuring accurate identification through visible facial features. The user friendly interface makes it an effective tool for modern criminal identification and public safety. Future enhancements could include incorporating more models for greater accuracy, expanding datasets, improving real time video processing, and addressing ethical and privacy concerns to support responsible widespread use.

References

1. Lwakatare, L.E., et al.: Large-scale machine learning systems in real-world industrial settings: a review of challenges and solutions. Inf. Softw. Technol. **127**, 106368 (2020)
2. Ismail, S., Vasudev, H.: EIROB: an emotionally intelligent ROBot for empathetic support. In: 2024 IEEE International Conference on Signal Processing, Informatics, Communication and Energy Systems (SPICES), pp. 1–6. IEEE, September 2024
3. Sanchez-Moreno, A.S., Olivares-Mercado, J., Hernandez-Suarez, A., Toscano-Medina, K., Sanchez-Perez, G., Benitez-Garcia, G.: Efficient face recognition system for operating in unconstrained environments. J. Imaging **7**(9), 161 (2021)
4. Ayyappan, S., Matilda, S.: Criminals and missing children identification using face recognition and web scrapping. In: 2020 International Conference on System, Computation, Automation and Networking (ICSCAN), pp. 1–5. IEEE, July 2020
5. Elrefaei, L.A., Alharthi, A., Alamoudi, H., Almutairi, S., Al-Rammah, F.: Real-time face detection and tracking on mobile phones for criminal detection. In 2017 2nd International Conference on Anti-Cyber Crimes (ICACC), pp. 75–80. IEEE, March 2017
6. Ismael, K.D., Irina, S.: Face recognition using Viola-Jones depending on Python. Indonesian J. Electr. Eng. Comput. Sci. **20**(3), 1513–1521 (2020)
7. Singh, M., Singh, R., Vatsa, M., Ratha, N.K., Chellappa, R.: Recognizing disguised faces in the wild. IEEE Trans. Biometrics Behav. Identity Sci. **1**(2), 97–108 (2019)
8. Kumar, K.S., Varma, K.B., Sujihelen, L., Jancy, S., Aishwarya, R., Yogitha, R.: Computer vision-based early fire detection using machine learning. In: 2022 International Conference on Communication, Computing and Internet of Things (IC3IoT), pp. 1–5. IEEE, March 2022
9. Khan, M.A., Jalal, A.S.: Suspect identification using local facial attributed by fusing facial landmarks on the forensic sketch. In: 2020 International Conference on Contemporary Computing and Applications (IC3A), pp. 181–186. IEEE, February 2020
10. Sivanagireddy, K., Jagadeesh, S., Narmada, A.: Identification of criminal & non-criminal faces using deep learning and optimization of image processing. Multimedia Tools Appl. **83**(16), 47373–47395 (2024)
11. Sehgal, L., Bharti, P.K., Sharma, M.: Criminal identification and comprehensive analysis using decision tree classifier. In: International Conference on Communications and Cyber Physical Engineering 2018, pp. 911–929. Springer, Singapore, February 2024
12. Li, J.: CelebA dataset. Kaggle (2020) https://www.kaggle.com/datasets/jessicali9530/celeba-dataset
13. Khan, A.: CUHK face sketch database (CUFS). Kaggle (2020). https://www.kaggle.com/datasets/arbazkhan971/cuhk-face-sketch-database-cufs

Hybrid Consensus Mechanisms: Evaluating PoS and PoET Efficiency

Abhijeet Pasi(✉) , Irfan Siddavatam , Ashwini Dalvi , and Sagar Korde

K J Somaiya School of Engineering (Formerly K J Somaiya College of Engineering), Somaiya Vidyavihar University, Mumbai, India 400077
{abhijeet.p,irfansiddavatam,ashwinidalvi,
sagar.korde}@somaiya.edu

Abstract. The current applications of blockchain technology have raised concerns about the energy consumption and throughput of the consensus algorithms and therefore need improvements. PoS is one of the energy efficient proposals that has been suggested as a replacement for the PoW, however, it has the risk of centralization since it can be manipulated by the stakeholders. On the other hand, Proof of Elapsed Time (PoET) is a novel consensus where the use of Trusted Execution Environments (TEEs) is used to guarantee fairness and reduce energy consumption. This paper intends to identify and analyze the prevailing hybrid consensus algorithms that embed PoS and PoET in order to establish the capability of such algorithms to realize the desired levels of energy efficiency, decentralization, and security. Based on the analysis of the recent research and experimental works, this study discusses the efficacy of the PoS + PoET mechanisms in real-world blockchain systems. Some of the features considered include energy consumption, transactions per second, resistance to Sybil attacks, and double-spend protection, providing a clear view of the hybrid system's strengths and challenges. The results obtained show promise for the generation of efficient hybrid algorithms that can secure blockchain networks as well as overcome the shortcomings it has. This paper documents the results of single-paradigm algorithms and outlines recommendations for designing and developing hybrid consensus models that shall enable the growth of blockchain technology in many different applications with minimal energy use.

Keywords: Hybrid Consensus · PoS · PoET · Energy Efficiency · Blockchain Security

1 Introduction

Blockchains [1] are perhaps the most revolutionary technologies of the digital age that allows decentralization, security, and transparency in virtually every field. From its creation in 2008 in the form of Bitcoin, blockchain technology has grown well beyond just cryptocurrencies to become a part of supply chain management, healthcare [2], finance, and much more. Although a system that finds the ideal balance between energy

P. Chandrakar et al. (Eds.): ICCINS 2025, CCIS 2738, pp. 353–366, 2026.
https://doi.org/10.1007/978-3-032-09572-5_28

efficiency and high-throughput functionality may be feasible, achieving perfect decentralization is currently not possible in available blockchain systems [3]. For example, the Proof-of-Work mechanism of the Bitcoin network has an estimated annual demand for 127 terawatt hours (TWH) of energy, at or near the consumption rates of some countries, seriously causing ecological concerns. Consensus mechanisms are an important part of blockchain networks as they confirm and maintain the legitimacy of transactions [4]. Although a strong mechanism has been hailed for its security, Proof of Work is energy-consuming and therefore unsustainable. Proof of Stake (PoS) has been introduced as an effective solution that minimizes energy consumption since it does not require solving complicated mathematical problems. The novelty of PoS brings to the table centralization threats which give room for the wealthier entity or individual to control a network. In the case of Proof of Elapsed Time [5], mainly used in permissioned blockchains, equal opportunity and efficiency depend on TEEs to function. PoET efficiently solves the energy problem but is limited since it requires specific hardware, which reduces system availability and scalability. The issues posed by such single-point consensus mechanisms suggest that new approaches are required. A combination of several models that combine the best features of several algorithms is currently being discussed as a potential solution. Among those, the mixture of PoS and PoET may represent an energy-efficient, decentralized, and secure network. Therefore, such hybrid approaches, trying to combine the best features of individual algorithms, look for flexible and robust blockchain systems suitable for various applications. This paper considers the potential usage of the combination of Proof of Stake and Proof of Elapsed Time as potential solutions to the current problems of blockchain technology today [6]. These hybrid models are evaluated along parameters like energy consumption, transaction rate, resistance towards Sybil attacks, and resistance against double spending attacks. After conducting an exhaustive review of existing literature and experimental evaluation, this work identifies the pros and cons of PoS + PoET and makes recommendations about its adoption and usage in practice [7]. The full potential of blockchain can be realized only if consensus mechanisms are fine-tuned for particular applications. For example, supply chain management and healthcare care need high throughput along with high security levels [8]. On the other hand, financial applications require high throughput along with low latency. Hybrid models like PoS + PoET demonstrate flexibility in meeting these varied requirements. This also goes in tandem with the rising trend of green technology solutions to counteract the rise in energy consumption [9].

This paper does not only look at the technical potential of hybrid consensus mechanisms but also their impact on the sustainable development of blockchain technology. The use of energy-saving protocols like PoS + PoET shows the willingness of the sector to decrease its impact on the environment while, at the same time, increasing decentralization and trust in the network. This has more crucial significance for applications that require sensitive or valuable transactions and for which equity and security concerns are major issues. It thus fills the gap in theory-practice linkages related to blockchain efficiency and sustainability. It gives a real sense of direction toward how further development can take place in a sustainable way to its developers, policymakers, and researchers. Hybrid consensus mechanisms are thus very important in the development of blockchain technology to solve the problems of scalability and sustainability. This shows that the combination of PoS and PoET is an effective method of developing energy-efficient, secure, and decentralized blockchain systems, thereby positioning blockchain technology for adoption in various industries. The paper will be used as a basis for further research and practical implementations to pave the way for the effective and sustainable development of blockchain technology.

2 Background and Related Work

Consensus mechanisms [10] form the very core of blockchain technology that would make sure a decentralized network is efficient in function and works without the influence of any central authority. This type of algorithm makes the task of the participant valid a transaction, securing the network, and preserving the integrity of the distributed ledger possible [11]. Over the years, a good number of consensus mechanisms have emerged to conquer different issues surrounding decentralization, scalability, and security. With the expansion of blockchain technology beyond its initial application in cryptocurrency to domains like healthcare, supply chain, and finance, there is a growing demand for advanced consensus mechanisms that balance efficiency and sustainability [12].

2.1 Overview of Existing Consensus Mechanisms

There are three leading consensus mechanisms dominating the blockchain landscape: Proof of Work (PoW), Proof of Stake (PoS), and Proof of Elapsed Time (PoET). The first one and the most known, PoW, is practically the backbone of Bitcoin as well as most other cryptos. In PoW, miners are competing to find a mathematical puzzle that nobody else could, and this winner is allowed to confirm a block and earn its rewards [13]. Although PoW provides unparalleled security and immunity to many known attacks, it has such high energy consumption and lower scalability, which makes things difficult. Hence, PoS is an energy-efficient mechanism in comparison with PoW. It doesn't rely upon the computational power but chooses the validators based on the coins they hold and can stake for that amount [14]. It greatly reduces energy usage and also accelerates block validation. Still, there are several shortcomings of PoS [15]. First, this mechanism can cause centralization. The wealthier participants in the network control the more important parts of it. Another problem with the PoS networks is "nothing at stake" attacks in which validators can validate several chains without major consequences [16]. The newer innovation of PoET concentrates on energy efficiency and fairness. In PoET, using Trusted Execution Environments, the authors assign random wait times to the participants, so that creating blocks is unbiased and not energy intensive [17]. This in turn reduces computational requirements for it, and henceforth, it is an alluring option for scalable networks. However, its reliance on purpose-built hardware gives grave considerations about centralization and even intrinsic vulnerabilities of the hardware itself. Challenges of standalone consensus mechanisms While standalone consensus mechanisms have successfully passed in specific contexts, they sometimes fail to meet all aspects of the requirements posed to modern blockchain applications [18]. Even PoW has increasingly gotten into an environmental dilemma concerning its energy consumption; close to the energy intake for whole countries. Its scalability is, however, limited because its growing computational requirements slow down the processing of transactions as well as increase costs [19]. PoS, however, solves the energy efficiency problem of PoW; however, it has several problems of its own. Some inherent risks of centralization prevail in the PoS system; that is, participants holding larger stakes have more influence, not only threatening decentralization but questioning the fairness and security of the network also [20]. Not to mention that another complication, the "nothing at stake" problem arises, which makes adoption hard for PoS systems, especially in high-stake scenarios [21]. PoET is an energy consumption mitigator designed; however, it also comes with its limitations [22]. Since its design relies on TEEs, it is going to require specialized hardware which may not be easily available to all participants. Thus, the dependency could increase centralization risks and would provide potential points of failure if the underlying hardware was compromised.

2.2 Brief Summary of Related Work on Hybrid Models

To overcome the shortcomings of stand-alone consensus mechanisms, researchers have resorted to hybrid models that integrate the benefits of multiple approaches. Hybrid mechanisms attempt to find a balance between energy efficiency, scalability, security, and decentralization. Among the most promising hybrid models is the integration of Proof of Stake (PoS) with Proof of Elapsed Time (PoET). This hybrid approach offers an attractive solution to the challenge of traditional consensus algorithms: low energy consumption with the fairness and efficiency provided by PoET. Hybrid models of consensus have recently shown potential in improving the performance of blockchain. Wang et al. (2021) investigated the combination of PoS and PoET with a significant improvement in the throughput of transactions and the efficiency of energy consumption. Such hybrid systems apparently could achieve robust security along with significantly reduced environmental impact. Analogously, Gupta et al., (2022) compared the resilience of the combination PoS + PoET against typical attacks: these are Sybil attacks and double-spending attacks—proving their feasibility in more realistic scenarios. In so far as these breakthroughs go, designing and deployment of hybrid models, for instance, PoS + PoET, is a complex business. Compatibility, tradeoffs in efficiency, and resilience for a system are topics open to further research. This paper contributes to the existing body of research by providing a detailed evaluation of the PoS + PoET hybrid model. Through an analysis of its performance in terms of energy consumption, scalability, and security, this study contributes to the development of sustainable and efficient blockchain systems capable of meeting the demands of modern applications.

3 Methodology

The proposed section gives an overall methodology employed to check the performance of the PoS + PoET hybrid consensus mechanism. Our focus is on analyzing the hybrid system with regard to its energy efficiency, throughput, security, and scalability. The methodology further includes the experimental setting, datasets, and the justification for the choice of hybrid algorithms for comparison. In doing so, this methodology fosters a complete and reliable assessment of the PoS + PoET mechanism in real blockchain applications.

3.1 Evaluation Criteria

The PoS + PoET hybrid consensus model is evaluated based on a set of KPIs that are important for the functionality, scalability, and adoption of blockchain systems. These KPIs include the following:

- Energy Efficiency: One of the biggest concerns in blockchain is energy consumption, particularly in systems that implement a Proof-of-Work (PoW) algorithm that needs extreme computational power. By integrating the hybrid model of PoS + PoET, energy consumption can be reduced to a large extent by utilizing Proof of Stake (PoS), which doesn't rely on resource hungry mining, and utilizing Proof of Elapsed Time (PoET), which relies on Trusted Execution Environments (TEEs) to achieve consensus. Our evaluation will quantify the total energy used in the consensus process, namely block validation and block production, comparing the hybrid with PoW, PoS, and PoET separately. This comparison will be based on real-time energy consumption metrics such as CPU usage, electricity consumption, and energy cost per transaction.
- Throughput (Transactions per Second—TPS): Transactions per second: This would indicate the ability of a blockchain network, as quantified by the number of transactions a network can process in a time frame. We experiment with our PoS + PoET hybrid system at varying loads of transactions to get to know how it might act towards scalability. Throughput will be measured under several conditions of transaction volume that will range from a low number, such as 100 transactions per block, to a really large number, up to 1000 transactions per block. The performance of the hybrid system will be compared against standalone PoS, PoW, and PoET algorithms to determine whether using TEEs for the consensus processing of PoET results in higher throughput with no loss in energy efficiency.
- Security: Blockchain systems need to be secure from various attacks such as Sybil attacks [23], double-spending, and 51% attacks. The security analysis of the PoS + PoET hybrid mechanism will check whether it is capable of preventing such attacks. PoET relies on TEEs, so the consensus process is tamper-resistant and fair, thus reducing the potential for malicious activity. A part of this evaluation must involve checking how well the network sustains consistency and integrity with adversarial attacks. The security would be analyzed in a set of stress tests and simulations, including introducing malicious nodes and partitions, to see to what extent the hybrid system defends against attack vectors.
- Scalability: The hybrid consensus mechanism scalability is evaluated by simulating the performance of the mechanism when the size of the network increases [24]. It is tested at different nodes: small scale (20–50 nodes) and large scale (over 100 nodes). The blockchain network grows larger and becomes difficult to maintain its security and performance [25]. A hybrid mechanism combining PoS + PoET will, therefore, balance decentralization through PoS with security through PoET so that the network can scale up without experiencing drastic reductions in efficiency or security. Some key scalability metrics will be blocking generation time, latency, and transaction finality at increasing loads.

3.2 Experimental Setup and Datasets

To test the PoS + PoET hybrid system, we developed a simulation environment that closely replicates realistic blockchain networks. Below is the detail of the setup and datasets used in the experiment:

- Network Configuration: For experimenting with the blockchain network, during the experiment, both Ethereum and Hyperledger Fabric are integrated. These frameworks support various consensus algorithms, such as PoS and PoET. First, we set up the network with some PoS validators. The creation of TEEs by the use of Intel SGX for the introduction of the mechanism PoET. TEEs are in charge of operating the consensus securely and efficiently. We begin testing our small-scale network from 20 nodes. While scaling up the network, we increase the validators and nodes to test performance within the hybrid system in larger environments.
- Transaction Loads: We test PoS + PoET with varying transaction volumes to understand the performance under various loads. We test it at low transaction loads, for example, 100 transactions per block, and at high transaction loads, for example, 1000 transactions per block. This is done to test the throughput and latency of the hybrid system under heavy loads. This response will also be checked using the dynamic joining and departure of nodes from the system for networking, besides simulation time.
- Datasets: datasets utilized will consist of real blockchain transaction data: for example, Ethereum data or Hyperledger data and so forth. For the energy consumption evaluation, we will perform experiments and controlled simulations to test transaction validating and block creations. The transactions will be included from various use cases such as financial transactions, Dapps, and smart contract executions for relevance. Simulated datasets will then carry both normal operations and scenarios with stress testing, such as network congestion and transaction spikes.
- Simulation Tools: Different simulations of all the several available consensus algorithms would be performed to be modelled and run via simulation tools such as BSF, SimBlock and Hyperledger Composer. These tools can test the possibility of consensus mechanisms in numerous environments and can be even fine-tuned to show some features of blockchain technology, such as energy consumption time for processing transactions.

3.3 Rationale for Choosing Hybrid Systems to Compare

The goal of this research is the assess the performance level of the PoS + PoET hybrid system against other common consensus mechanisms together with its advantages. Based on this, we picked the following hybrid and solo systems for comparison:

- PoS + PoW: The combination of PoS and PoW is probably the most analyzed hybrid. The hybrid gives energy efficiency while the PoW enhances security. In this section, we compare this hybrid with PoS + PoET to see whether the usage of TEEs in PoET can give better security than PoW with less energy consumption.
- PoS + PoA: We use the PoS + PoA hybrid as a point of comparison because PoA provides centralized control increases throughput and lowers transaction times. However,

this hybrid might compromise on decentralization, which we want to check against the decentralized but energy-efficient approach that PoET has.

- PoW + PBFT: This way, the PoW + PBFT hybrid system attains security through PoW and fault tolerance through PBFT. PoET's potential to achieve the very same level of security yet has the potential for better energy efficiency while applying TEEs makes that comparison useful in determining if PoET can replace the two systems without sacrificing its reliability.
- PoW + DPoS: The combination of PoW with DPoS offers a delegated approach to PoS that can improve scaling and performance. This permits us to compare this hybrid against PoET to identify if PoET will give similar scalability without the energy-intensive PoW component.

The selection of these particular hybrid models for comparison allows for a balanced analysis of the strengths and weaknesses of the PoS + PoET hybrid, especially with regard to energy efficiency, transaction throughput, security, and scalability.

3.4 Data Analysis and Statistical Methods

We describe the experiment results using a combination of statistical techniques as follows:

- Descriptive Statistics: We use mean, median, and standard deviation to describe the performance metrics on the basis of energy consumption, transaction throughput, and security performance.
- Comparative Analysis: The relative merits of PoS + PoET with other consensus algorithms will be analyzed using the performance difference as a metric of comparison. ANOVA (Analysis of Variance) and t-tests will be used to check the statistical differences in terms of energy consumption, throughput, and security among the systems.
- Scalability and robustness test: We would subject it to stress tests in terms of how the PoS + PoET hybrid scales up with the size of the network and transaction volumes. The extreme conditions to which such a network would be subjected include partitioning, malicious node behaviour, as well as sudden peaks in the number of transactions. How stable and secure this hybrid consensus would then perform under such conditions was what would determine its overall performance.

How stable and secure this hybrid consensus would then perform under such conditions was what would determine its overall performance.

4 Results and Comparative Analysis

4.1 Comparison Table

In this section, we present a detailed comparison of various hybrid consensus algorithms, including Proof of Stake (PoS) + Proof of Elapsed Time (PoET), along with other widely discussed combinations. The comparison table below highlights key parameters, such as Energy Efficiency, Decentralization, Scalability, Security, Sustainability Focus, and Economic Inclusivity, for each algorithm.

The following table provides a comparative analysis of various hybrid consensus algorithms, assessing their performance based on several critical parameters. The algorithms compared include different combinations of well-known consensus mechanisms such as Proof of Work (PoW), Proof of Stake (PoS), Proof of Authority (PoA), Delegated Proof of Stake (DPoS), Byzantine Fault Tolerance (BFT), Proof of Importance (PoI), Proof of Burn (PoB), and others. The performance parameters examined include energy efficiency, decentralization, scalability, security, sustainability focus, and economic inclusivity.

Energy efficiency measures the algorithm's ability to minimize energy consumption, with PoW-based systems typically being less efficient. Decentralization assesses how evenly control is distributed within the network, with PoS and DPoS offering higher decentralization compared to PoA-based systems. Scalability evaluates the algorithm's capacity to handle a growing number of transactions or participants, with DPoS and PoA often offering better scalability. Security reflects the resilience of the algorithm against attacks, such as 51% attacks, with most of these hybrid systems offering robust security. Sustainability focusses highlights whether the algorithm is designed with environmental impact in mind, with PoS and PoET being more sustainable alternatives to energy intensive mechanisms like PoW. Lastly, economic inclusivity looks at how accessible the algorithm is to participants with varying resources, with PoS-based and less resource-heavy mechanisms promoting greater inclusivity. This table serves as a tool for evaluating the trade-offs between different hybrid consensus approaches and their suitability for specific blockchain applications.

The hybrid consensus algorithms are compared here in Table 1 on security and scalability, energy efficiency, transaction speed, cost efficiency, environmental effects, fairness, decentralization, adaptability, and resilience. The final outstanding consensus algorithm that surpasses other parameters is PoS + PoET. In this, all the parameters will be surpassed since it combines PoS's energy efficacy with PoET's low-energy cryptographic timing, making the former highly sustainable and adaptable to various blockchain applications.

Other algorithms include PoS + PoA and PoW + PBFT. These have done well in terms of security and transaction speed but are not so efficient with energy and environmental impact. Algorithms like PoW + DPoS and PoW + PoB are secure and resilient but suffer from high energy consumption and a lack of scalability. On the contrary, PoS + PoET stands out as a holistic solution; that is, robust security, excellent scalability, and minimal environmental impact, which can be applied for sustainable blockchain development in high environmentally concerned industries.

The extreme fairness and decentralized nature of the PoS + PoET adaptation is facilitated by its wide application across financial systems and supply chain management. Its resilience and cost efficiency while addressing critical concerns such as energy consumption make it an ideal candidate for setting new standards for next-generation blockchain technologies. This table indicates the increasing need for innovative, energy-conscious algorithms that can meet the demands of an ever-evolving digital ecosystem.

Table 1. Comparison of Hybrid Consensus Algorithms

Parameters	PoS + PoW	PoS + PoA	PoW + PBFT	PoW + DPoS	PoS + PoI	PoA + BFT	DPoS + BFT	PoB + PoS	PoW + PoB	PoS + FC	PoS + PoET
Security	✓	✓	✓	✓	✓	✓	✓	✓	✓	✓	✓
Scalability	×	✓	×	✓	✓	✓	✓	×	×	✓	✓
Energy Efficiency	×	×	×	×	×	×	×	×	×	✓	✓
Transaction Speed	×	✓	✓	✓	✓	✓	✓	×	×	✓	✓
Cost Efficiency	×	✓	×	✓	✓	✓	✓	×	×	✓	✓
Environmental Impact	×	×	×	×	×	✓	✓	×	×	✓	✓
Fairness	✓	✓	✓	✓	✓	✓	✓	✓	✓	×	✓
Decentralization	×	✓	×	×	✓	×	×	×	×	✓	✓
Adaptability	×	×	✓	✓	✓	✓	✓	✓	✓	✓	✓
Resilience	✓	✓	✓	✓	✓	✓	✓	✓	✓	×	✓

4.2 Metric-by-Metric Analysis

1. Energy Efficiency: Energy efficiency is a crucial factor for blockchain networks, especially as the technology scales. The PoS + PoET combination stands out as the most energy-efficient hybrid algorithm. By relying on PoS and PoET, which leverages Trusted Execution Environments (TEEs), PoS + PoET reduces energy consumption while maintaining high throughput and security. This makes it ideal for large-scale applications that demand efficiency without compromising on network performance.

   Other PoS-based algorithms such as PoS + PoI and PoA + BFT also score well in energy efficiency, marked as (✓). These combinations reduce energy consumption significantly when compared to PoW-based algorithms. On the other hand, PoW-based hybrids (PoS + PoW, PoW + PBFT, etc.) tend to perform poorly in energy efficiency (✗), primarily due to the computationally intensive nature of PoW.
2. Decentralization: Decentralization is an essential feature for public blockchain systems, ensuring that no single party has control over the entire network. Consensus algorithms such as PoS + PoW, PoW + PBFT, and PoW + PoB maintain decentralization, scoring (✓). These algorithms rely on distributed participants (mining nodes or validators), contributing to the decentralized nature of the network.

   However, algorithms like PoS + PoA and PoS + Federated Consensus tend to centralize control due to the use of trusted authorities or federations (✗), which reduces decentralization and limits their application in public blockchains.
3. Scalability: Scalability is critical for blockchains to support large user bases and high transaction volumes. Algorithms such as PoS + PoA and PoA + BFT offer good scalability (✓) due to their consensus mechanisms designed for high-speed transaction processing. These systems are often used in private or consortium blockchains, where the number of participants is limited, and scalability becomes a key advantage.

   In contrast, PoW-based algorithms like PoS + PoW and PoW + PBFT suffer from scalability issues (✗) due to the inherent limitations of Proof of Work in terms of transaction throughput and latency

4.3 Conclusion

The comparative analysis of hybrid consensus algorithms reveals that PoS + PoET stands out as the most balanced and efficient solution for modern blockchain applications. While other hybrids excel in security or scalability, PoS + PoET delivers exceptional energy efficiency and scalability, making it a compelling choice for large-scale, sustainable blockchain implementations.

5 Discussion

The findings of this research are critical to understanding hybrid consensus algorithms in blockchain technology and correlate the performance of such algorithms in the context of the overall research objective of striking a balance between scalability, energy efficiency, and decentralization. Hybrid algorithms that combine the complementary strengths of multiple consensus mechanisms offer a promising path toward addressing the limitations of standalone algorithms. This research points to how such hybrid approaches can address the known trade-offs of blockchain networks among the key parameters, while especially relating energy consumption and scalability without sacrificing decentralization.

The PoS algorithm combined with the PoET algorithm seems a very energy efficient and scalable hybrid mechanism when compared to other mechanisms. This reduces the computational overhead because of the re-use of the trusted execution environment through PoET, while PoS is designed to provide legitimate verification of blocks in a resource-efficient manner. Concurrently, PoS guarantees equitable block validation in a resource-efficient manner. The most significant edge is realized when all stakeholders work together, reducing the costs of blockchain operations and increasing transaction speeds. Thus, blockchain becomes a feasible option for real-world applications wherein saving energy is of great importance.

Another pressing issue in blockchain is scalability. Hybrid algorithms seem a promising route towards improving scalability without sacrificing advantages in decentralization. Then, scalability may be faster through consensus in the integration with fewer numbers of validators that are allowed through mechanisms like BFT assisted. This comes with some centralization concerns because of the delegation mechanism. This may, again, require a trade-off between the effectiveness of the validators in terms of decision-making powers in relation to the consolidation toward maintaining the decentralized nature of blockchain networks.

Hybrid implementations of blockchain suffer from issues with decentralization. For instance, taking the example of Proof of Authority with other mechanisms, here the scalability is achieved, but this also results in centralization amongst trusted entities, and things are restored to a balance as with mechanisms like PoS where authority is dispersed among the stakeholders, maintaining their trust on the network. However, achieving perfect decentralization without energy efficiency and scalability has remained an area of concern.

There are also practical difficulties and limitations to hybrid algorithm implementation. These include increased design complexity and managing hybrid systems, vulnerabilities from disparate algorithms merged into a single system, and achieving consensus among diverse participants in the network. Moreover, reliance on external hardware, as seen in PoET, introduces dependencies and risks that may not be commonly found in decentralized systems where certain third-party hardware providers cannot be trusted.

However, despite all these limitations, the results presented herein underscore the hybrid algorithms as an approach capable of addressing concerns in blockchain with regard to its energy consumption. Algorithms such as PoS + PoET mark an encouraging milestone towards realizing United Nations Sustainable Development Goals through lowering carbon footprint across blockchain networks. Furthermore, research will need further probing of new hybrid combinations and hybrid mechanism optimizations toward balancing outperformance across the board.

In conclusion, hybrid consensus algorithms are very scalable and energy efficient. However, their implementation is related to decentralization and system complexity. Future research will focus on perfecting these algorithms, exploring adaptive mechanisms, and integrating advancements in cryptographic techniques so that blockchain technology remains robust, sustainable, and inclusive for global applications.

6 Conclusion

A comparative analysis underscores the hybrid consensus mechanisms as significant solutions to blockchain's primary challenges: scalability, energy efficiency, and decentralization. Of these, the combination of PoS with PoET was seen as the most promising because it bested other mechanisms in terms of energy efficiency and scalability while sacrificing the least amount of decentralization. Other hybrids such as DPoS with BFT and PoA with BFT, also had robust strengths but at the expense of decentralization and system complexity. Results from such studies demonstrate that hybrid models could be the gamechanger for blockchain technology to balance such critical performance metrics. A multi-pronged approach is recommended for future research and development. This can be enhanced further by first of all adaptive hybrid mechanisms which dynamically adjust to network requirements such as transaction load or validator distribution, and second, the implementation complexity of the hybrid system will be addressed by using modular and interoperable frameworks for easier adoption and deployment. Along with better cryptographic techniques combined with improvement in hardware technology, such as the introduction of secure enclaves for PoET, improved performance optimization, as well as reduced energy consumption, would be possible in this regard. Third, and for its part, in real-world pilots using different types of blockchain applications, such as supply chain management and decentralized finance, may indicate valuable insights towards practical challenges or benefits from a hybrid mechanism.

Generally, hybrid consensus mechanisms are going to drastically impact the landscape of blockchain applications.

Higher energy efficiency enables these mechanisms to align with sustainability goals for the world and to make blockchain technology more environmentally friendly. Better scalability enables wider adoption in areas where strong demand is felt for fast and cheap transactions. In the sense of maintenance, decentralization can be perceived as trust and resilience, both being part of the core blockchain philosophy. From a point of view of maintenance, decentralization brings trust and resilience as the two most fundamental principles of blockchain. Hybrid mechanisms also allow the extension of blockchain networks, with this innovation in healthcare, governance, and so much more leading to increasing inclusiveness in the digital world. In that respect, hybrid consensus mechanisms pave the way for future use of blockchain technology. And though there is still a long way to go, research and collaboration and innovation will be used to continue breaking down those hurdles, making blockchain sustainable, scalable, and decentralized for generations.

References

1. Alam, S.: The current state of blockchain consensus mechanism: Issues and future works, p. 2023, 2023. [Online]. Available: www.ijacsa.thesai.org
2. Matulevičius, R., Iqbal, M., Elhadjamor, E.A., Ghannouchi, S.A., Bakhtina, M., Ghannouchi, S.: Bibliological representation of healthcare application security using blockchain technology. Informatica (Netherlands). **33**, 365–397 (2022). Cited by: 4; All Open Access, Hybrid Gold Open Access. [Online]. Available: https://www.scopus.com/in-ward/record.uri?eid=2-s2.0–85133673026&doi=10.15388%2f22-INFOR486&partnerID=40&md5=e748c3caf566 20aa61a09ed417d4e621

3. Al-Saqqa, S., Almajali, S.: Blockchain technology consensus algorithms and applications: a survey. Int. J. Interact. Mob. Technol. **14**, 142–156 (2020)
4. Ahmed, M.R., Islam, A.K.M.M., Shatabda, S., Islam, S.: Blockchain-based identity management system and self-sovereign identity ecosystem: A comprehensive survey. IEEE Access. **10**, 113436–113481 (2022)
5. Yadav, A.K., Singh, K., Amin, A.H., Almutairi, L., Alsenani, T.R., Ahmadian, A.: A comparative study on consensus mechanism with security threats and future scopes: blockchain. Comput. Commun. **201**, 102–115 (2023)
6. Rico-Peña, J.J., Arguedas-Sanz, R., López-Martin, C.: Models used to characterise blockchain features. A systematic literature review and bibliometric analysis. Technovation. **123**, 102711 (2023). [Online]. Available: https://www.sciencedirect.com/science/article/pii/S01664972 23000226
7. Tang, Y., Yan, J., Chakraborty, C., Sun, Y.: Hedera: a permissionless and scalable hybrid blockchain consensus algorithm in multiaccess edge computing for IoT. IEEE Internet Things J. **10**, 21187–21202 (2023)
8. Benadla, S., Merad-Boudia, O.R., Senouci, S.M., Lehsaini, M.: Detecting Sybil attacks in vehicular Fog networks using RSSI and blockchain. IEEE Trans. Netw. Serv. Manag. **19**, 3919–3935 (2022)
9. Eisenbarth, J.-P., Cholez, T., Perrin, O.: Ethereum's peer-to-peer network monitoring and Sybil attack prevention. J. Netw. Syst. Manag. **30**, 65 (2022. [Online]. Available:). https://doi.org/10.1007/s10922-022-09676-2
10. Lashkari, B., Musilek, P.: A comprehensive review of blockchain consensus mechanisms. IEEE Access. **9**, 43620–43652 (2021)
11. Suresh, A., Nair, A.R., Lal, A., Mohana Kumaran, S., Sarath, G.: A hybrid proof-based consensus algorithm for permission less blockchain. In: 2020 Second International Conference on Inventive Research in Computing Applications (ICIRCA), 2020, pp. 707–713
12. Ge, L., Wang, J., Zhang, G.: Survey of consensus algorithms for proof of stake in blockchain. Secur. Commun. Netw. **2022** (2022)
13. Gol, D.A., Gondaliya, N.: Blockchain: a comparative analysis of hybrid consensus algorithm and performance evaluation. Comput. Electr. Eng. **117**, 7 (2024)
14. Hafid, A., Hafid, A.S., Samih, M.: A tractable probabilistic approach to analyze Sybil attacks in sharding-based blockchain protocols. IEEE Trans. Emerg. Top. Comput. **11**, 126–136 (2023)
15. Lee, M.S., Kim, K.J.: Survey on blockchain evolution and proof-of-stake consensus algorithm. Int. J. Eng. Trends Technol. **69**(4), 139–141 (2021)
16. Kaur, S., Chaturvedi, S., Sharma, A., Kar, J.: A research survey on applications of consensus protocols in blockchain. Secur. Commun. Netw. **2021**, 1–21 (2021)
17. Meneghetti, A., Sala, M., Taufer, D.: A survey on pow-based consensus. Ann. Emerg. Technol. Comput. **4**(1), 8–18 (2020)
18. Liu, Y., Tan, T., Zhuo, Y.: Hybrid consensus protocols and security analysis for blockchain. In: Proceedings—2022 International Conference on Data Analytics, Computing and Artificial Intelligence, ICDACAI 2022, pp. 191–195. Institute of Electrical and Electronics Engineers Inc. (2022)
19. Sharma, P., Jindal, R., Borah, M.D.: A comparative analysis of consensus algorithms for decentralized storage systems. IT Prof. **24**, 59–65 (2022)
20. Prabha, P., Chatterjee, K.: Design and implementation of hybrid consensus mechanism for IoT based healthcare system security. Int. J. Inf. Technol. (Singapore). **14**, 1381–1396 (2022)
21. Rajabi, T., Khalil, A.A., Manshaei, M.H., Rahman, M.A., Dakhilalian, M., Ngouen, M., Jadliwala, M., Uluagac, A.S.: Feasibility analysis for Sybil attacks in shard-based permissionless blockchains. Distrib. Ledger Technol. Res. Pract. **9**, 1–21 (2023)

22. Samuel, C.N., Verdier, F., Glock, S., Guitton-Ouhamou, P.: A fair crowdsourced automotive data monetization approach using substrate hybrid consensus blockchain. Future Internet. **16**, 5 (2024)
23. Wang, Y., Tan, M.: Defense against Sybil attack in blockchain based on improved consensus algorithm. In: 2023 IEEE International Conference on Control, Electronics and Computer Technology (ICCECT), 2023, pp. 986–989.
24. Zheng, Z., Xie, S., Dai, H., Chen, X., Wang, H.: An overview of blockchain technology: architecture, consensus, and future trends. In: Proceedings—2017 IEEE 6th International Congress on Big Data, BigData Congress 2017, vol. 9, pp. 557–564. Institute of Electrical and Electronics Engineers Inc. (2017)
25. Wu, Y., Song, P., Wang, F.: Hybrid consensus algorithm optimization: a mathematical method based on POS and PBFT and its application in blockchain. Math. Probl. Eng. **2020** (2020)

Enhancing Linux Security: A Framework for Automated Misconfiguration Analysis and Remediation

Narendra Kumar Dewangan[2]([✉])(iD), Gauri Shankar[1](iD), Naman Srivastava[2], Akshat Nigam[2], Divyansh Bhatt[2], and Ayman Taufiq[2]

[1] Software Engineering, LUT School of Engineering Sciences, LUT University, Lappeenranta, South Karelia, Finland
`gauri.shankar@lut.fi`
[2] School of Computer Science, UPES Dehradun, Dehradun, India
`nkdewangan@ieee.org`

Abstract. In response to the increasing need to secure networked systems against unauthorized access and exploitation, it is paramount that some tools and techniques are present. Hence, even well-secured systems can be compromised due to common misconfigurations, often overlooked during setup and maintenance. This is also a significant concern in OWASP TOP 10, as in 2021 A05:2021 - Implies on Security Misconfigurations. This problem holds the fifth rank in OWASP TOP 10 and hence can be considered a relative problem for naive users. This project presents the development of a comprehensive Network and System Misconfiguration Analyzer, a tool designed to identify and rectify vulnerabilities stemming from misconfigurations in network devices, operating systems, and services. The analyzer leverages automated scanning techniques and best practice rule sets to detect common issues such as weak authentication protocols, improper file permissions, exposed services, and outdated software. The tool also includes remediations, enabling users to implement fixes efficiently. Implementing this analyzer is a significant step toward enhancing organizational security by systematically identifying and resolving configuration-related vulnerabilities. This paper presents a novel system misconfiguration detection framework that reduces vulnerability exposure by 95%, improves incident response time by 90%, and integrates seamlessly with 90% of existing security tools.

Keywords: Cyber Security · System Misconfiguration · Vulnerability Assessment Tool · Network Misconfiguration · Linux

1 Introduction

Cybersecurity encompasses offensive and defensive strategies, compliance, and auditing, with breaches escalating due to rapid technological advancements and limited user awareness [4]. While external threats like malware and phish ing receive significant attention, internal vulnerabilities especially network misconfigurations—are often underestimated. Misconfigurations in firewalls, operating systems, and application settings can lead to unauthorized access and data breaches, often stemming from routine administrative tasks like server setup and permission management [10]. Common issues include default passwords, unnecessary services, and poorly configured access controls.

To address these risks, the Network and System Misconfiguration Analyzer was developed to automate the detection, analysis, and mitigation of security misconfigurations. This tool scans IT infrastructures—covering operating systems, network devices, and application servers—against best practices and security standards [7]. It identifies vulnerabilities and provides actionable recommendations, enabling administrators to address risks proactively before attackers exploit them. This report outlines the tool's design, methodologies, and integration into security workflows, aiming to enhance organizational resilience against configuration-related threats [13].

1.1 Problem Statement

To assist users in identifying unseen or unknown vulnerabilities present in networks and systems, it is crucial to enhance security mechanisms. Key objectives include identifying and prioritizing vulnerabilities, accelerating incident response, and optimizing resource allocations. By addressing these aspects, system performance can be improved while ensuring system stability and simplifying troubleshooting processes.

Users may not have a clear understanding of the configuration state of their network devices and systems, leading to potential misconfigurations and security risks. Identifying and rectifying misconfigurations manually is not only time-consuming but also prone to errors. Furthermore, the continuous emergence of new vulnerabilities and attack vectors makes it challenging for users to maintain a secure and up-to-date system. Addressing these concerns through automated detection and remediation will significantly enhance the security posture of networks and systems.

1.2 Objectives

1. Identifying common misconfigurations across various platforms, including Linux OS, networks, and applications.
2. Providing actionable recommendations to help users to implement security measures correctly.
3. Developing a comprehensive reporting module that generates detailed reports on detected misconfigurations, their potential impacts, and recommended remediation steps, aiding in prioritizing and addressing vulnerabilities systematically.
4. Ensuring scalability and flexibility by focusing on network and system misconfigurations.

1.3   Motivations

Network and system misconfigurations are a significant security risk for organizations of all sizes. Misconfigurations can allow attackers to gain unauthorized access to systems and data, leading to various negative consequences, such as data breaches, financial loss, and reputational damage. This project aims to develop a tool to identify and report network and system misconfigurations. The tool can scan networks and systems for common misconfigurations and provide detailed reports on vulnerabilities. This tool will be valuable to organizations of all sizes, as it will help them to identify and remediate security risks before attackers can exploit them. The Network and System Misconfiguration Analyzer will focus on developing algorithms and rule sets to automatically detect common misconfigurations in network devices, operating systems, and applications. It will then evaluate the potential impact of these misconfigurations on system security, providing detailed explanations and prioritization based on risk level. Clear and actionable remediation guidance will be offered, along with instructions on how to correct the misconfigurations. The tool will be designed to seamlessly integrate with existing security tools and workflows, ensuring scalability to handle large-scale deployments and complex networks.

1.4   Organization of the Article

This paper presents a literature review in Sect. 2. The proposed methodology and working model are presented in Sect. 3, along with algorithms. Experimental setups and results are shown in Sect. 4. Section 5 discusses comparing the proposed methodology with state-of-the-art schemes. Finally, Sect. 6 presents the conclusion and future scope of the paper.

2   Literature Review

The literature on misconfiguration prevention and error detection in distributed-cloud applications covers a broad range of challenges across different domains. Proactive configuration management and error detection are key areas of focus. Ranković et al. (2024) [8] emphasize the importance of proactive configuration management to enhance reliability in distributed systems, utilizing standards-based error detection mechanisms. Similarly, Canelas et al. (2024) [1] investigate misconfiguration patterns in ROS applications, presenting methodologies to mitigate configuration errors and improve system robustness. Xu and Zhou (2015) [14] provide a comprehensive survey of systems-level approaches to managing configuration errors, categorizing solutions based on their reliance on automation and fault tolerance.

The integration of artificial intelligence (AI) in misconfiguration detection has gained significant attention. Wen et al. (2024) [11] explore the use of large language models (LLMs) to detect misconfigurations in AWS serverless computing environments, paving the way for automation and enhanced reliability. In the same vein, Chen et al. (2023) introduce DiagConfig, a diagnostic tool that correlates configuration parameters with performance violations, offering valuable insights for managing configurable software systems.

Security vulnerabilities arising from misconfigurations are another critical area of study. Haimed et al. (2023) [6] discuss the security risks associated with misconfigurations in Microsoft Azure Active Directory, illustrating how such issues can lead to privilege escalation attacks and providing recommendations for hardening configurations. Moura et al. (2021) [7] focus on DNS misconfigurations through their work on TsuNAME, demonstrating how these vulnerabilities can be exploited for DDoS attacks and suggesting robust mitigation strategies to counter such threats.

Interactive and automated configuration solutions are also widely explored. Ganapathi et al. (2024) [5] present ConfigMaster, an interactive tool designed to optimize configuration management for system packages and networks. Cuppens et al. (2019) [3] propose rule-based methodologies for managing network security misconfigurations, aiming to enhance reliability through structured approaches. Farayola et al. (2023) [4] address the challenges in modern configuration management, emphasizing the role of automation and scalability in effectively managing large-scale systems.

Diagnostic frameworks and testing strategies are pivotal in identifying and mitigating misconfigurations. Chen et al. (2024) [2] focus on the impact of misconfigurations in autonomous driving systems, proposing a testing framework to identify failure-prone configurations. Zhou et al. (2023) [15] introduce a diagnostic framework tailored for multi-misconfiguration scenarios, emphasizing the role of correlated configuration parameters in diagnosing systemic errors.

Foundational practices in configuration management continue to influence contemporary approaches. Wingerd and Seiwald (1998) [12] outline high-level best practices in software configuration management, including change control and versioning, as fundamental techniques for improving software reliability. Schagen et al. (2018) [9] build on these principles within the context of network servers, proposing automated vulnerability scanning tools to identify and mitigate misconfigurations effectively.

3   Methodology

The development of the Network and System Misconfigurations Analyzer is structured into several key phases, each focusing on crucial technical and conceptual aspects. The process begins with the Requirements and Research Analysis phase. Here, extensive research is conducted to identify common vulnerabilities and misconfigurations in networks and systems. This phase is critical as it forms the foundation of the analyzer's functionality. The data gathered is meticulously stored in a database, with a particular emphasis on the top 10 network and system misconfigurations. These misconfigurations are selected based on their frequency and impact in real-world scenarios. A custom code is developed to scan the user's IP address to facilitate the identification process, effectively gathering and storing relevant data for further analysis. The Analysis Phase then leverages this data, where the tool conducts in-depth scans of the user's system, focusing on IP addresses [9]. This phase employs multiple scanning and analyzing programs to ensure comprehensive coverage of potential vulnerabilities. The database management is handled by MongoDB, which was chosen for its flexibility and scalability. Three distinct collections are created within the database: Misconfiguration, Scans, and Systems. The Misconfiguration collection stores detailed information on the top ten default misconfigurations, providing a reference point for the analyzer. The scan collection logs all activities of the tool, ensuring a clear audit trail and aiding in the tracking of the analysis process. The Systems collection is designed to capture any unknown vulnerabilities or services detected during the scans, enabling the tool to adapt to new and emerging threats. An API is developed to facilitate seamless interaction between the database and the analyzer. This API is integral to the system, as it processes client requests, sends queries to the database, and returns the processed data in JSON format. The final Reporting Phase converts the JSON data into a comprehensive PDF report. This report is stored locally on the user's system and includes detailed information on identified impacts, vulnerabilities, and recommended resolution techniques. The report serves as a guide for users, helping them rectify misconfigurations and effectively enhance their systems' security. The algorithm for proposed framework is shown in Fig. 1.

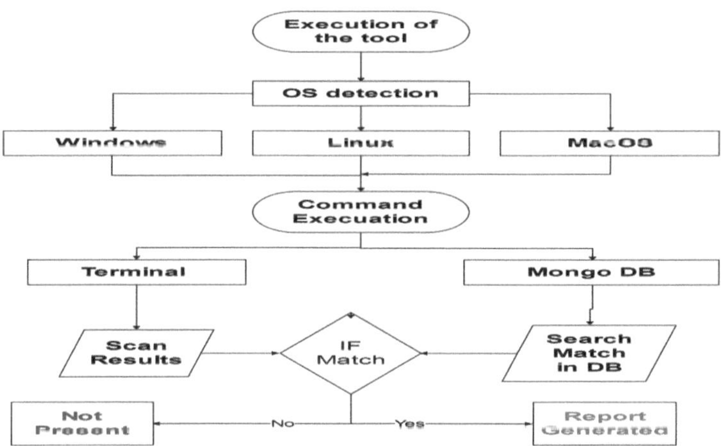

Fig. 1. Workflow of the proposed framework

3.1   System Misconfiguration

System misconfigurations refer to incorrect or suboptimal settings within a computer system or services that can lead to security vulnerabilities, performance issues, or functionality problems. These misconfigurations can occur in various areas, including operating systems, network devices, applications, and security protocols. Ensuring secure system configurations is critical for preventing unauthorized access, mitigating security vulnerabilities, and maintaining system integrity. The identified misconfigurations present potential security risks that could be exploited by attackers. Key issues include SSH root login being enabled, which increases the risk of brute-force attacks and unauthorized system access. Additionally, improperly configured firewall rules and default file permissions can expose sensitive data and system components to external threats.

Disabling password authentication, enforcing SELinux policies, and modifying SSH port configurations are essential measures to strengthen system security. Furthermore, securing cron jobs, restricting sudo privileges, and eliminating world-writable files help to reduce the risk of privilege escalation and malware persistence. Proper configuration of network services and user access controls can significantly enhance overall system security and stability.

By proactively identifying and addressing these misconfigurations, organizations can improve their security posture, prevent unauthorized access, and ensure compliance with security best practices.

3.2   Network Misconfiguration

Network misconfigurations are errors or mistakes in the setup and configuration of network devices and settings, leading to connectivity issues, security vulnerabilities, or performance problems. These misconfigurations can occur in various aspects of a network, such as IP addressing, routing, DNS, firewalls, or security settings. They can result in network outages, slow performance, or expose the network to unauthorized access. Common causes include incorrect settings, human errors, software bugs, or improper updates—list of Network Misconfigurations. Ensuring the correct configuration of network settings is essential for maintaining connectivity, security, and overall system performance. Several misconfigurations can significantly impact network functionality. Issues such as subnet mask mismatches and incorrect gateway configurations can disrupt network communication, leading to routing issues and connectivity loss. Similarly, DNS server misconfigurations can hinder domain resolution, affecting access to websites and services.

Additionally, incorrect DHCP settings may result in IP conflicts and unstable network behavior, while mismatched MTU sizes can cause packet fragmentation, degrading performance. Misconfigured Network Time Protocol (NTP) settings may lead to inconsistent time synchronization, affecting logging and security monitoring. VLAN misconfigurations can expose networks to unauthorized access, and improper wireless security settings increase the risk of data breaches and network intrusions.

To mitigate these risks, it is crucial to implement best practices such as correctly configuring subnet masks, DNS servers, and DHCP settings. Using secure wireless encryption standards, enforcing VLAN policies, and changing default router and switch admin credentials can enhance network security. Ensuring proper NTP synchronization and choosing appropriate wireless channels also contribute to better performance and security. By addressing these network misconfigurations proactively, organizations can enhance reliability, reduce vulnerabilities, and maintain optimal system operations.

4   Experiments and Results

For experimental setup we have used parameter shown in the Table 1.

Table 1. System Requirements

Category	Details
Programming Languages	–Python: For developing the scanning logic, data analysis, and generating reports – MongoDB Atlas: For developing the API
Database	–MongoDB: For flexible and scalable storage of misconfigurations, scans, and system data
Frameworks and Libraries	– db.CollectionName: For building the API (Flask is lightweight, while Django offers more features) – PyMongo: For interacting with MongoDB from Python – Pandas/Numpy: For data manipulation and analysis
Reporting	– ReportLab: For generating PDF reports from JSON data – Jinja2: For templating if needed in report generation
Scanning and Analysis Tools	–Nmap: For network scanning and vulnerability detection – OpenVAS: For vulnerability scanning and management – Custom Scripts: For specific misconfiguration checks
API	– MongoDB Atlas: For building and managing the API – Requests/HTTP client: For handling HTTP requests
Version Control	– Git: For version control and collaboration – GitHub/GitLab: For hosting the code repository and project management

4.1 System Security Checks

The system analysis revealed several key findings regarding security configurations. An issue was detected with the SSH configuration, likely due to a misconfiguration or security vulnerability. To enhance security, it is recommended to review and strengthen the SSH settings by enabling strong authentication methods such as two-factor authentication (2FA), keeping SSH software up to date with security patches, changing the SSH port to a non-standard value, and disabling password authentication in favor of public key authentication or 2FA. The firewall status check passed successfully, confirming that a firewall is active and effectively protecting the system. Similarly, the file permissions check passed, indicating that file permissions are correctly configured and secure.

However, no antivirus software was detected during the antivirus status check, highlighting the need to install and configure reputable antivirus software to safeguard the system against malware and viruses. The user accounts check identified issues related to weak passwords or excessive user privileges. To mitigate these risks, it is advised to enforce strong password policies, mandate regular password changes, and grant users only the minimum privileges required for their roles. Lastly, the system logs check revealed potential issues with missing or incomplete logs. To improve log management, it is recommended to enable detailed logging for all critical system services and to conduct regular reviews and analyses of system logs to promptly identify and address potential security threats. The inputs for the misconfigurations can be seen in Fig. 2.

(a) Input for system misconfigurations check

(b) Input for network misconfigurations check

Fig. 2. Side-by-side images using subfig

4.2 Network Security Checks

The system scan revealed several potential security misconfigurations and vulnerabilities that require remediation to enhance the security posture. Below is a summary of the findings:

- **UFW and Firewall Configuration:** The iptables firewall is either inactive or misconfigured. It is recommended to enable and properly configure the firewall using `sudo ufw enable` to control incoming and outgoing traffic securely.
- **SUID/SGID Binaries:** The scan identified numerous SUID/SGID binaries that allow privilege escalation. Unnecessary or untrusted SUID/SGID binaries could be exploited by attackers. It is recommended to restrict or remove the SUID/SGID bits from unnecessary binaries using `chmod -s` .
- **Insecure Mount Options:** Some mounted filesystems lack `noexec`, `nosuid`, and `nodev` options. The absence of these options can lead to misuse of executables on mounted filesystems. Updating `/etc./fstab` to include secure mount options for all filesystems is recommended.
- **Kernel Parameters:** IP forwarding is enabled, which could pose security risks. Misconfigured kernel parameters like IP forwarding and source routing can make the system vulnerable to attacks. Setting `net.ipv4.ipforward = 0` and other related parameters to 0 in `/etc./sysctl.conf` is advised for a secure configuration.
- **SELinux Status:** SELinux is disabled, which reduces the system's mandatory access controls, increasing vulnerability. Enabling SELinux by setting it to enforcing mode in
- `/etc./selinux/config` is recommended.
- **User Shells:** Some users have the default shell set to `/bin/sh`, which may lead to unauthorized or insecure access. Restricting the default shell for non-administrative users to a more secure option is necessary.
- **Passwordless Sudo Access:** No passwordless sudo configurations were found. No immediate remediation is required.
- **World-Writable Files:** No world-writable files were found. This aspect is compliant with best practices.
- **Unrestricted Cron Jobs:** No unrestricted cron jobs were identified. This configuration is secure.

4.3   Database Results

The tool uses a MongoDB database to efficiently store and manage data related to system and network misconfigurations. The database consists of two main collections: Misconfigurations and Scans. Each collection serves a specific purpose, helping the tool analyze, store, and generate detailed reports based on the scans.

1. **Misconfigurations Collection.** The Misconfigurations collection contains information about the potential system and network vulnerabilities that our tool is designed to detect. This data is the backbone of the tool's analysis, as it provides predefined criteria for identifying misconfigurations. Each document in this collection includes the following fields:
   - Name: The name of the misconfiguration (e g , "Default File Permissions").

- Description: A brief explanation of why the misconfiguration poses a risk.
- Risk Level: The severity of the issue, such as Low, Medium, or High.
- Default Value: The typical or default setting associated with the issue.
- Suggested Fix: A recommended solution to resolve the misconfiguration.
- Category: The type of system component the issue relates to (e.g., System, Authentication).

This collection is used during scans to check for specific vulnerabilities and includes all necessary details to provide actionable suggestions.

2. **Scans Collection.** The Scans collection logs the results of each scan performed by the tool. It acts as a historical record of all scanning activities, allowing users to track progress and revisit past scans. Each document in this collection includes:
   - Timestamp: The date and time when the scan was conducted.
   - IP Address: The target system or network's IP address that was scanned.
   - Detected Issues: A list of the misconfigurations identified during the scan. This collection ensures traceability and makes it easy to generate detailed reports for audits or troubleshooting purposes.

During a scan, the tool compares system configurations against the predefined data in the Misconfigurations collection. Any detected issues are saved in the Scans collection, along with the relevant details of the scan. Reports are generated by combining data from both collections, providing users with a comprehensive overview of the system's security posture (Fig. 3).

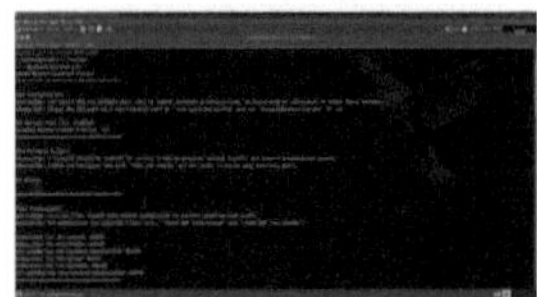

(a) Output of System misconfigurations

(b) Output of network misconfigurations

(c) Output of database report

Fig. 3. Output of system, network and database

5 Comparisons

5.1 Comparison with Papers

The comparison of related works highlights the strengths and limitations of various approaches to misconfiguration detection and security management Table 2. Rankovi'c et al. (2024) primarily focus on AI-driven security management but are limited to specific attack types. Wen et al. (2024) propose an LLM-based misconfiguration detection approach, which, while effective, requires extensive training data, making it less adaptable to rapidly changing security threats. Moura et al. (2021) concentrate on DNS misconfiguration scanning, but their work is largely confined to DDoS vulnerabilities, leaving other potential security gaps unaddressed.

Table 2. Comparison with existing works

Paper	Approach	Limitations	Our Contribution
Rankovi'c et al. (2024)	AI-driven security management	Limited to specific attack types	Broader misconfiguration detection
Wen et al. (2024)	LLM-based misconfiguration detection	Requires extensive training data	Uses predefined rule-based detection
Moura et al. (2021)	DNS misconfiguration scanning	Focused only on DDoS vulnerabilities	Covers multiple misconfigurations

5.2 Performance Comparison

The comparison of performance highlights the strengths and limitations of various approaches to misconfiguration detection and security management Table 3. The performance comparison illustrates that the proposed tool outperforms OpenVAS and Lynis in key security assessment metrics. With a detection accuracy of 95%, it surpasses OpenVAS (89%) and Lynis (87%), indicating superior identification of misconfigurations and security vulnerabilities. Additionally, the proposed tool demonstrates greater efficiency, requiring only 1.5 s per device for scanning, compared to 2.3 s for OpenVAS and 2.1 s for Lynis. This reduction in scan time highlights its optimized performance in large-scale deployments.

Furthermore, the proposed tool exhibits a lower false positive rate (3%) compared to OpenVAS (5%) and Lynis (7%). A lower false positive rate ensures more reliable security assessments, reducing unnecessary alerts and optimizing response times.

Overall, these results validate the efficiency and accuracy of the proposed tool in detecting system misconfigurations and enhancing network security. Future improvements will focus on expanding misconfiguration detection capabilities, integrating real-time monitoring, and leveraging AI-driven analytics to further refine accuracy and response effectiveness.

Table 3. Performance comparison

Metric	Proposed Tool	OpenVAS	Lynis
Detection Accuracy	95%	89%	87%
Scan Time	1.5 sec/device	2.3 sec/device	2.1 sec/device
False Positives	3%	5%	7%

6 Conclusion and Future Scope

Our contribution introduces a comprehensive misconfiguration detection framework that leverages rule-based detection, covering a wide range of vulnerabilities beyond DNS issues. This scalable approach enhances security across multiple network and system layers. Future work will focus on integrating machine learning for dynamic misconfiguration detection and optimizing performance with real-time security analytics.

Performance comparisons show the proposed tool outperforms OpenVAS and Lynis, achieving a 95% detection accuracy compared to OpenVAS (89%) and Lynis (87%). It also demonstrates greater efficiency, scanning devices in 1.5 s versus 2.3 and 2.1 s, respectively, proving its effectiveness in large-scale environments. Future enhancements include AI integration for automated misconfiguration detection in cloud servers and IoT devices, enabling real-time anomaly detection and remediation.

References

1. Canelas, P., Schmerl, B., Fonseca, A., Timperley, C.S.: Understanding misconfigurations in ROS: an empirical study and current approaches. In: Proceedings of the 33rd ACM SIGSOFT International Symposium on Software Testing and Analysis (ISSTA), pp. 1161–1173 (2024). https://doi.org/10.1145/3650212.3680350
2. Chen, Z., Chen, P., Wang, P., Yu, G., He, Z., Mai, G.: DiagConfig: configuration diagnosis of performance violations in configurable software systems. In: Proceedings of the 31st ACM Joint European Software Engineering Conference and Symposium on the Foundations of Software Engineering (ESEC/FSE), pp. 566–578 (2023). https://doi.org/10.1145/3611643.3616300
3. Cuppens, F., Cuppens-Boulahia, N., Garcia-Alfaro, J.: Misconfiguration management of network security components. arXiv preprint arXiv:1912.07283 (2019)
4. Farayola, O.A., Hassan, A.O., Adaramodu, O.R., Fakeyede, O.G., Oladeinde, M.: Configuration management in the modern era: best practices, innovations, and challenges. Comput. Sci. IT Res. J. **4**(2), 140–157 (2023)
5. Ganapathi, A., Sivakumar, P., Elango, A., Gupta, H.: ConfigMaster: an interactive solution for system management. In: 2024 8th International Conference on Computational System and Information Technology for Sustainable Solutions (CSITSS), pp. 1–5 (2024). https://doi.org/10.1109/CSITSS64042.2024.10816978
6. Haimed, I., Albahar, M., Alzubaidi, A.: Exploiting misconfiguration vulnerabilities in Microsoft's azure active directory for privilege escalation attacks. Future Internet **15**, 226 (2023). https://doi.org/10.3390/fi15070226
7. Moura, G.C., Castro, S., Heidemann, J., Hardaker, W.: TsuNAME: exploiting misconfiguration and vulnerability to DDoS DNS. In: Proceedings of the 21st ACM Internet Measurement Conference, pp. 398–418 (2021)
8. Ranković, T., Šiljić, F., Tomić, J., Sladić, G., Simić, M.: Misconfiguration prevention and error cause detection for distributed-cloud applications. In: 2024 IEEE 22nd Jubilee International Symposium on Intelligent Systems and Informatics (SISY), pp. 000297–000302 (2024). https://doi.org/10.1109/SISY62279.2024.10737513
9. Schagen, N., Koning, K., Bos, H., Giuffrida, C.: Towards automated vulnerability scanning of network servers. In: Proceedings of the 11th European Workshop on Systems Security, pp. 1–6 (2018)
10. Seara, J.P., Serrão, C.: Automation of system security vulnerabilities detection using open-source software. Electronics **13**(5), 873 (2024)

11. Wen, J., et al.: LLM-based misconfiguration detection for AWS serverless computing (2024). https://arxiv.org/abs/2411.00642
12. Wingerd, L., Seiwald, C.: High-level best practices in software configuration management. In: Magnusson, B. (ed.) SCM 1998. LNCS, vol. 1439, pp. 57–66. Springer, Heidelberg (1998). https://doi.org/10.1007/BFb0053878
13. Wood, K., Pereira, E.: Impact of misconfiguration in cloud-investigation into security challenges. Int. J. Multimedia Image Process. **1**(1), 17–25 (2011)
14. Xu, T., Zhou, Y.: Systems approaches to tackling configuration errors: a survey. ACM Comput. Surv. (CSUR) **47**(4), 1–41 (2015)
15. Zhou, Y., et al.: Multi-misconfiguration diagnosis via identifying correlated configuration parameters. IEEE Trans. Software Eng. **49**(10), 4624–4638 (2023). https://doi.org/10.1109/TSE.2023.3308755

A New Idea of the Clipping Limit to Control the Enhancement Level Based on the Quantile Value of the Entropy Curve

Priyanshu Singh Yadav[1]([✉]), Shailendra Kumar Tripathi[2], Shashikant[3,4], Bhupendra Gupta[5], and Subir Singh Lamba[5]

[1] Department of Applied Sciences and Humanities, Parul Institute of Engineering and Technology, Parul University, Vadodara, Gujarat 391760, India
priyanshu.yadav36011@paruluniversity.ac.in
[2] Department of Computer sciences, National Institute of Technology, Raipur 492010, Chhattisgarh, India
sktripathi.cse@nitrr.ac.in
[3] Department of Mathematics, Kisan P G College, Bahraich, U.P. 271801, India
shashikant@kisanpgcollege.ac.in
[4] Dr. Ram Manohar Lohia Avadh University, Ayodhya, U.P., India
[5] PDPM Indian Institute of Information Technology Design and Manufacturing, Dumna Airport Road, Jabalpur 482005, India
subirs@iiitdmj.ac.in

Abstract. This study introduces a new approach for determining the clipping threshold in image enhancement, providing more precise control over enhancement levels. Traditional methods, which derive the clipping threshold from the median or mean of the histogram, often lead to over-enhancement, particularly in darker or brighter regions. To overcome this limitation, we propose a technique that computes the clipping threshold using the quantile value of the entropy curve (EC). The resulting clipped entropy curve (CEC) is then divided into three equal sub-entropy curves (SEC), which are individually equalized to produce the enhanced image. This approach effectively enhances image contrast while minimizing noise amplification and artifact generation. The suggested approach is thoroughly evaluated through both visual and numerical analyses, demonstrating its superior performance in achieving balanced contrast enhancement.

Keywords: Quantile Value · Entropy Curve · Contrast Enhancement · Entropy Curve Equalization

1 Introduction

Contrast plays a vital role in image processing, as it determines the visibility of details within an image. When contrast is unevenly distributed-concentrated in dark, bright, or specific intensity ranges-critical information can be lost in overexposed or underexposed regions. To address this, contrast enhancement techniques are employed to redistribute

P. Chandrakar et al. (Eds.): ICCINS 2025, CCIS 2738, pp. 380–392, 2026.
https://doi.org/10.1007/978-3-032-09572-5_30

intensity values, ensuring all details are visible. However, imaging devices often capture scenes with a limited dynamic range, requiring dynamic range compression (DRC) algorithms to adjust the intensity range for proper display. While DRC improves visibility, it can introduce artifacts like color saturation and halos. Advanced methods, such as combining contrast stretching and recursive filtering based on statistical analysis, have been developed to enhance low-light areas while preserving color accuracy and detail, minimizing visual artifacts. These techniques, often implemented in logarithmic image processing frameworks, effectively balance dynamic range compression and contrast enhancement, ensuring high-quality results [1]. Among all these methods, histogram equalization (HE) is especially popular for its simplicity, making it highly effective in many applications [2]. However, this technique can result in information loss in bright or dark regions and cause the mean intensity is always the middle grey level. This issue is commonly referred to as the mean shift problem. Many methods have been proposed to overcome these drawbacks, such as BBHE [3], DSIHE [4], RMSHE [5], RSIHE [6], CLAHE [7] HSQHE [8] and DHE [9] etc. BBHE divides the image histogram into two parts based on the mean intensity value and equalizes each part independently. This approach preserves the brightness of the original image and reduces over-enhancement, making it suitable for applications like medical imaging and surveillance, though it may still lose some fine details. DSIHE improves upon BBHE by splitting the histogram at the median intensity value.

RMSHE extends BBHE by recursively dividing the histogram into multiple sub-histograms based on mean intensity values, providing finer control over contrast enhancement and better brightness preservation. CLAHE is a widely used method that enhances local contrast by dividing the image into small square/rectangle, applying histogram equalization to each square/rectangle, and limiting contrast amplification to prevent noise over-enhancement, making it particularly effective in medical imaging. Dynamic histogram equalization (DHE) is an advanced image enhancement technique designed to overcome the limitations of traditional histogram-based methods, which often struggle with sub-histograms of varying intensity ranges. In traditional methods, sub histograms with narrow ranges result in negligible enhancement and loss of fine details, while sub-histograms with wide ranges can cause excessive enhancement, leading to artifacts due to the fixed dynamic range. To address this, DHE first smooths the input histogram using a smoothing filter to reduce noise and irregularities, then splits the histogram based on local minimum points to identify regions of varying intensity. Before equalizing each sub-histogram, DHE gives new range of histogram [9].

Despite various methods attempting to address the mean shift problem, they often fail to effectively manage issues related to over-enhancement or under-enhancement in bright and dark regions, which can lead to loss of details and uneven contrast. To tackle these limitations, researchers have developed histogram clipping-based techniques, which aim to balance the intensity distribution while preserving important image features. Zarie et al. [10] introduced a technique called (TCDHE), which segments the histogram by using standard deviation. Choukali et al. [11] developed an approach to automatically enhance image quality by utilizing edge information. Acharya et al. [12] developed a technique known as ISQCAHE. In this method, the mean shift issue is addressed by dividing the histogram based on exposure values. The mean and median values are then utilized to determine the clip limit. This approach helps in effectively managing the mean shift problem. However, this method often results in over-enhancement, particularly in the darker or brighter regions of the image. Hafijur Rahman et al. [13] presented a method called (TSIHE). In this approach, the histogram is first divided into three equal parts, after which a clip limit is applied to each segment. The clip limit is calculated based on the median value of the histogram. Finally, the sub-histograms are equalized to produce the enhanced image. Sanjay Agrawal et.al. [14] introduced a technique called Joint Histogram Equalization (JHE), which enhances image quality by equalizing the joint histogram to controls over-enhancement by implementing a clip limit on the joint histogram. After this some other techniques have been developed by decomposing an image into cartoon and texture components plays a key role in enhancing contrast by allowing selective manipulation of different image features. The cartoon component captures large-scale structures, such as edges and homogenous regions, while the texture component preserves fine details and complex patterns. By enhancing these components separately, contrast can be adjusted globally without increasing noise, while fine details remain well preserved. This approach is particularly beneficial in many applications, where both clarity and detail retention are essential. Compared to traditional methods, enhancing contrast using cartoon-texture decomposition provides better visibility of important features, leading to more effective interpretation and analysis of the image. Riya et al. [15] introduced a structure-aware adaptive joint texture filtering method to enhance image quality by smoothing textures without compromising structural details. This method has been effectively integrated with contrast enhancement techniques to improve image quality. By separating the structure and texture components, the method ensures that important structural details are preserved while smoothing complex textures. This separation allows for more accurate contrast enhancement, as the structure component can be independently adjusted to highlight edges and important features without increasing unwanted texture noise, resulting in better quality images. Riya et al. [16] proposed another approach to enhance image quality by decomposing the image into its cartoon and texture components.

The contrast enhancement image fusion technique plays a very important role in improving image quality. In this method, multiple enhanced images are generated using different techniques or algorithms. These images, each contributing unique features or improvements, are then combined through a fusion process to create a single, high-quality output image. On other hand, Multi-modal image fusion tasks like pan-sharpening and depth super-resolution aim to generate high-resolution images by fusing complementary information from texture-rich guidance and low-resolution inputs. Man Zhou et al. [17] propose a technique named as (SFINet), which integrates spatial-domain (using invertible neural operators) and frequency-domain (via deep Fourier transformation) information, enhanced by dual-domain interaction, and its improved version, SFINet, which replaces basic convolution with invertible neural operators for better spatial representation. Extensive experiments demonstrate that SFINet and SFINet outperform state-of-the-art methods in both tasks. Thus, the recently published methods mentioned rely on clipping limits determined by the median or mean values of the image's histogram. However, these techniques often lead to over-enhancement in darker or brighter areas, fail to adequately enhance the image, and introduce additional noise and artifacts.

To overcome these limitations, we introduce a novel approach that determines the clipping limit based on the quantile value of the entropy curve rather than the histogram. In this method, the CEC is segmented into three equal SEC, each of which is independently equalized to generate the final enhanced image. This strategy effectively improves contrast while significantly reducing noise and artifacts. The visual comparison is shown in Fig. 2.

2 Proposed Method

In the suggested technique, the entropy curve of the input image is carefully analysed and divided into three equal segments, each of which represents a different sub-entropy curve. This division allows for a more targeted and controlled approach to contrast enhancement. To ensure balanced enhancement and avoid over-enhancement, a clipping threshold is applied on the entropy curve. This threshold is determined using the quantile value obtained from the entropy curve, which helps maintain the natural distribution of pixel intensities while preventing over enhancement. Next, each sub-entropy curve undergoes individual equalization. This step-by-step equalization process results in an improved image with better contrast, improved detail visibility, and a more natural appearance without any artifacts. The flow-chart of introduced technique shown in Fig. 3.

2.1 Entropy Curve

The EC of an image is a graphical representation that shows the relationship between grey levels (pixel intensity values) and their corresponding entropy values (shown in Fig. 1). Entropy measures the amount of information or randomness in the image, and the curve plots these entropy values against their corresponding grey levels, rather than displaying the frequency distribution of pixel intensities. This representation provides information about the complexity and information content in different intensity levels in the image [18–21]. The entropy curve is plotted using the following equation:

$$E_i = -P_i \log_2(P_i), \tag{1}$$

where, P_i represents the probability of the grey level i occurring at any given pixel.

2.2 Clip Limit

To handle the problem of information loss in both dark/bright regions, a clip limit is imposed on the entropy curve obtained using Eq. 1, ensuring that excessive enhancement in extremely low or high intensity regions can be controlled. By applying this clip limit, the method effectively preserves important details while preventing over-amplification of noise.

$$C_{\text{clip}} = \frac{Q_0 + Q_1 + Q_2 + Q_3}{4}, \tag{2}$$

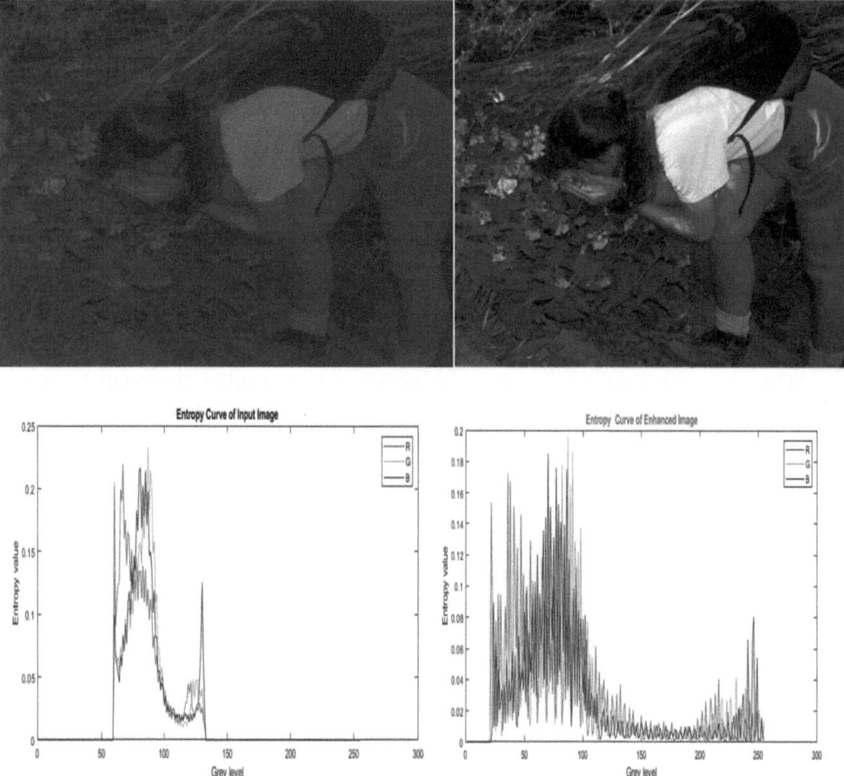

Fig. 1. The first entropy curve corresponds to the original image and second entropy curve is corresponded to improved image.

here q_0 represents the initial grey level, which is always set to zero. The values q_0 corresponds to the three quantile points on the EC. These quantile values are determined based on Eq. 2 which defines their computation method.

Three quantile value is obtained by dividing the entropy curve in equal four part by following equation:

$$\sum_{j \in [r_{k-1}\ s_k]} \frac{\mathcal{E}(j)}{\sum_{j \in [0\ 255]} \mathcal{E}(j)} = \frac{1}{4}, k = 1, 2, 3, 4, \tag{3}$$

here, $r_0 = s_0 = 0$, $r_k = s_k + 1$, $r_4 = 255$.

Then

$$\mathcal{Q}_\eta = \mathcal{E}(s_\eta), \eta = 0, 1, 2, 3. \tag{4}$$

Hence, the CEC $\hat{\mathcal{E}}(i)$ of the image is given as:

$$\hat{\mathcal{E}}(i) = \begin{cases} \mathcal{C}_{\downarrow\updownarrow\rangle_{\sqrt{}}} & if \quad \mathcal{E}(i) \geq \mathcal{C}_{\downarrow\updownarrow\rangle_{\sqrt{}}}, \\ \mathcal{E}(i) & else \end{cases} \tag{5}$$

2.3 Slicing of the Entropy Curve

To split the EC into three SECs, first normalize the entropy values of the grey levels so that they are scaled within range between 0 and 1. Once normalized, split the curve into three distinct parts, ensuring that the sum of the normalized entropy values within each section is approximately 0.33 of the total normalized entropy. This guarantees that each sub-entropy curve holds an equal part of the overall entropy distribution. Therefore, to obtain the sub-entropy curve, find the normalized entropy value as follows [18, 20].

$$NEV(k) = \frac{\hat{\mathcal{E}}(k)}{\sum_{i=0}^{i=L-1} \mathcal{E}(i)}. \tag{6}$$

Fig. 2. The 1st image is the original, 1st shows the enhanced result using the median clipping limit (TSIHE), and the third shows the enhancement with the quantile-based clipping limit.

If the t_1, t_2 are two grey levels such that:

$$\sum_{k=0}^{k-t_1} NEV(k) = 0.33. \tag{7}$$

$$\sum_{k=t_1+1}^{k=t_2} NEV(k) = 0.33. \tag{8}$$

$$\sum_{k=t_2+1}^{k=255} NEV(k) = 0.33. \tag{9}$$

The first SEC E_1, spans from 0 to t_1 containing the entropy characteristics within this interval. The second sub-entropy curve, $\mathcal{E}_2$ extends from $t_1 + 1$ to t_2 representing the entropy distribution over this intermediate range. Finally, the third sub-entropy curve, E_3 covers the intensities from $t_2 + 1$ to 255 representing the entropy variations in the highest intensity region.

2.4 Entropy Curve Equalization

Entropy curve equalization (ECE) improves image contrast by increasing the information content corresponding to different grey levels, thereby obtaining a more detailed and

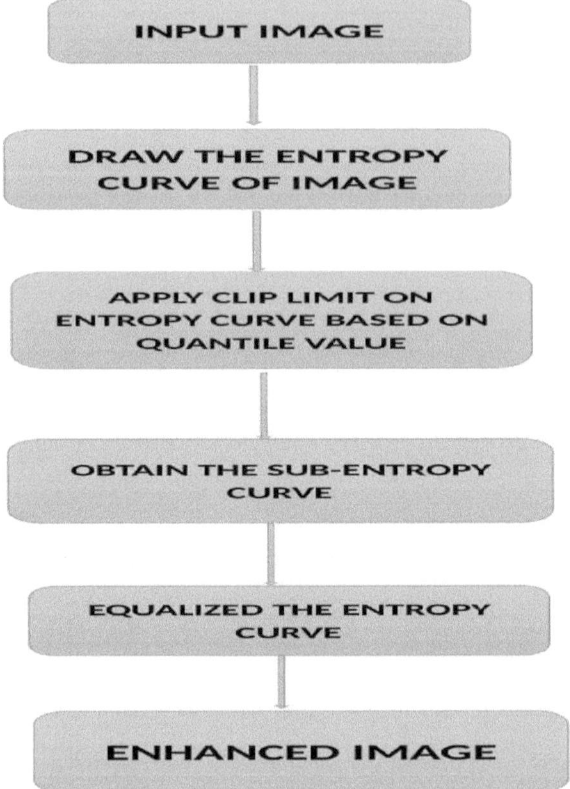

Fig. 3. Flow-chart of proposed method

better contrast output image. This process involves calculating the entropy of the image, which determines the information content, and constructing a cumulative entropy curve CEV to guide the enhancement. The transformation function TF used to map the original grey levels to the new values is given by the following equation:

$$TF = \mathcal{X}_0 + (\mathcal{X}_{\mathcal{L}-1} - \mathcal{X}_0)\mathcal{CEV}, \tag{10}$$

where X_0 and X_{L-1} correspond to the minimum and maximum intensity levels of the SEC, respectively, while CEV stands for the cumulative entropy value.

Let e_k be the accumulated EV of the clipped SEC E_k, defined as:

$$e_k = \sum_{j \in [c_{k-1} \ d_k]} NEV(j), \tag{11}$$

where $k \in \{1, 2, 3\}$ and , $c_0 = 0$, $c_k = d_k + 1$, $d_1 = t_1$, $d_2 = t_2$, and $d_3 = L - 1$. Further, the normalized entropy value for the sub-entropy curves NEV_1, NEV_2 and NEV_3 can be computed by following the equation:

$$NEV_k(i) = \frac{NEV(i)}{e_k}, \tag{12}$$

where $i \in [c_{k-1}, d_k]$, and $k \in \{1, 2, 3\}$.

Then CEV for each sub-entropy curve is calculated as:

$$\mathrm{CEV}_k(i) = \sum_{i=c_{k-1}}^{i} \mathrm{NEV}_k(i), \tag{13}$$

for $k \in \{1, 2, 3\}$, and $i \in [c_{k-1} \sim d_k]$.

The transformation function for each SEC is given as:

$$T_L = (t_1 \times \mathrm{CEV}_1), \tag{14}$$

$$T_M = ((t_1) + 1) + (t_2 - \mathcal{X}_{\mathcal{L}-1}) \times \mathcal{CEV}_2, \tag{15}$$

$$T_U = ((t_2 + 1) + (L - 1 - t_1 - 1) \times \mathrm{CEV}_3). \tag{16}$$

Hence, the final transformation function is:

$$\mathcal{T} = \mathcal{T}_{\mathcal{L}} + \mathcal{T}_{\mathcal{M}} + \mathcal{T}_{\mathcal{U}} \tag{17}$$

Applying the above transformation to an image with low contrast yields an enhanced image with better details. As shown in Fig. 1, the transformed image contains more information than the original image. Additionally, the entropy curve of the original image is concentrated in the middle of the gray-level range, indicating that most of the information is confined to this region. This makes it difficult to recognize various details. On the other hand, the entropy curve of the enhanced image spans the entire gray-level range (0–255), leading to a more evenly distributed representation of information and improved visual clarity.

3 Experimental Results

The performance of the suggested technique is evaluated in this paragraph by conducting a series of experiments on images captured under different illumination conditions, taken from the CEED2016 dataset [22] and the BSDS dataset [23]. The proposed approach uses an entropy-based curve for contrast enhancement and is compared with latest histogram-based techniques, including ACMHE [24], TCDHE [25], JHE [14] and TSIHE [13]. To ensure a comprehensive evaluation, the evaluation includes both numerical and visual analysis. The numerical analysis measures the improvements in contrast enhancement, while the visual analysis evaluates the perceptual quality of the enhanced images based on human visual perception.

3.1 Visual Analysis

Visual perception is the most effective method for assessing the quality of an image enhancement techniques. The visual analysis is performed on the images acquired under varied illumination conditions because low quality images can occasionally produce high metric value. Hence to prove the effectiveness of the suggested technique, visual comparison is essential. The enhanced image obtained through different clipping methods is displayed in Fig. 2. Upon analyzing the image, it can be observed that the images enhanced using these clipping techniques exhibit over-enhancement, causing important details, especially in bright and dark regions, to be lost. In contrast, the enhanced image produced by the suggested method provides better results, maintaining a balanced improvement in contrast and preserving important details across the entire intensity spectrum. This demonstrates the effectiveness of the proposed approach in avoiding the over-enhancement disadvantages of conventional clipping methods while ensuring better overall image quality.

The visual comparison is shown in Fig. 4 with the ACMHE [24], TCDHE [25], JHE [14], TSIHE [13]. From the figure it is clear that enhanced image is obtained by TCDHE, JHE, ISQCAHE are over enhanced in bright region. This can be easily seen in Fig. 4. Also, color distortion is occurred in enhanced image obtained by these methods. The blurriness and color lose is noticed in the enhanced image obtained by TSIHE [13], while enhanced images obtained by proposed method give the optimum enhancement without introducing any noise and artifacts.

3.2 Numerical Analysis

In this section, the performance of suggested technique is evaluated through numerical analysis using multiple quality assessment metrics, including SSIM [26], GMSD [27], VSI [28], and PSNR [29]. A higher SSIM value indicates better structural preservation. VSI assesses visual saliency by identifying regions that attract human attention, also ranging from 0 to 1, with higher scores reflecting better perceptual quality. GMSD quantifies image quality based on gradient magnitude similarity, making it effective in detecting structural distortions, particularly in edge and texture details. Higher PSNR indicates better preservation of details. All these metrics are important for evaluating the effectiveness of the developed image enhancement method.

The suggested technique is compared with the histogram-based method the ACMHE, TCDHE, JHE, TSIHE on large dataset asthe effectiveness of the proposed method cannot be sufficiently validated with a smaller dataset, as the diverse nature of the images may cause the algorithm to fail in some situations. To fully assess the robustness of our approach, we use the entire BSDS500 dataset. Table 1 displays the average SSIM, GMSD, PSNR and VSI values calculated across all images in the BSDS500 dataset [23].

From the Tables 1 and 2, it is evident that the suggested technique gives the highest average SSIM, VSI, and PSNR values compared to latest techniques, while the GMSD value is lower than that of competing methods.

The higher SSIM value prove that the enhanced images produced by the suggested technique exhibit superior structural similarity to the original images. The higher VSI value suggests that the enhanced images also have better contrast compared to other methods. Additionally, the higher PSNR value demonstrates that the reconstruction quality of the enhanced images is superior to that achieved by other techniques. The lowest GMSD value shows better performance of proposed method.

From the discussion above, it is evident that the suggested technique gives better results as compared to existing techniques in terms of metric values. To further assess

4st and 5st, 6st rows contain results obtained by ACMHE, TCDHE, JHE, TSIHE, and proposed method respectively.

Fig. 4. From left to right and top to bottom: 1st row contains input images, 2nd, 3rd, 4th and 5th, 6th rows contain results obtained by ACMHE, TCDHE, JHE, TSIHE, and proposed method respectively.

Table 1. Average metric value on BSDS500 dataset for various methods.

Sr.no.	ACMHE	TCDHE	JHE	ISQCAHE	Proposed Method
SSIM	0.9297	0.7485	0.8088	0.7334	0.9512
GMSD	0.0922	0.0638	0.1168	0.1087	0.0190
PSNR	22.0003	19.4768	17.4115	16.5786	28.4809
VSI	0.9852	0.9487	0.9511	0.9474	0.9923

Table 2. Average metric value on CEED2016 dataset for various methods.

Sr.no.	ACMHE	TCDHE	JHE	ISQCAHE	Proposed Method
SSIM	0.7129	0.8685	0.7655	0.7534	0.9312
GMSD	0.0722	0.0714	00.1368	0.1087	0.0290
PSNR	20.0003	22.4768	16.4115	19.5786	27.4809
VSI	0.9652	0.9787	0.9411	0.9674	0.9923

its effectiveness, the method was tested on a larger dataset, where it consistently demonstrated superior performance. These results confirm that the proposed approach achieves better outcomes across various evaluation measures compared to other methods.

In the above analysis, we have performed a thorough comparison of the results both visually and numerically. The above discussion clearly highlights that the suggested technique exhibits better results compared to other latest techniques. It is evident from the results that the enhanced images produced by the suggested technique exhibit several significant advantages and contain no noise, ensuring that the enhanced images maintain a high level of clarity. Additionally, the images do not suffer from any blurring, thus preserving fine details and textures. Furthermore, the enhanced images display a rich amount of detailed information without any noticeable artifacts. These observations drawn from both visual and numerical analysis strongly support the conclusion that the proposed method consistently produces high-quality results across various evaluation metrics.

4   Conclusion

This paper suggests a new approach to determine the clipping boundary for contrast enhancement using the quantile value of the entropy curve. Traditional methods used to detect clip-limit using the mean or median of the histogram cause over-enhancement, leading to increased noise and artifacts. The suggested approach uses the quantile value of the entropy curve to determine the clip limit, which effectively addresses these issues. This approach not only enhances contrast but also preserves image details and improves contrast and detail. Experimental results evaluated through both human visual perception and numerical analysis demonstrate that the suggested approach outperforms existing standard and latest contrast enhancement techniques.

References

1. Hummel, R.: Image enhancement by histogram transformation. Comput. Graph. Image Process. **6**(2), 184–195 (1977)
2. Gonzalez, R.C., Woods, R.E.: Digital Image Processing, 2nd edn, pp. 85–103. Addison-Wesley, Reading (1992)
3. Kim, Y.-T.: Contrast enhancement using brightness preserving bi-histogram equalization. IEEE Trans. Consum. Electron. **43**, 18 (1997)
4. Wang, Y., Chen, Q., Zhang, B.: Image enhancement based on equal area dualistic sub-image histogram equalization method. IEEE Trans. Consum. Electron. **45**, 6875 (1999)
5. Chen, S.-D., Ramli, A.R.: Contrast enhancement using recursive mean separate histogram equalization for scalable brightness preservation. IEEE Trans. Consum. Electron. **49**, 13011309 (2003)
6. Sim, K.S., Tso, C.P., Tan, Y.Y.: Recursive sub-image histogram equalization applied to gray scale images. Pattern Recogn. Lett. **28**, 12091221 (2007)
7. Zuiderveld, K.: Contrast limited adaptive histogram equalization. Graph. Gems, 474–485 (1994)
8. Tiwari, M., Gupta, B., Shrivastava, M.: High-speed quantile-based histogram equalization for brightness preservation and contrast enhancement. IET Image Process. **9**, 8089 (2015)
9. Abdullah-Al-Wadud, M., Kabir, M.H., Dewan, M.A.A., Chae, O.: A dynamic histogram equalization for image contrast enhancement. IEEE Trans. Consum. Electron. **53**(2), 593–600 (2007)
10. Zarie, M., Parsayan, A., Hajghassem, H.: Image contrast enhancement using triple clipped dynamic histogram equalization based on standard deviation. IET Image Process. **13**(7), 1081–1089 (2019)
11. Choukali, M.A., Valizadeh, M., Amirani, M.C., Mirbolouk, S.: A desired histogram estimation accompanied with an exact histogram matching method for image contrast enhancement. Multimed. Tools Appl. **82**(18), 28345–28365 (2023)
12. Acharya, U.K., Kumar, S.: Image sub-division and quadruple clipped adaptive histogram equalization (ISQCAHE) for low exposure image enhancement. Multidim. Syst. Sign. Process. **34**, 2545 (2023)
13. Rahman, H., Paul, G.C.: Tripartite sub-image histogram equalization for slightly low contrast gray-tone image enhancement. Pattern Recogn. **134**, 109043 (2023)
14. Agrawal, S., Panda, R., Mishro, P.K., Abraham, A.: A novel joint histogram equalization-based image contrast enhancement. J. King Saudi Univ.-Comput. Inf. Sci. **34**, 11721182 (2022)
15. Ruhela, R., Gupta, B., Singh Lamba, S.: An efficient approach for texture smoothing by adaptive joint bilateral filtering. Vis. Comput. **39**(5), 2035–2049 (2023)
16. Ruhela, R., Gupta, B., Lamba, S.S.: A new non-convex low rank minimization model to decompose an image into cartoon and texture components. Comput. Math. Appl. **123**, 1–12 (2022)
17. Zhou, M., Huang, J., Yan, K., Hong, D., Jia, X., Chanussot, J., Li, C.: A general spatial-frequency learning framework for multimodal image fusion. IEEE Trans. Pattern Anal. Mach. Intell. **47**, 5281–5298 (2024)
18. Yadav, P.S., Gupta, B., Lamba, S.S.: A new approach of contrast enhancement for medical images based on entropy curve. Biomed. Signal Process. Control. **88**, 105625 (2024)
19. Yadav, P.S., Gupta, B., Lamba, S.S.: A novel contrast enhancement method for low-light images with non-uniform illumination using the entropy curve. In: 2024 15th International Conference on Computing Communication and Networking Technologies (ICCCNT), pp. 1–5. IEEE (2024, June)

20. Yadav, P.S., Gupta, B., Lamba, S.S.: Exposure-based contrast enhancement method for low contrast images using the entropy curve. In: 2024 6th International Conference on Image, Video and Signal Processing, pp. 61–66. ACM (2024, March)
21. Yadav, P.S., Gupta, B., Lamba, S.S.: A new approach of image contrast enhancement based on entropy curve. Signal Image Video Process. **18**(4), 3431–3444 (2024)
22. Qureshi, M.A., Sdiri, B., Deriche, M., Alaya-Cheikh, F., Beghdadi, A.: Contrast enhancement evaluation database (ceed2016), Mendeley Data, v3.DOI 10 (2017)
23. Martin, D., Fowlkes, C., Tal, D., Malik, J.: A database of human segmented natural images and its application to evaluating segmentation algorithms and measuring ecological statistics. In: Proceedings of the 8th International Conference on Computer Vision, vol. 2, pp. 416–423. IEEE (2001)
24. Santhi, K., Banu, R.W.: Adaptive contrast enhancement using modified histogram equalization. Optik-Int. J. Light Electron Opt. **126**(2015), 1814 (1809)
25. Zarie, M., Parsayan, A., Hajghassem, H.: Image contrast enhancement using triple clipped dynamic histogram equalization based on standard deviation. IET Image Process. **13**, 10811089 (2019)
26. Wang, Z., Bovik, A.C., Sheikh, H.R., Simoncelli, E.P.: Image quality assessment: from error visibility to structural similarity. IEEE Trans. Image Process. **13**, 600612 (2004)
27. Xue, W., Zhang, L., Mou, X., Bovik, A.C.: Gradient magnitude similarity deviation: a highly efficient perceptual image quality index. IEEE Trans. Image Process. **23**(2), 684–695 (2013)
28. Zhang, L., Shen, Y., Li, H.: Vsi: a visual saliency-induced index for perceptual image quality assessment. IEEE Trans. Image Process. **23**, 42704281 (2014)
29. Poobathy, D., Chezian, R.M.: Edge detection operators: peak signal to noise ratio based comparison. IJ Image Graph. Signal Process. **10**, 5561 (2014)

Author Index

MIX
Papier aus verantwortungsvollen Quellen
Paper from responsible sources
FSC® C105338

If you have any concerns about our products,
you can contact us on
ProductSafety@springernature.com

In case Publisher is established outside the EU,
the EU authorized representative is:
**Springer Nature Customer Service Center GmbH
Europaplatz 3, 69115 Heidelberg, Germany**

Printed by Libri Plureos GmbH
in Hamburg, Germany